A SHORT HISTORY OF
TWENTIETH-CENTURY
TECHNOLOGY

c.1900–*c*.1950

A SHORT HISTORY OF
TWENTIETH-CENTURY TECHNOLOGY

c.1900 – *c*.1950

Trevor I. Williams

CLARENDON PRESS · OXFORD
OXFORD UNIVERSITY PRESS · NEW YORK
1982

Oxford University Press, Walton Street, Oxford OX2 6DP

London Glasgow New York Toronto
Delhi Bombay Calcutta Madras Karachi
Kuala Lumpur Singapore Hong Kong Tokyo
Nairobi Dar es Salaam Cape Town
Melbourne Auckland

and associate companies in
Beirut Berlin Ibadan Mexico City

Published in the United States
by Oxford University Press, New York

British Library Cataloguing in Publication Data

Williams, Trevor I.
A short history of twentieth-century
technology c. 1900-c. 1950
1. Technology—History—20th century
I. Title
609'.04   120
ISBN 0-19-858159-9

Library of Congress Cataloging in Publication Data

Williams, Trevor Illtyd.
A short history of twentieth-century
technology c. 1900-c. 1950.

Sequel to: A short history of technology from the earliest times to A.D. 1900 /
by T. K. Derry and Trevor I. Williams.
Includes bibliographies and index.
1. Technology — History — 20th century.
I. Derry, T. K. (Thomas Kingston), 1905-
Short history of technology from the earliest
times to A.D. 1900. II. Title.
T20.W55   609'.04   82-4362
ISBN 0-19-858159-9   AACR2

Set by Macmillan India Ltd
Printed in Great Britain at the University Press, Oxford,
by Eric Buckley,
Printer to the University

For Sylvia;
patience rewarded

PREFACE

In 1973 the Oxford University Press invited me to edit two further volumes (VI and VII) of *A History of Technology*, covering roughly the period 1900 to 1950. In 1978, the Press's quincentennial year, these two volumes were published, thus bringing to a conclusion a project on which work had begun in 1949. The first volume was published in 1954 and covered the period from the dawn of civilization to the fall of the ancient empires. The fifth, bringing the *History* up to the end of the nineteenth century, appeared in 1958. At that stage it was thought appropriate to pause and prepare a smaller work covering much the same ground but designed to introduce the subject to a wider readership; this led to the publication in 1960 of *A Short History of Technology*, which I wrote with Dr T. K. Derry.

The success of that *Short History*, which has since been published in a paperback edition and in several foreign-language translations, suggested that the same formula might be acceptable in respect of the two most recent volumes covering the first half of this century. The present *Short History of Technology in the Twentieth Century* thus bears much the same relationship to the last two volumes of the main work as the original *Short History* did to the first five. While it falls mainly within the confines of the larger work, and draws heavily on it for factual information, it is not a mere abridgement of it. This would in any event have been impracticable, partly because of the degree of compression necessary and partly because the plan of the major work did not lend itself to such treatment.

This book is, therefore, an independent work designed, within a fairly modest compass, to give the general reader a broad conspectus of the way in which technology developed in the first half of this century and a glimpse of the social, economic, and political factors that influenced it. Without subscribing wholeheartedly to the doctrine that the history of science and technology need not concern itself overmuch with the attribution of discoveries and inventions to individuals – which indeed is often very difficult – I have tacitly assumed that for many general readers inventions became interesting when they first make a discernible social impact.

The planning of a book of this kind presents considerable difficulties both because of the wide range of activities to be considered and because the period covered – although relatively brief in the historical sense – was one of exceptionally rapid change, punctuated by two world wars which gave a powerful but somewhat artificial stimulus to certain branches of technology. These punctuations were not, however, of a kind that seemed to call for a division of the book in a chronological sense.

To deal adequately with each major aspect of technology, without undue repetition, is another problem. To take a random example, agriculture demands consideration of such diverse topics as industrial chemistry, a source of artificial fertilizers, insecticides, and herbicides; the internal combustion engine, resulting in a virtually total change from draught animals to tractors as a source of motive power; the development of refrigerated shipping to export the sheep and cattle raised on the vast new grazing areas of Australasia and South America; and the further development of canning as a large-scale preservative process. At the same time, the chemical, agricultural, shipbuilding, and food preservation industries clearly demand consideration in their own right. While cross-referencing can reduce repetition it cannot eradicate it. Indeed, a degree of repetition is perhaps no bad thing, in that it reminds the reader that modern technologies are highly interdependent.

Nominally, this book covers the first half of this century but the remit has been fairly elastically interpreted, as in the main work. Some subjects are most sensibly treated by picking up the threads somewhat before 1900; others would have terminated too abruptly if not extended a little beyond 1950. It is difficult to explain computers, for example, without some reference to the mechanical calculating-machines of the nineteenth century and readers would surely have been disappointed had there been no mention of space flight or the beginning of atomic power.

In science, and in much of the technology derived from it, the turn of the century was something of a watershed. Up to that time the immense developments in the physical sciences had been largely the natural consequence of the Scientific Revolution of the seventeenth century. This was based on an essentially mechanistic view of the workings of nature, and it provided a technique of discovery which was immensely fruitful for more than two centuries. In the latter half of the nineteenth century, however, a whole range of new phenomena, mainly in the electrical field, were emerging which were not explicable in simple and generally understandable terms. In the public mind the transition is closely identified with Einstein's special theory of relativity, published in 1905, but in fact this reflected a change in the direction of thought that began some thirty years earlier. This new line of development had far-reaching technological consequences, the most dramatic of which were the atomic bomb and nuclear energy. The exposition of such developments for a general readership presents difficulties more considerable than those encountered in describing developments stemming from Newtonian physics, for they involve fundamental conceptions that are not only not generally familiar but in some cases seem also to be at variance with common experience. Electronic devices, too, such as radio and television are not easily explained to those with no relevant technical knowledge. In a book of this kind emphasis must, therefore, necessarily be on what they do rather than on how they do it. However, these problems are to some extent diminished by the existence of a new and increasing generation of readers to whom some science has been taught, even if not at an advanced level, in a way that takes account of these new developments.

The choice of what to include and what to omit is necessarily subjective. In planning the book I have tried to ensure that, while all the major developments are described, some minor, but possibly not less generally interesting ones, are not omitted. For many, domestic appliances will be at least as appealing as coal mining or turbines. For those who wish to read further, the appropriate chapters of Volumes VI and VII of the main *History* will, of course, be directly relevant; in addition, each chapter in this book is followed by an extensive bibliography. Most of the works listed there will themselves provide further references. Many of the illustrations appeared in the original work, but the opportunity has been taken of including some new ones, more appropriate to the present treatment of the subject.

In its published form, this book is a tribute to the mastery of printing technology that the Oxford University Press has acquired since 1478. My own personal technology is still limited to pushing a pen over paper, and I am greatly indebted to Mrs Yvonne Rue for her patience in deciphering the result and preparing the typescript for press.

Oxford, March 1980   TREVOR I. WILLIAMS

CONTENTS

Contents

LIST OF ILLUSTRATIONS

List of Illustrations

1

INTRODUCTION

The first half of the twentieth century was in a sense a period of transition in western civilization. For better or for worse, an essentially rural economy had been transformed into an urban one based on the technological developments ushered in by the Industrial Revolution. That these were capable of generating great wealth was apparent, but it was insufficiently recognized that industrialization brings penalties as well as prizes. Up to the end of the nineteenth century the emphasis had been on technology: if this could be got right the resulting prosperity would surely suffice to cure the ills of society. It was not sufficiently realized that rapid industrialization creates social, economic, and political problems not capable of solution by the dispassionate methods that had served so well to advance science and technology. Within industry the human element began to assert itself: on the one hand through the growth of the trade union movement and on the other by formal studies in the techniques of management. The people at large began to question whether technology was taking them where they really wanted to go. The educational systems of the western world began slowly to adapt themselves to the needs of the new society in which its younger generation was growing up. Governments increasingly involved themselves in industrial affairs, sometimes by promoting developments that seemed in the national interest, at others by restricting those which seemed undesirable.

These somewhat tardy reactions to the consequences of nineteenth-century developments were complicated by the fact that technology itself was entering a new phase. Hitherto it had been based largely – though by no means exclusively – on progress in mechanical engineering, but a new force was making itself felt before the turn of the century. This new factor was electricity. It was not, of course, by then a newly discovered phenomenon, having been familiar to philosophers of the seventeenth and eighteenth centuries. Nevertheless, it is true to say that, save in telegraphy, electricity made little social impact until the 1880s, when the first public supply systems began to appear in the larger cities. From that time onward electricity became increasingly important as a novel source of power and, especially, of light. More important, the allied science of electronics, then in its infancy, was destined to bring about great changes before the middle of this century. The thermionic valve, of paramount importance in the early days of radio, paved the way to the transistor and the microprocessor, of profound significance in the development of computers and automatic control systems. By the 1970s the consequences of the microprocessor revolution had become a matter of great concern to governments throughout the western world.

While the nature of technology changed greatly during the first half of the twentieth century, so too did its geographical pattern. In 1900 the major centres of industrial production were still Britain and Germany, but during the next fifty years powerful and successful competitors appeared. The United States, for example, not only established a commanding position in traditional industries, such as iron and steel, but was a pioneer in wholly new industries, such as the manufacture of motor cars and aeroplanes. After the Revolution in 1918, Russia embarked on a programme of massive industrialization to support the new socialist state. In 1961 she was to astonish the entire world by achieving the first manned flight in space. Japan, who had developed in virtually complete isolation from the western world, did not introduce modern industry until the 1870s, but then proceeded to dispel disparaging beliefs that she was no better than an imitator. By the end of the period of our present concern she had emerged decisively as an innovator *par excellence*, taking the lead in the highly sophisticated new industries – based mainly on electronics, optics, and precision engineering – that developed after the Second World War. In agriculture, too, trends that began in the nineteenth century continued in the twentieth. Improved transport, advances in farm machinery, and better methods of preservation encouraged the production of grain and meat in the vast open spaces of Australia and North and South America.

Various factors can be discerned as having influenced these great changes in the nature and location of industry. Among the most important were the educational establishments; the two sides of industry, management and workers respectively; and government. To discern them is not difficult: to evaluate their relative importance and disentangle their strong, sometimes violent, interactions is an altogether different matter. For convenience we will consider them separately, but will try at the same time to draw attention to the many cross-linkages.

I. EDUCATION

It has often been pointed out that the Industrial Revolution in Britain was initiated and carried through by men with little or no formal education, certainly not at university level. We are less often reminded that the universities of those days had little, except mathematics, to offer the new pioneers, even had they attended them. Science was viewed with suspicion and engineering as beneath the dignity of a gentleman; on the other hand, a sound grounding in the classics had little relevance to the new demands of industry. Not until the latter part of the nineteenth century (1875) was a Chair of Applied Mechanism and Applied Mechanics established at Cambridge; there was no chair of engineering at Oxford until 1907. It was left largely to the civic universities, in the new centres of industry and commerce, such as Liverpool and Manchester, to provide for the needs of manufacturers in the early years of this century. Even there, however, the academic atmosphere was more sympathetic to the pure sciences than to the applied: in the university world, and in the nation at large, engineers lacked the prestige of their arts colleagues. The relationship between academic departments of applied science and industry was a tenuous one. Only a few chairs – such as the Livesey Professorship of Coal Gas and Fuel Industries at Leeds (1908) – were directly related to the needs of any particular industry. The pure sciences fared better, but there was a reluctance on the part of the best graduates to go into industry. Those who did, found that their universities had done little to prepare them for the practical problems of industrial life and their employers often failed to use them to the best advantage. Even at mid-century, scientists and engineers complained that they were on tap and not on top.

Elsewhere, a more positive attitude prevailed towards vocational training in science and engineering. In Germany, for example, the old universities showed the same sort of antipathy towards technology as was evinced in Britain, but this was countered by the founding of separate institutions for higher education in technology, the *Technische Hochschulen*; these had the status of the older universities and conferred their own degrees. The earliest university of this type was the Universität Fridericiana at Karlsruhe (1825) but most were founded a good deal later: the Technical University of Munich was founded in 1868 and the Berlin Technische Hochschule in 1879. It was customary for the professors to maintain strong links with industry. Thus, by the start of the twentieth century, German industry had a source of appropriately trained graduates independent of the traditional universities and a link with centres of research relevant to its own activities. The wisdom of this policy became apparent in the Second World War, which Germany entered well prepared technically for a long siege, whereas Britain faced immediate shortages of such basic materials as dyes, optical devices, special grades of steel, and pharmaceutical products. France, with her splendid but rather irrelevant *lycées*, found herself in like straits. The old engineering foundations, such as the École Nationale des Ponts et Chaussées (1747) and the École Polytechnique (1794), were inadequate for new needs. At the other end of Europe, Russia had a number of important centres for basic and applied science, but she did not become a major industrial power until after the Revolution, when Lenin propounded his famous equation: Electrification plus Soviet power equals Communism.

Beyond Europe, the importance of education with an emphasis on science and technology was widely recognized. Japan, coming very late into the field, was well aware of the importance of scientific and technological education, which was emphasized in the five imperial universities founded between 1877 (Tokyo) and 1911 (Tohoku). The practical success of this policy was apparent when Japan defeated Russia in 1905: in less than forty years she had emerged from obscurity to become a world power. In the United States, where there were no inhibitions about the pursuit of material wealth, the need for men properly qualified to manage the rising new industries and commercial enterprises was recognized. The word 'manage' should be noted: the initiation and direction of the new industries was still largely in the hands of men, such as Andrew Carnegie, Sam Goldwyn, and Henry Ford, who had little formal education of any sort, let alone in science and technology. But the sorts of industries such men created could flourish only with the assistance of well-qualified employees at all levels. One of their main assets was the ability to spot and utilize talent.

In the United States, early colleges supported by state or private funds were reinforced by the land-grant colleges and universities founded under the Morrill Act of 1862. This provided land in each State 'for the endowment, support, and maintenance of at least one college where the leading object shall be . . . to teach such branches of learning as are related to agriculture and the mechanic arts'. Each State received 30 000 acres for each senator and representative then in Congress. This land was gradually sold off and the proceeds invested. After a rather slow start, the fund totalled nearly $20 million in 1927, with land valued at $6 million: this yielded an annual income of over $1 million. Over the years much non-government support was forthcoming, and as early as 1892 only one-third of the total income of the land-grant colleges came from Federal sources: by the middle of the twentieth century this had fallen to one-tenth.

The American universities generally were well disposed towards science and technology, and developed in the tradition of such independent foundations as the Massachusetts Institute of Technology (1861) and Stevens Institute of Technology (1870).

Thus the industrialized world entered the twentieth century in varying states of preparation with regard to its educational systems. Perhaps the most fully seized of the importance of a wide understanding of science and technology were Germany and the United States, and it is significant that these were the two principal rivals for Britain's industrial supremacy. At the beginning of the twentieth century the total state grant to universities and colleges in Britain was a meagre £27 000: this compared with a state grant of £130 000 to the University of Berlin alone and nearly £1 million to the universities of New York State. In 1913 there were 40 000 students of science and technology in the United States, 17 000 in Germany, and 6500 in Britain. While these figures certainly represent substantial differences, they should not be regarded as strictly comparable owing to differences in curricula, duration of courses, quality of teaching staff, entrance requirements, and so on.

The First World War underlined the strategic importance of science and technology and the need to have adequately trained practitioners at all levels. To some extent deficiencies were made good as the war progressed, but this was of necessity largely a matter of improvisation; not until peace was restored could proper programmes of development be planned and implemented. In Britain, the emphasis remained on pure rather than applied science. While the number of university places nearly doubled between 1919 (20 500) and 1938 (37 200), those allocated to technology remained virtually static at 4200, with the result that the proportion of technological students fell from 20 per cent to 11 per cent. Relatively speaking, Germany supported twice as many students of technology. Further east, the USSR was putting into effect a programme of industrialization even more intensive than that which had drawn Japan so swiftly into the western camp. The Five-year Plans, beginning in 1928, included massive provision for the expansion of higher education, with a heavy emphasis on science and technology; an unusual feature was the high proportion of women students, which by 1940 amounted to more than 40 per cent in the industrial field generally.

The Second World War stressed still further the strategic importance of educational systems with a strong emphasis on science and technology. Far more than in any previous major conflict, victory was the reward of technological superiority – superiority not only in the traditional industries such as steel-making, shipbuilding, and the manufacture of motor cars, but in novel developments such as radar, penicillin, and, of course, all the very advanced technology necessary to develop the ultimately decisive atomic bomb. Again major errors of judgement were revealed. Britain, for example, lost penicillin manufacture to the United States largely because of inadequate facilities for carrying out long-term fermentations under sterile conditions. Germany suffered severely from an ideological obsession with a pure Aryan science as opposed to an allegedly corrupt Jewish one. The expulsion of the Jews was to provide the Allies with a highly intelligent and knowledgeable accretion of scientific manpower in the very fields where it was most necessary.

With the return of peace in 1945, there seemed no limit to what science and technology could do for the benefit of mankind. Young people, encouraged by parents and teachers alike, sought qualifications in these fields, and the number of places in universities and technical colleges was substantially enlarged. Later, as environmentalists and sociologists increasingly drew attention to the destructive concomitants of technology, this enthusiasm was to wane, but it is undeniable that the immediate post-war years saw the tide running much more strongly in favour of technologically relevant education: numeracy became as much esteemed as literacy. For a few brief years it seemed that without a knowledge of science and its applications life's battle was lost before it was begun.

It would be satisfying to record that these changes in the educational pattern kept in step with the changing demands of industry, but such was not the case. Even in the most progressive countries there remained a considerable disparity. The reasons for this are too complex to analyse here, but undoubtedly one of the most important was the difference between the academic and the industrial cycles. If need be, industry can respond very rapidly to new situations such as the deterioration of economic circumstances or the appearance of profitable new technologies, either of which has immediate repercussions on the educational world. In the first situation recruiting is curtailed; in the other, new knowledge and skills are demanded. The universities, however, cannot respond anything like as rapidly. Many of their students will have made their general choice of subject two or three years before entry and last-minute changes are not easy. Would-be engineers, for example, are handicapped if they lack mathematics. For students committed to the arts, it is in practice very difficult to change to a course in science or technology. Once in the university system, students are committed to a three- or four-year course which cannot easily be modified. Universities therefore had to proceed cautiously, aiming at producing generally well qualified graduates rather than responding to the rapidly changing demands of industry. In so doing, they attracted a good deal of criticism, much of it ill-informed.

While the development and management of the new industries depended on the availability of a nucleus of well and appropriately trained staff, this was not sufficient in itself. The new equipment was dependent upon a better educated work-force, with emphasis on science, engineering, and mathematics. Although they could not be studied in any depth, of course, these subjects were increasingly introduced into secondary school curricula.

Here, as in the universities, they presented difficulties because – unlike some of the traditional subjects where texts were literally handed down from father to son – their dynamic nature demanded constant revision of the syllabus. Science in 1950 had to include many topics that were virtually unknown in 1900.

No less important, the public at large had to acquire in one way or another the sort of background knowledge and experience necessary to accept and utilize the new products of industry. How far this was a consequence of more broadly based education, as opposed to the natural adaptability of the human race, is debatable. We cannot ignore the fact that the oldest of our citizens have taken in their stride major developments that occurred long after their formal education was over: as examples – taking account of the date of social impact rather than actual invention – we may cite the cinema, telephone, radio, and television; the motor car and the aeroplane; electric light; refrigeration and frozen foods; computers and automation. All these became normal parts of daily life in the western world, utilized without difficulty by millions of people who – regardless of the nature of their education – had no real understanding of the way in which they work. Motoring, for example, was at first regarded as an art that could be mastered only with difficulty by men and virtually beyond the capacity of women. Sophisticated products of modern technology – such as transistor radios and earth-moving equipment – are worked by operators with no formal qualifications whatever. This is, of course, largely a consequence of deliberate design innovations on the part of manufacturers: the larger the market to be developed, the less must be the demand on the user.

This opens up very important considerations that cannot be pursued at any length here. In particular, there is the question of the transfer of technology. How far can modern technology – as distinct from the technological products of advanced countries – be transferred to developing countries, and what are the principal determining factors? How far can a backward country rely on its own resources and how far, and for how long, must it rely on foreign advice and supervision? To quantify such considerations is impossible, but it is clear that education is of paramount importance, and that in this context there are two aspects to be considered. On the one hand, there is the purely technological aspect: for example, oil refineries cannot be designed, built, and operated without the services of highly qualified engineers, even though the latter may be able to make use of a surprising amount of unskilled and ignorant labour. No less important, however, is the creation of an educational background favourable to technological changes desired for political and economic reasons. Within this century there are countless examples of relatively backward countries in which a policy of industrialization has been thwarted not so much by technical difficulties as by a violent popular reaction deeply rooted in a dislike of foreign intervention and respect for traditional religious and social observances. In such situations we should not be too ready to assume that the popular instinct is not right.

II. INDUSTRIAL ORGANIZATION

For industrial organization, like education, the first half of the twentieth century was a period of transition. In so brief an account as this it is possible to do no more than generalize, recognizing that changes took place at different rates and in different ways in different places and in different industries. At the beginning of this century employers were dominant, though not necessarily oppressive; unions were small and weak. By mid-century a professional managerial class had appeared and the unions, developing from the old guilds of skilled craftsmen, had penetrated deeply into the ranks of the unskilled. With this polarization of management and unions there arose complex negotiating machinery in place of arbitrary decisions based simply on authority. As a further complicating factor inter-union disputes arose, mainly over demarcation of areas of work and the maintenance of wage differentials. Over and above all this, all countries developed complex and constantly changing legislation – often purely politically motivated – to control wages and relations between management and labour. The question of the right to strike, for example, was one of perennial importance. In such a complex situation it is impossible to distinguish in any quantitative way how the changing organization of industry influenced the development of technology as a whole: historians have found it difficult to reach firm conclusions even in respect of single industries in a particular country.

Nevertheless, despite these reservations, certain fundamental developments can be clearly seen, and in the context of management the first of these was F. W. Taylor's Theory of Scientific Management, which he first publicized in 1895. Briefly, Taylor's doctrine was that productivity could be increased if every stage of a worker's task was carefully analysed with a view to shortening or eliminating it. In effect, he sought to apply to the human worker the same principles as had proved so successful in improving the design of machines. He assumed that the worker's only motivation was economic and that increased wages would be sufficient reward for greater productivity.

Taylor had some conspicuous successes and his methods were widely acclaimed in American industry and later abroad. Thus in one steel works he increased the capacity of a pig-iron handler from $12\frac{1}{2}$ to $47\frac{1}{2}$ tons per day:

as a result, the workman received 60 per cent more wages than his colleagues. The weakness of Taylorism was that it ignored the worker's feelings and motivation. In reality he did not want his employer to know the limit of his capacity. The point was succinctly put (1914) by N. P. A. Alifas, a spokesman for the International Association of Machinists:

The only way that the workman has been able to retain time enough in which to do the work with the speed with which he thinks he ought to do it, has been to keep the employer somewhat in ignorance of exactly the time needed. . . . Most people walk to work in the morning, if it isn't too far. If somebody should discover they could run to work in one third the time, they might have no objection to have that fact ascertained, but if the man who ascertained it has the power to make them run, they might object to having him find it out.

Not surprisingly, Taylorism fell into disrepute and was replaced by new worker-oriented doctrines. The scientific managers were succeeded by the industrial psychologists, initially in Germany, who believed that the right approach was to match the individual's personality to the task he was to perform. These in turn were followed in the 1920s by the industrial sociologists, particularly identified with Elton Mayo in Australia, who believed that productivity was closely related to the individual's relationship with his fellow workers and his personal life. This was, of course, the very antithesis of Taylorism.

While these rival theories of management technique were being argued about, the nature of the enterprises to which they were applied was changing steadily. The small family business – in which authority lay firmly with the owner and his relations, who were rarely much bothered about management theory – was giving way to big corporations, some multinational, in which a large measure of authority had necessarily to be delegated to professional managers, who had no direct personal involvement beyond the retention of their own jobs and the prospects of promotion. Increasingly, first in the United States and then in Europe, professional managers were trained in the various business schools attached to universities and other institutes of higher education. By 1925 virtually every American university had a school of business studies, but in Britain this was essentially a post-war development. The British Institute of Management was established in London in 1947 but not until 1965 were schools of business studies established in Manchester, London, and Oxford. By the end of our period higher management was calling to its assistance a wealth of disciplines ranging from psychology to statistics and the theory of games.

While management had been adapting itself to the changing industrial situation so also had the unions. Here, too, a remarkable transition took place: in Britain, for example, total trade union membership in 1889 was not more than 750 000, representing about 5 per cent of a working population of about 16 million. By 1948 the British Trade Unions Congress had an affiliated membership of nearly eight million; much of this tenfold increase took place after the First World War. In general, the growth of the trade union movement in Europe paralleled that in Britain: total membership rose from about eight million in 1910 to around 26 million immediately after the end of the First World War. Unlike Britain, however, the continental unions embodied a strong religious element. They were also more prone to destructive internal political dissension between communists and non-communists, as in France after the First World War. Germany, too, suffered from disputes between social democrats and communists: these were abruptly ended in 1933 when the Nazis dissolved the movement. Not until after the Second World War did trade unionism revive in Germany.

In America, where there was much legislative restriction, the movement grew relatively slowly. In 1910, membership there scarcely exceeded two million, a quarter of that in Europe, and even in 1939 it was no more than about six million. Most of its growth followed the foundation of the Congress of Industrial Organisations in 1935, on the initiative of John L. Lewis, of the United Mine Workers, in the wake of the New Deal. This opened the way to the unionization of many industries hitherto unorganized, such as the steel, textile, and motor-car industries.

While these changes in the attitudes of management and in the organization of labour substantially altered the pattern of industry, the sort of figures quoted above make it clear that much industrial production was still effected by non-union labour managed by men brought up in the old authoritarian tradition. In Britain, as late as 1945, Imperial Chemical Industries – one of the country's major companies, with an excellent record as an employer – still based its labour relations on benevolent paternalism or, as some would have it, benevolent authoritarianism. At that time one of the directors of the Company, Sir Frederick Bain, had to remind the Board that:

The worker of today, even on the docks, is not the illiterate clodhopper of the last generation. He has that little knowledge which is a dangerous thing, but it is enough to convince him that there was something wrong with his old conditions of employment and in particular, with the system under which his own livelihood and that of his family were subject to the threat of one hour or one week's notice.

On balance, this is a fair statement of the general change of attitude on the part of workers that occurred during the first half of this century. They did not know clearly what they wanted, or even where their true interests lay, but they were not satisfied with what they had and they knew that change could best be effected by collective action.

The consequences for the history of technology are difficult to measure. On the one hand, innovation lay increasingly – but by no means exclusively – with the large corporations, and unionization was a natural response to the growth of these. In this sense, the unions may be said to have promoted technological progress. On the other hand, their direct response to new developments was almost wholly based on the narrow issue of whether there was an immediate threat to the employment or remuneration of their members; long-term issues were of little or no interest, an attitude still widely prevalent. The long-standing attitude of union officials was epitomized by Moss Evans, general secretary of the powerful Transport and General Workers' Union in Britain, in 1979: 'I am not bothered by percentages. It is not my responsibility to manage the economy. We are concerned about getting the rate for the job.' In consequence, it is very difficult to find any example of technological advance that was initiated by union activity; the instinctive response was obstructive. To an extent, this reflects the fact, sometimes ignored, that the organizations of management and labour are diametrically opposite. In management, power is concentrated at the top and becomes more diffuse as one descends lower into the organization. In the unions, by contrast, power resides with the workers, who elect their officers to do their bidding. In practice, of course, it is not so simple as this, for much depends on personalities. A determined and persuasive union leader may carry his members with him, just as an ambitious manager may persuade his superiors of the value of his own ideas.

That unions should see their role as protectors of the short-term interests of their members is natural, if rather disappointing, but at least this served as a counterbalance to sometimes inhumane authoritarianism. In the event new industries did not, overall, create unemployment: in 1950, millions of workers found employment in industries that did not even exist in 1900. In 1950, for example, ten million motor vehicles were manufactured and twenty million radio receivers: neither of these great and growing industries was economically significant at the beginning of the century. Today the world market in tyres alone amounts to £2000 million annually. Nevertheless, it is indisputable that technological innovation was constantly changing the pattern of industry, and displaced workers did not always have the skill necessary for new trades, nor could they always readily move to the centres of new employment. What could be good for society might be disastrous for the individual.

Thus the reorganization of industry, with its polarization into professional management and unionization, presents us with somewhat the same

dilemma as the changes in education. We can discern certain clear trends –
running with different force in different circumstances – but it is not easy to
distinguish between cause and effect.

III. GOVERNMENT INTERVENTION

Government intervention in technological affairs has a long history. As
examples from earlier ages we may recall the great irrigation systems of
Egypt and Mesopotamia; the Athenian silver-mines at Laurion; the Roman
system of roads; and the Grand Canal constructed by the Chinese to
transport the grain tribute to Pekin. These were all great enterprises
initiated and controlled by the state in the national interest. They were,
however, no more than particularly striking manifestations of a system of
state intervention that was accepted in industry throughout the western
world long before the start of the Industrial Revolution and had become
highly complex by the start of the twentieth century.

Broadly speaking, the role of government may be direct or indirect.
Further examples of direct involvement are provided by government
arsenals and naval dockyards. In the same category must be placed the vast
Manhattan Project, which produced the first atomic bombs in 1945. Others
arise when governments place orders for manufacture, or for research and
development, with independent contractors. In some instances the involvement may be very close indeed, as in the modern aerospace industry, where
the governments concerned are the principal customers.

In the extreme case, of course, governments may acquire industries
outright. This happened in Russia on a grand scale after the Revolution and
in the new Communist states of eastern Europe after the Second World War.
Elsewhere, in mixed economies, governments were content to control key
industries, as in Britain after the Second World War. The industries so
nationalized were predominantly public utilities such as transport and
power.

A very great deal of government influence on the development of
technology has, however, been exercised indirectly, through legislation of
one kind or another. This might take the form of taxation or of restrictive
regulation designed to favour or discourage some particular development,
or to protect the consumer. New industries have been helped to establish
themselves by the restriction or heavy taxing of imports. Needless to say, the
results were not always what the legislators intended. In the course of this
work we shall encounter many instances of the influence of legislation on
technology; it suffices at this point merely to cite one or two examples. In
Britain, the early development of the motor car was seriously impeded by
the Red Flag Act which required that a man carrying a red flag should walk

in front of every vehicle: this was not repealed until 1896. In the British chemical industry the development of an oil-from-coal process in the 1930s hinged upon the readiness of the government to give a margin of preference to home-produced oil. After much wrangling, this was finally conceded in 1934. The plant subsequently erected by ICI at Billingham, together with a second one opened at Heysham, was to give Britain millions of gallons of urgently needed aviation fuel during the Second World War. In the United States, the Tennessee Valley Authority was one of the early consequences of the New Deal of 1933: this engendered a vast enterprise for the production of electricity, the manufacture of fertilizer, and the provision of flood control and irrigation.

A crucial factor in the development of most inventions is the protection of the inventors' rights by patents, giving him a statutory period of time in which to reap the benefit of his discoveries. By the beginning of this century all the principal countries had well-established patent offices and a new specialist profession – that of patent agent – had been created. In modern technology patents serve a dual role: to promote one's own inventions and to block those of rivals. Their growing complexity often gave grounds for dispute, however, and many major inventions have been the subject of long and sometimes bitter litigation. As an example within our present period we may mention the rival claims of Lee de Forest and J. A. Fleming to the invention of the thermionic valve; the dispute substantially delayed the development of the radio industry. While patents still have an essential role to play, there is some evidence that, at least in advanced technology, they have become relatively less important. This is perhaps related to the diminishing significance of the individual inventor and the small firm. The success of many processes depends less upon availability of precise directions for carrying them out (which in any case patent specifications rarely provide) as upon what is expressively called know-how. Recipes for the great dishes of the world are freely available, but only a master chef with the proper materials and facilities can make them to perfection. An additional factor in a trend towards protection by secrecy and know-how was the rising cost of patents. Although as early as 1883 an international convention had been agreed facilitating the patenting of an invention in several countries, the cost was still considerable, especially as, with increasing complexity, a whole series of patents might need to be filed. The cost of patenting, like that of insurance, had to be carefully watched lest the price of premiums was disproportionate to the risk. In this context, however, we must recall that know-how is often as much a marketable commodity as a patent.

By 1900 industry was inured to a mass of regulating legislation but up to that time the emphasis was basically economic. Much national revenue accrued from taxation on such varied products as alcoholic drinks, tobacco, and matches. Price regulation, and dividend control, as in the coal-gas industry for example, were designed to protect the purse of the consumer. In the opening years of this century the sliding scale was incorporated in the Acts governing many of the British gas companies. This provided that dividends might exceed 10 per cent if the price of gas was reduced or, conversely, could be reduced if the price was increased.

More often, however, government control was directed at protecting the consumer not only from commercial exploitation but also from the possible harmful effects of industrial products. By mid-century, and particularly in the following decades, consumerism had reached levels that seriously affected industrial innovation. The manufacturer had not only to satisfy many requirements, which might vary substantially from country to country but also to face severe penalties for adverse qualities of his product that might elude even the most conscientious pre-marketing evaluation. Perhaps the industries most seriously affected were the chemical and the very closely allied pharmaceutical industries. That early synthetic insecticides of the chlorinated hydrocarbon type had unforeseen toxic properties, magnified by careless use, is indisputable. Nevertheless, the number of human deaths attributable to these agents was minute compared with those saved by their use, whether by the control of food-destroying pests or of the insect vectors of disease, such as mosquitoes, tsetse fly, and lice. We cannot afford to ignore the fact that in 1971 the World Health Organization reported that three-quarters of the estimated 1800 million people living in formerly malarious areas of the world then lived in areas where the disease had been virtually eliminated by eradication of mosquitoes.

Undoubtedly a substantial part of this success can be attributed to another aspect of technology, namely land drainage schemes, but it is a matter of fact that tens of millions of people are alive today who would otherwise be dead, and tens of millions more have been saved from lasting ill health. Paradoxically, many of these eradication programmes were carried out by the same governments as were in due course to condemn the use of the very agents they so successfully employed. That more stringent regulations should have gone hand in hand with advancing knowledge was as inevitable as it was desirable, but many of the environmentalists who exerted a growing influence in the years after the end of the Second World War gave insufficient thought to the consequences of the restrictions they advocated so passionately. As so often in history, governments tended to over-react to popular emotional demands. In the 1970s more pragmatic decisions had to be taken, as when the statutory elimination of lead from petrol – undoubtedly harmful and a serious cause of atmospheric pollution, especially in urban areas – had to be postponed in face of the energy crisis and the need to extract the maximum value from every gallon of petrol burned.

Technology is commonly regarded as being synonymous with applied science, but this is not so. Many technologies were highly developed long before science as we now know it was conceived: textiles, pottery, and shipbuilding are important and familiar examples. Nevertheless, by 1900 technology had reached a stage of development at which the main road to progress was by the application of science. To some extent this was possible simply by utilizing the accumulated backlog of knowledge, but more and more it became necessary to acquire new knowledge by research. Governments, therefore, increasingly directed their attention to this, though they did so in different ways and with different attitudes of mind. At one extreme was the somewhat naïve assumption that to support research was a good thing *per se*, because useful applications would follow as a matter of course, even though their nature might be unpredictable. At the other was the hard-headed belief that expenditure on research – or at least public expenditure – had to be directed towards specific objectives in order to justify it. In practice, neither argument is tenable. Modern research is very heavily dependent on the products of modern technology, as a visit to any laboratory will effectively demonstrate. Equally, practical objectives are not worth pursuing until the basic knowledge is either available or can confidently be expected to be made available. This basic truism is well illustrated by the Manhattan Project and by the space-flight projects that culminated in the moon landings. Each started with recognition that there were major problems to be solved and that their solution was not at all clear: nevertheless, the state of science and technology was such that the problems appeared soluble provided sufficient resources were available and the right general approach was adopted. The last is an important qualification. In space research, for example, a moon landing based on combustion rocketry was clearly feasible. On the other hand, a landing based on the development of some kind of anti-gravity device held out no promise whatever of success: the present state of knowledge indicates that such a device – if it is, indeed, ultimately realizable at all – could not be produced now, however much effort and money were spent.

In allocating money for research, governments necessarily had to compromise, and give support to both pure and applied work. In our consideration of education we have already seen one way in which this was done: research is possible only when there is a sufficiency of trained people to carry it out. Even in countries, such as Britain, in which there was a considerable private education sector, vast and increasing sums were spent on education and, as we have seen, the tendency was to introduce more science and technology. In this sense, therefore, governments supported research by improving education at the secondary and tertiary levels. In countries such as Germany, which had founded technological universities, a relatively high proportion of funds allocated to education went to the support of research in science and technology.

In most countries the universities were the chief centres of research but, even when they were not unduly addicted to fundamental work selected without regard to its potential usefulness, there was need for direct government research in other ways. Much research is relevant to industry as a whole, or at least to a large section of it, but it is not of such specific interest to any particular firm as to command its support. As the adage goes – what is everybody's business is nobody's business. A typical example is corrosion: worldwide, damage caused by corrosion runs into thousands of millions of pounds annually but it afflicts everybody, so that the individual burden is small. Another is the maintenance of standards, of length, mass, and time so that all work is referred to the same base-lines.

For such reasons, governments have long encouraged institutions for research providing general support for industry or for the country as a whole. An early example was the Royal Observatory at Greenwich, founded in 1675 for the benefit of navigation in general. From about the turn of this century, however, these responsibilities were taken more seriously. In Britain the office of Government Chemist was created in 1894, primarily to control analytical procedures necessary to the enforcement of a growing volume of technical legislation, much of it concerned with the collection of revenue. The next decade saw the foundation of the National Physical Laboratory (1900) in England, the National Bureau of Standards (1901) in the USA, and the Kaiser Wilhelm (later Max Planck) Institute (1910) in Germany. Subsequently, various countries established major research organizations for research and development work in co-operation with industry: such, for example, was the Department of Scientific and Industrial Research in Britain (1917), which spawned a series of Research Associations identified with the major industries, and the TNO organiz-ation (Nederlandse Centrale Organisatie voor Toegepast Natuur, 1932) in the Netherlands. In the United States, with its predilection for private enterprise, a number of independently incorporated research establish-ments were founded whose profits were ploughed back to improve their facilities and research activities. The first of these was the Mellon Institute (1913) in Pittsburgh; the Battelle Memorial Institute was founded in 1925 and after the Second World War (1952) it opened outstations in Europe at

Geneva and Frankfurt to do sponsored research for industry. Although independent, such institutes depended so heavily on government contracts ($224 million in 1969) that they must be regarded, for practical purposes, as a form of government support for research and development.

In various countries, governments sought to establish 'centres of excellence' to encourage particular fields of research regarded as of special national importance. These could not be expected to arise spontaneously. In Britain they were established through such agencies as the Medical Research Council (1920) and the Agricultural Research Council (1931), which had their own research establishments and also sponsored research in appropriate university departments. In the United States, Federal Contract Research Centers (later Federally Funded Research and Development Centers) were established shortly after the Second World War. They were administered by universities; industrial consortia; or non-profit research and development institutes of the kind already described.

Space permits no more than a cursory description of the complex web of institutes and associations by which governments involved themselves in research and development activities. As we have seen, this involvement was well established at the beginning of the century and became steadily more elaborate as the years passed. The Second World War, and its aftermath, with its demand for the rapid development of new technologies of strategic importance – as in atomic energy and the aerospace industries –, accentuated the process and led to the creation of many more highly specialized government laboratories. Examples are to be found at Harwell in Britain; at Saclay and Marcoule in France; and at Houston in the United States. These fulfilled their required functions well but later presented problems which proved very intractable. Briefly, what is to be done with very large, very specialized research establishments when their main purpose has been wholly or largely achieved? For the answer we must turn to the writings of future historians.

IV. INNOVATION

To the historian, whatever his persuasion, static situations are of little interest, save perhaps in seeking for the reasons for stagnation. It is change and its causes that really concern him. In the history of technology the symptom of change is innovation, the introduction of new products or processes, and this introductory chapter can appropriately be concluded with some consideration of innovation and its causes, with particular reference to the present century. In earlier sections we have considered some of the relevant factors – changes in the educational system, changes in the

organization of industry, and increasing intervention by government – but
these relate to the general milieu of innovation and we have not considered
the underlying causes.

We must begin by making it clear that this is not a subject on which there
is a clear consensus of opinion, for it lends itself to subjective rather than
strictly logical reasoning. To ask what makes a man invent is akin to asking
somebody what made him a creative painter or writer. Very often, he does
not know himself: how, then, can others tell him? The best we can hope to do
is to identify the sort of milieu that seems conducive to innovation, bearing
in mind that this line of inquiry may sometimes be sterile and also that
inventions may be made in widely differing circumstances, including some
that are seemingly unfavourable.

As a first step we can classify innovation in various ways, beginning with
radical and minor innovation. In the first category we may include wholly
new products not directly descended from existing ones: as examples within
our period we may cite the thermionic valve and the transistor, the powered
aeroplane, colour photography, xerography, and atomic energy. In the
same category are novel processes for making existing products: typical
examples are low-pressure processes for making polyethylene, freeze-drying
of food, and the catalytic cracking of petroleum. In this context we must, of
course, take a sensible view of what we mean by direct descent. Purists could
argue that nothing arises spontaneously, that the aeroplane was implicit in
George Cayley's experiments with gliders in the first half of the nineteenth
century or even in the drawings of Leonardo; that the transistor was no more
than an obvious development of the cat's whisker detector in early crystal
radio receivers. Such tenuous threads exist in innumerable instances: our
concern here must be with developments that triggered off major new
industries or led to major changes in existing ones. Many inventions fail to
be accepted because they are, as it were, ahead of their time: there may be
insufficient demand for them, their price may be unacceptable, or current
technology may be inadequate for their realization. Re-invention is a
common occurrence in technology.

Striking developments are, however, not necessarily the most important;
minor improvements in existing products and processes may be economically no less rewarding. For example, a new catalyst that improves the yield
of a chemical process by as little as 1 per cent can earn £1 million per year
if – as is by no means uncommon – annual turnover is £100 million. The
innovator can, therefore, very adequately earn his keep even if the real
block-busters of new technology elude him.

Various studies have been made of the sources of new products and new
processes. On balance, the tendency is for new products to originate outside

the industries concerned and for new processes to be internally generated. This is not altogether surprising. Existing firms on the whole prefer to expand by steadily enlarging their frontiers rather than by jumping into new territory, and they tend to see all too clearly difficulties of which the outsider is happily unaware. Process improvement, by contrast, is a natural target for existing firms, who already have a great deal of background knowledge and experience not easily available to outsiders, who are, therefore, at a disadvantage. As an example of this general thesis we may cite the experience of du Pont, a leading American chemical company, between 1920 and 1949. During that period their own staff were responsible for five out of seven major innovations in manufacturing processes, but for only five out of eighteen new products.

We must recognize that invention is, in more senses than one, very far removed from marketing a new product or putting a new process into operation. Unless circumstances are exceptional – as with the Manhattan Project, where expenditure and the commitment of resources took little account of economics – the time-lag between invention and marketing is commonly around ten years. During this period much money has to be spent without any return at all (negative cash flow, in the language of economists) and with the clear recognition that, in the end, success may not be achieved, and the expenditure will have to be written off. An example is provided by the development of catalytic cracking processes which became very important in the petroleum industry from the 1920s onward. One of the most successful of these, the Houdry process, was developed by the French engineer E. J. Houdry over the period 1925–36 (illustrating the ten-year lag) and cost the consortium of companies concerned $11 million. The later fluid process (1938–41) demanded a capital outlay of over $30 million.

Expenditure of this scale is possible only for large firms or – in the most costly projects, such as Manhattan – governments. It has been argued that this has been a major factor in the growth of large industries during this century. While accepting this argument, we must recognize that many other factors have encouraged the concentration of industry. Mergers may, for example, be arranged for no worthier reason than defence against more progressive rivals, as when the United Alkali Company was formed in Britain in 1890 as a consortium of some forty-five Leblanc soda firms threatened by the basically more efficient Solvay process: despite technological inferiority, they contrived to hold the fort until absorbed into ICI when that combine was formed in 1926. Other mergers were motivated by a desire to acquire a rival's assets or cash, or both.

In a free economy governed by market forces the normal incentive to innovation is, of course, to gain a temporary advantage over rivals but innovation *per se* is not necessarily profitable. In the case of governments the advantage sought may not be financial at all but strategic, as with such innovations as radar. As we have noted, profitable successes have been matched by expensive failures. Moreover, the innovating firm does not necessarily score a clear advantage over rivals who buy their way into the field later in the day when teething troubles have been overcome. For example, the first firms to manufacture penicillin in Britain did not find it a profitable venture, because their technology was quickly superseded.

Finally, we must direct our attention to innovators themselves. The supposition that personal profit is a prime incentive is not tenable. There were, of course, innovators such as T. A. Edison and Henry Ford who made no bones about their regard for financial returns. But they can be contrasted with their pioneers with no such motivation, such as J. K. Northrop whose chief objective seems to have been to make increasingly efficient aircraft and who finally spent himself on an unsuccessful attempt to develop an aircraft consisting entirely of wing, with no fuselage. Other aircraft pioneers tacitly demonstrated their detachment from the profit motive by literally risking – and sometimes losing – their lives in their own enterprises long before there was any substantial market for aeroplanes. This market did not develop until the Second World War was well under way.

The same conclusion must be drawn if we look at innovation within the great international companies that came increasingly to dominate the industrial scene during this century. Although these concerns often had to turn to the outside world for novel products, and to some extent for novel processes, their total innovative capacity became formidable. Yet this was provided by executives who knew full well that in even the most liberal companies their reward would be insignificant in relation to the resulting profits. The manufacture of man-made fibres, for example, is now an enormous new business, largely developed after the Second World War, with powerful influences on the development of the giant chemical and textile industries. Yet the principal inventors – W. H. Carothers (nylon) and J. R. Whinfield and J. T. Dickson (Terylene/Dacron) – gained neither personal profit nor even any significant public acclaim. Though half the world goes clad in man-made fibres, few would even know their names. Carothers was elected to the National Academy of Sciences in the United States – the first industrial organic chemist to be so honoured – but the Royal Society in London, though founded in the seventeenth century to 'promote . . . the science of natural things and of useful arts', totally ignored Terylene, one of the world's great inventions. As we have remarked, in Britain it was the pure scientist who was esteemed; a hint of usefulness in a man's work made him an object of suspicion in intellectual circles. This attitude has proved, and still proves, a grave national weakness.

As for monetary reward, the position of the individual inventor in industry is undeniably ambivalent. On the one hand, it is arguable that a man whose genius demonstrably makes large profits for his company (and its shareholders) deserves not only recognition but some share of the proceeds. On the other, it can be said that he is no more than a member of a team, making and developing his invention in a milieu created by his fellow workers. To hold out the promise of individual gain would encourage secrecy and destroy the very basis of the teamwork on which progress now increasingly depends. Nevertheless, there have been moves to make it obligatory for employers to reward the individual inventor, but these largely fall outside the period of our present concern. In the first half of the twentieth century many major innovations were made by men who neither sought nor expected greater reward than their fellows and in many cases devoted themselves a great deal more assiduously to their work, sometimes undermining their health in the process. Clearly, much innovation was, and is, the outward manifestation of an inward compulsion.

BIBLIOGRAPHY

Aitken, H. G. J. *Taylorism at Watertown Arsenal: scientific management in action, 1908–1915*. Harvard University Press, Cambridge, Mass. (1960).

Allen, J. A. *Studies in innovation in the steel and chemical industries*. Manchester University Press, Manchester (1967).

Anderman, S. D. (ed.) *Trade unions and technological change*. Allen and Unwin, London (1967).

Argles, M. *South Kensington to Robbins: an account of English technical and scientific education since 1851*. Longmans, Green and Co., London (1964).

Argyle, M. *The social psychology of work*. Penguin, Harmondsworth (1972).

Bain, J. O. E. S. *Industrial organization*. Wiley, New York (1968).

Bennett, J. W. and Ishino, I. *Paternalism in the Japanese economy*. University of Minnesota Press, Minneapolis (1963).

Bowe, C. (ed.). *Industrial efficiency and the role of government*. HMSO, London (1977).

Boyle, S. E. *Industrial organization*. Holt Rinehart and Winston., New York (1972).

Bright, J. R. *Automation and management*. Harvard Graduate School of Business Studies (1958).

Burns, T. and Stalker, G. M. *The management of innovation*. Tavistock London (1961).

Bush, V. *Science, the endless frontier: a report to the President on a program for post-war scientific research*. US Government Printing Office, Washington, DC (1945).

Cochrane, Raymond C. *The National Academy of Sciences: the first hundred years, 1863–1963*. National Academy of Sciences, New York (1978).

Cofer, C. N. and Appley, M. H. *Motivation: theory and research*. Wiley, New York (1964).

Copley, F. B. *Frederick Winslow Taylor, father of scientific management*. 2 vols. Harper, New York (1923).

Cotgrove, S. and Box, S. *Science, industry and society: studies in the sociology of science*. Allen and Unwin, London (1970).

Denison, E. F. *The sources of economic growth in the United States and the alternatives*. Report to the US Committee for Economic Development, New York (1962).

De Witt, N. *Eduction and professional employment in the U.S.S.R* National Science Foundation, Washington, DC (1961).

Ellul, J. *The technological society*. A. K. Knopf, New York (1964).

Evely, R. and Little, I.M.D. *Concentration in British industry*. Cambridge University Press, Cambridge (1960).

Freeman, C. *The economics of industrial innovation*. Penguin, Harmondsworth (1974).

Gabor, Dennis. *Innovations: scientific, technological and social*. Oxford University Press, New York (1970).

George, C. S., Jr. *The history of management thought*. (2nd edn.) Prentice-Hall, Englewood Cliffs, NJ (1963).

Giedion, S. *Mechanization takes command*. Oxford University Press, New York (1948).

Herzberg, F., Mausner, B., and Snyderman, B. B. *The motivation to work*. Wiley, New York (1959).

Hoxie, R. F. *Scientific management and labor*. Appleton, New York and London (1915).

Jewkes, J., Sawers, D., Stillerman, R. *The sources of invention* (2nd edn.) Macmillan London (1968).

Kakar, S. *Frederick Taylor: a study in personality and innovation*. MIT Press, Cambridge, Mass. (1970).

Korol, Alexander G. *Soviet education for science and technology*. Chapman and Hall, London (1957).

Kranzberg, Melvin and Pursell, Carroll, W., Jr. *Technology in Western civilization: Vol. II, the twentieth century*. Oxford University Press, New York (1967).

Langrish, J. *et al. Wealth from knowledge*. Macmillan, London (1972).

McCrensky, Edward. *Scientific manpower in Europe*. Pergamon Press, London (1958).

Mann, F. C. and Hoffman, L. R. *Automation and the worker*. Holt, Rinehart, and Winston, New York (1960).

Mansfield, E. *The economics of technological change*. Norton, New York (1968).

Marquand, H. M. *et al. Organised labour in four continents*. Longmans, London (1939).

Mayo, E. *The human problems of an industrial civilization*. Macmillan, New York (1933).

Mortimer, J. E. *Trade Unions and technological change*. Clarendon Press, Oxford (1971).

Mumford, Lewis. *Technics and civilization*. Harcourt, New York (1934).
    The myth of the machine. Secker and Warburg, London (1967).

Nadworny, J. *Scientific management and the unions, 1900–1932: a historical analysis*. Harvard University Press, Cambridge, Mass. (1955).

Nakayama, Shigeru, Swain, D. L., and Yagi Eri (ed.). *Science and society in modern Japan*. MIT Press, Cambridge, Mass. (1974).

Nelson, D. *Managers and workers: origins of the New Factory System in the United States 1880–1920*. University of Wisconsin Press, Madison (1975).

Nelson, R. R. (ed.). *The rate and direction of inventive activity*. Princeton University Press, Princeton, NJ (1962).

—Peck, M. J., and Kalachek, E. D. *Technology, economic growth and public policy*. The Brookings Institution, Washington, DC (1967).

OECD. *The conditions for success in technological innovation*. Paris (1971).

Penrose, E. *The theory of the growth of the firm*. Blackwell, Oxford (1966).

Pollard, S. *The genesis of modern management*. Arnold, London (1965).

Porter, G. *The rise of big business, 1860–1910*. Crowell, New York (1973).

Reader, W. J. *Imperial Chemical Industries – A history*. 2 vols. Oxford University Press, London (1970, 1975).

Roethlisberger, F. J. and Dickson, W. J. *Management and the worker*. Harvard University Press, Cambridge, Mass. (1949).

Rosenberg, N. *The economics of technological change*. Penguin, Harmondsworth (1971).

Salter, W. E. G. *Productivity and technical change*. Cambridge University Press, Cambridge (1969).

Sanderson, M. *The universities and British industry: 1850–1970*. Routledge and Kegan Paul, London (1972).

Schumpeter, J. *The theory of economic development*. Oxford University Press, London (1961).

Scott, W. H. *et al*. *Technical change and industrial relations*. Liverpool University Press, Liverpool (1956).

Slichter, S. H. *Union policies and industrial management*. The Brookings Institution, Washington, DC (1941).

Taylor, F. W. *Scientific management*. Harper, New York (1947).

Taylor James. *The scientific community*. Clarendon Press, Oxford (1973).

Townsend, H. *Scale, innovation, merger and monopoly*. Pergamon Press, Oxford (1968).

Vroom, V. H. *Work and motivation*. Wiley, New York (1964).

Walker, C. R. *Toward the automatic factory: a case study of men and machines*. Yale University Press, New Haven, Conn. (1957).

Whyte, W. F. *Money and motivation*. Harper, New York (1955).

Williamson, H. F. (ed.). *Evolution of international management structures*. University of Delaware Press, Newark (1975).

Zweig, F. *Productivity and trade unions*. Basil Blackwell, Oxford (1951).

2

THE CHANGING PATTERN OF ENERGY SOURCES

Changes in the source and utilization of energy again reveal the first half of the twentieth century as a period of transition. In 1900 industrial and domestic demands for energy were growing rapidly, but lack of energy was not seen as any bar to technological advance. There were vast reserves of coal cheaply available, and the petroleum industry, though still in its infancy, clearly represented a major new source of fossil fuel. Coal had long been the basis of a flourishing and expanding gas industry. Electricity represented a new and flexible kind of energy whose potential value was already becoming clear. Although its production depended largely on generating steam in coal-fired boilers – so that it was an indirect way of using fossil fuel rather than a new source of energy –, by the 1890s the Niagara project had demonstrated the possibilities of hydroelectric power in favourable situations. While there would inevitably be problems of continuously balancing supply and demand, there seemed no reason to doubt that, overall, ample energy would be available for all industrial developments.

By mid-century the situation was rather different. Coal was no longer pre-eminent, not because of an absolute decline but because of the very rapid growth of alternative energy sources, notably petroleum. World coal production (in millions of tons) was 1300 in 1913; 1200 in 1928; 1100 in 1938; and 1100 in 1946. In contrast to this virtually static position, petroleum production multiplied many times over. In 1900, world production, largely in the USA, was a mere 150 million barrels. In 1919 it was 550 million barrels; thereafter production had doubled by 1926, and doubled again by 1937. By 1950 it had soared to more than 4000 million barrels, equivalent (in weight) to more than half the world's coal production. A great part of this vast increase represented, of course, the demands of road vehicles and, later, the aeroplane. Shipping, too, increasingly switched from coal to oil. At the same time, there had been a great expansion, again largely in the USA, in the use of natural gas as a fuel. This was destined, in due course, to oust gas manufactured from coal.

Development in electricity was comparable. In Britain, for example, consumption rose from 125 million units in 1900 to 35 000 million in 1946. The world picture was similar: from modest beginnings, global production had risen to over 800 000 million Kwh in 1950 – compared with about 380 000 million pre-war – with the USA by far the largest single consumer. Three-fifths of this electricity was generated by burning fossil fuel and virtually all the remainder was hydroelectricity. These figures are important, in that consumption of energy has long been regarded as a very significant index of industrial activity.

As we have noted, although electricity was an important new form of energy it had its origin in conventional sources – fossil fuels or, in the case of hydroelectricity, sophisticated versions of the ancient water-wheel. Not until after the Second World War did a wholly new source of energy appear. This was atomic energy, first developed, in the wartime Manhattan Project, for military purposes. Only after the war was attention directed to nuclear power production, and interest was then greatest in Britain and France. In the USA, with seemingly ample and cheap reserves of petroleum, natural gas, and coal, attention was directed mainly to propulsion units for submarines and large sea-going vessels, the main attraction being the possibility of remaining at sea for long periods, without need for refuelling. In Britain, research and development was directed primarily to commercial power-producing reactors, and the first nuclear power station, at Calder Hall, was commissioned in the autumn of 1956. From that time onwards, nuclear power was to make a steadily growing, but not unopposed, contribution to the energy needs of the western world.

By the 1950s, however, there was already widespread concern at the rapid depletion of fossil fuels. It had, of course, always been clear that as these are now being generated in nature at a negligible rate – if at all – the supply must as a matter of arithmetic ultimately be exhausted. Hitherto, however, new fields had been discovered so steadily that it seemed as though, for practical purposes, the world could carry on as though the supply were limitless. This complacency was shaken by the Suez crisis of 1956, and earlier words of warning began to be heeded. The consequences of the short interruption of supply from the Middle East – petrol rationing was briefly introduced in Britain in December – brought home to the public their dependence on petroleum to sustain their way of life. It reminded them, too, that the mere existence of vast petroleum deposits was not sufficient in itself; access to them was also essential. The strategic significance of petroleum was further emphasized by the massive OPEC price increases in 1973. The resulting energy crisis had far-reaching consequences, of which the most important were measures to conserve fuel and the direction of increased attention to the development of other sources of energy.

This was a conventional reaction to shortage and a heavy price increase in an essential commodity: the customer must temporarily accept the situation but extricates himself from it as best and as quickly as he can, by developing alternative sources of supply. In this instance, the sheer magnitude of the western world's dependence on Middle East oil precluded any rapid solution but a new energy strategy began to emerge. It had three main features. Firstly, more attention was paid to the exploitation of petroleum and natural-gas fields outside the OPEC area; apart from new fields, the sharp increase in the price of oil made it feasible to reopen existing fields that had become uneconomic. Two major developments were the Alaska oilfields and those of the North Sea. Secondly, there was renewed interest in coal both as a raw fuel and as a source of gas made by processes much more efficient than the old method of carbonization. Thirdly, other natural sources of power – the sun, tides, wind, the heat found deep in the earth, and so on – were looked at afresh as a source of electrical, rather than mechanical, energy. These had, in fact, long been exploited in a small way where local circumstances were particularly favourable, but in the new economic climate their utilization on a large scale began to look more attractive.

Before we consider the way in which the technology of natural sources of energy developed, one fundamental point deserves consideration. This is that the demand for energy is not constant. The domestic demand, for example, is less in the long warm days of summer than in winter; the industrial demand is highest during normal working hours. Ideally, the power industry ought to be able to adapt itself to these variations and satisfy the consumers fully at all times, but in practice this is not possible, for a variety of reasons. Most importantly, if capacity is sufficient to meet the maximum predictable demand – such as might arise exceptionally, as in Europe in 1947, as the result of an unusually long spell of exceptionally cold and severe weather in winter – then in normal times there must be excess capacity standing idle, which is uneconomic. Even if total production capacity is sufficient, there may be problems of distribution. Part of the cause of Britain's fuel crisis in 1947, for example, was that coal stocks were frozen solid and the icy conditions made transport of what little could be broken out very difficult. Moreover, while fuel – such as oil or coal – can be stockpiled against a time of need, this is not normally true of energy, which must be generated as it is needed. A surge in demand for electricity, for example, must be almost instantly reflected in increased production in the power-stations. If this cannot be effected, power cuts and all their consequences are inevitable. There are, of course, exceptions to this generalization about availability, but they do not affect its basic truth. Thus electrical energy can be stored, in chemical form, in an accumulator; mechanical energy can be stored in flywheels; surplus electricity can be used to pump water up to storage reservoirs to be released to drive turbo-generators when demand increases; night-storage heaters enable electricity available at cheap night rates to be used during the day. Nevertheless, one of the unsolved technological problems of this century is that of storing energy when it is abundant, against a time of need. Vast amounts of precious energy continuously go to waste simply because there is no practicable means of storing it; an example is the huge quantity of natural gas flared off at oilfields.

Finally, if we are to understand the basic problems of power technology, we must consider the unavoidably high levels of wastage that prevail. While energy is indestructible, in that if it disappears in one form an exactly equivalent amount appears in another, energy conversion is potentially wasteful. Thus even at mid-century the best of Britain's power-stations could convert only about 40 per cent of the energy in coal into electricity; some of the older ones could do no better than 30 per cent. The remaining 60 – 70 per cent of energy, in the form of heat, was simply wasted at source and more was lost in transmission of electricity to the consumer. As the century progressed, and particularly after the First World War, the mounting cost of fuel encouraged the use of more efficient appliances and greater care in their maintenance, and the OPEC crisis naturally intensified this: more effective use of available supplies is, of course, equivalent to increasing the supply at the existing level of efficiency. The 1970s saw moves, for example, towards industrial units in which surplus heat from power-stations is used on the spot for complementary manufacturing purposes.

The new petrol engine, which came to dominate road transport, was not conspicuously efficient. At best its efficiency was raised to around 25 per cent, though diesel engines, used mainly for heavy vehicles, and the massive power units of ships and locomotives, might attain 40 per cent. Nevertheless, in America, with huge native reserves of cheap petroleum, the emphasis was on the development of large cars – demanding roads and other facilities to scale – with heavy fuel consumption. In Europe, where fuel had to be imported and traffic was more congested, for geographical reasons, smaller cars were popular. This exemplifies the influence of economic and social factors on the development of technology. The same field can be used to illustrate the impact of another factor – the conservation of the environment – that was to become increasingly important during this century. In 1921 it was discovered that the efficiency of petrol engines could be substantially increased by addition of a little lead tetraethyl to the fuel as an 'antiknock' agent. This was hailed as a major technological advance, which indeed it was, but only half a century later many countries were contemplating legislation to forbid its use on the ground of atmospheric pollution by lead.

In conclusion, it is interesting to note that the heavy inroads into resources of fossil fuels; the exhaustion of the most easily accessible deposits; and the sharp increase in price all seemed to point inexorably towards increasing dependence on nuclear power. As we shall see, this presented very great technological difficulties but they were matched by comparable sociological ones. Doubtless because of its fearful military potential, nuclear power, even in a wholly peaceful context, became the centre of controversy so highly emotive and so well organized that technological progress was severely hindered.

BIBLIOGRAPHY

Brame, J. S. S. and King, J. G. *Fuel: solid, liquid and gaseous*. (6th edn.) St. Martin's Press, New York (1967). (Also earlier editions: Brame, 1914, 1917, 1924; Brame and King 1935, 1955.)

Crabbe, David and McBride, Richard. *The world energy book: an A-Z, atlas, and statistical source book*. Kogan Page, London (1978).

Forrester, C. (ed.) *The efficient use of fuel*. HMSO, London (1958).

Jensen, W. G. *Energy in Europe 1945–1980*. Foulis, London (1967).

Little, I. M. D. *The price of fuel*. Clarendon Press, Oxford (1953).

Parker, A. *World energy resources and their utilization*. World Power Conference Statistical Year Book (1949).

Posner, Michael V. *Fuel policy: a study in applied economics*. Macmillan, London (1973).

World Power Conference Proceedings. London (1924).

3

FOSSIL FUELS: COAL, GAS, AND PETROLEUM

I. COAL

Of the fossil fuels coal, as we have seen, retained a commanding position throughout the first half of this century, though world output after the First World War remained fairly static. If we look at individual countries, however, the situation is rather different: there were considerable differences in the uses to which coal was put and in the economics of its production.

Within the confines of this chapter we must regard coal as an entity, but in fact it shows considerable variations which affect its use. In some countries, such as Germany and Austria, the variety known as lignite was extensively mined. This is a brown coal, with a low calorific value, that is intermediate between peat and the hard black bituminous coals familiar in Britain. Again, not all coals are suitable for the manufacture of metallurgical coke, essential for steel-making. Thus a particular region may be rich in coal, but nevertheless have to import varieties to satisfy local requirements. As we shall see, a similar position pertains in the case of petroleum. Although the North Sea was to make Britain self-sufficient in oil, it did not make her independent of imports. For certain purposes Middle East crudes were necessary, though the cost of these could be offset by export of North Sea oil to countries, such as the USA, where it was in demand.

Coal production during this period was influenced by various factors. Energy consumption, as we remarked in the previous chapter, is an important index of industrial activity. Inevitably, therefore, the demand for coal, as the dominant fuel, reflected the world economic situation. Up to the First World War, with industry expanding generally, the demand rose steadily, from about 750 million tons in 1900 to nearly twice as much in 1914. After the war, growth resumed, but at a very much lower rate – about $\frac{3}{4}$ per cent annually. The business depression of the 1930s cut output by 40 per cent between 1929 and 1932 and it had not entirely recovered even at the outbreak of the Second World War in 1939. Nevertheless, some countries fared better than others. Whereas Britain and the United States lagged, Germany had fully recovered by 1939. These three countries, incidentally, produced 90 per cent of the world's coal in 1900, and in 1950 the figure was still 60 per cent. Russia, by contrast, was a very small producer in 1900 (15 million tons), and was scarcely affected by the depression that afflicted the western world, but by 1950 was the second-largest coal producer in the world. At that time China entered the lists to create a new industrial economy based on coal: her annual production of 40 million tons had increased to 400 million tons in 1960.

Another factor influencing the demand for coal was the adoption of new kinds of power. Electricity and coal gas were not really rivals, because they both required coal for their generation. Petroleum was a different matter, however; while its main use was as a fuel for road transport, a development which was novel to the twentieth century and in which coal had never had a real foothold, oil became increasingly used at sea. In this connection we must distinguish between the use of oil as a substitute for coal to raise steam to drive either reciprocating engines or turbines and its use in internal combustion engines of the diesel type. The first important vessel to be fitted with diesel engines was the British *Selandia* (1912); by 1950 a third of the world's merchant vessels were propelled by diesel engines.

There were, in fact, two aspects to fuel and shipping. On the one hand, growing use of oil for propulsion correspondingly diminished the demand for coal. No less important was the fact that countries dependent on imported fuel turned increasingly to oil which, by the use of tankers, could be more cheaply transported than coal. Thus in the first half of this century countries such as Norway and Argentina halved their coal imports. In Britain, traditionally an exporting country, coal exports fell from 73 million tonnes in 1913 to 13 million tonnes in 1950.

Although fuel economy inspired by exhaustion of resources was not to become a major international preoccupation until the 1970s, in the first half of this century rising costs alone gave a strong incentive to economy, and the use of alternative fuels. In the United States, for example, an over-all increase of one-third in the efficiency of coal utilization was effected between 1909 and 1929. In the British steel industry 25 cwt of coke was required for every ton of pig-iron made in 1900: by 1950 this had been reduced to 13 cwt and in the 1970s the most efficient producers required only 10 cwt.

II. COAL GAS

An important use for coal was in the manufacture of gas by carbonization. By 1900 this was a well-established industry, although its pattern was changing. Originally, the main use of gas had been in lighting, but towards the end of the nineteenth century electric lighting appeared, which was to supersede it almost entirely within a few decades. The invention of the Welsbach mantle in 1885 gave a new lease of life to gas lighting but by the turn of the century gas was being used more and more for heating, both in the house – where the gas cooker lightened work in the kitchen – and in industry. This change vastly enlarged the market for gas: in Britain, for example, demand rose eightfold between 1887 and 1946, although up to then the domestic consumption continued to exceed industrial and commercial combined.

At the beginning of this century most coal gas was made by methods basically little different from those employed nearly a century earlier. Coal was strongly heated in banks of sealed refractory retorts until all gas had been expelled; the residual coke was then mechanically expelled. This coke was a valuable domestic fuel, and its sale was a vital factor in the economics of gas manufacture. This method was a batch process. The introduction of vertical retorts, which required very much less ground space, was linked with methods of continuous operation, with its attendant advantages. Disadvantages were a much higher capital cost and greater maintenance problems. Nevertheless by 1934, 60 per cent of the gas manufactured in Britain came from vertical retorts.

Gas coke could not be used for smelting purposes in the iron and steel industries, because its mechanical properties were unsuitable. These industries, therefore, made their own coke in coke ovens, which simultaneously generated a gas almost identical with that produced by the gas industry. In theory, therefore, gas companies in the vicinity of coke-oven installations could utilize coke-oven gas; this could be purchased on advantageous terms because, so far as the iron and steel industry was concerned, it was largely a waste product. In practice, however, there were drawbacks. Firstly, the production of coke-oven gas rose and fell with the demand for iron and steel: the supply was, therefore, not reliable and the gas industry was obliged to have stand-by plant of its own. Secondly, coke-oven gas did not provide the by-products – notably domestic coke, tar, and ammonia – which were important for the economic viability of the gas industry. For these reasons, and because the long-distance transmission of manufactured gas was not economically attractive during the greater part of the period of our present concern – except between large sources of supply and areas of concentrated demand – less use was made of coke-oven gas than might be expected. In Britain, for example, it amounted to only 7 per cent of total gas supplied even in 1935, and most of this was in the Sheffield area. Elsewhere, however, the situation was somewhat different. In Germany, for example, the iron and steel industry of the Ruhr had been systematically reorganized on the basis that surplus coke-oven gas would be largely absorbed by gas undertakings. The very large quantities of gas produced within a relatively small area made it feasible to pump the gas much greater distances than could be contemplated in Britain. In the 1930s economic transmission up to 35 miles was practicable only for quantities up to 25 million cubic feet daily; this approximated to the consumption of a city the size of Glasgow. In the Ruhr, quantities ten times greater than this were in question.

While coke was an important fuel in its own right, it could also be used to provide, at short notice, additional supplies of gas. If steam is blown through red-hot coke a combustible mixture of carbon monoxide and hydrogen is formed: this is known as blue water-gas, because it burns with a pale blue flame. The steam, however, rapidly cools the coke, but this can be made incandescent again by blowing air through it: this causes production of producer gas, a mixture consisting largely of inert nitrogen (from the air) and carbon monoxide. The process can be kept going by blowing steam and air alternately or by continuously blowing through an appropriate mixture of air and steam, to give semi-water-gas. Although all these gases burn readily, their calorific value is much lower than that of the coal gas to which, therefore, they could be added in only small proportions. This defect was overcome by treating water-gas with vaporized oil in a carburetter: the carburetted water-gas has a much higher calorific value. Although it began to be manufactured in the nineteenth century – France (1865), United States (1873), Britain (1890) – carburetted water-gas made no great contribution to total production until the twentieth century. Apart from other considerations, its high content of toxic carbon monoxide presented problems. By the 1930s, however, many British gasworks generated around one-third of their total make in this form.

In all the above processes the whole of the coke was eventually turned into burnable gas: that is to say, the coal had been totally gasified except for its mineral content (ash). Not surprisingly, attempts were made to achieve this result in a single process. Of these, the most important – though not the first – was the Lurgi process, developed in Germany in the 1930s. In this, low-grade brown coal – as opposed to coke – was gasified by a mixture of superheated steam and oxygen at high pressure (20–30 atmospheres).

Although the Lurgi process made it possible to use low-grade fuel unsuitable for carbonization and substantially reduced transmission costs by generating gas at high pressure, it had serious disadvantages. In particular, it needed a great deal of steam – which produced correspondingly large volumes of dilute waste products, difficult to dispose of – and the oxygen required was expensive. Nevertheless, the Lurgi process did have some operational success in Germany and elsewhere: for example, a plant with a daily capacity of 40 million cubic feet of gas was operated in Scotland from 1960 to 1974.

By the middle of the century, supplies of coals suitable for gas-making were dwindling and their price was becoming prohibitive. At the same time, however, the rapidly expanding petroleum-refining industry could offer, at competitive prices, oil fractions suitable for gasification and capable of producing gas similar to that made by traditional processes of coal carbonization. Typical of such processes was the catalytic reforming process – originally developed by ICI in Britain to make synthesis gas for ammonia manufacture – and the catalytic rich gas process developed by the British Gas Council at Solihull. In the USA, too, plants to generate high-pressure gas from oil were developed after the Second World War, mainly to supplement supplies of natural gas at times of peak demand.

III. NATURAL GAS

In America the gas industry developed very differently from that in Europe. Initially, it was quick to adopt the coal carbonization process: the streets of Baltimore were lit by coal gas as early as 1816. But in areas where natural gas was available – it was often encountered at quite shallow levels in the search for petroleum – it was piped away for domestic and industrial use. Although attempts to utilize natural gas in America can be traced back at least to 1821, the first commercial success seems to have been at Titusville in 1873. Thereafter, the natural-gas industry expanded enormously and transmission over long distances became commonplace. Between 1935 and 1950 sales of natural gas nearly quadrupled and represented more than 90 per cent of total gas sales. In 1955 one new gas well was said to be being brought into production every 23 minutes. At the same time there were 115 000 miles of transmission mains to convey natural gas to gas companies which, of course, had their own distribution system for their consumers. An outstanding technological achievement was that of the Trans-Continental Gas Pipeline Corporation, which in 1951 completed a 30-inch pipeline 1840 miles long to transport gas from the coast of the Texas–Louisiana Gulf to the fuel-hungry areas around New York, New Jersey, and Philadelphia.

Natural-gas exploitation in Europe is essentially a post-war development and thus falls largely outside the scope of the present work. Nevertheless, it is too important to dismiss without brief mention. In Russia, the Nobel Brothers had established their Naphtha Company in 1878 to exploit oil-fields around Baku in Azerbaijan, on the Caspian. As late as 1939 this area represented about three-quarters of Russian oil production, and some use was made of associated natural gas. After the Second World War a tremendous expansion took place: in 1950 the gas trunk-line system extended to 2273 kilometres. By 1971 this had increased to 68 000 kilometres and Russian production of natural gas equalled that of the USA in 1955. So much gas was available that Russia could export a surplus not only to the

Eastern Bloc allies but also to western nations such as Italy and West Germany. At the same time, Russia imported gas from Iran and Afghanistan, countries whose resources were not directly accessible to the western world. In France a gas field was discovered at Saint-Marcet, near Toulouse, in 1939 but the much more important field at Lacq was not exploited until 1957. Even Britain was not wholly without hope for the future. In 1937 a well sunk at Eskdale, in Yorkshire, produced 2.5 million cubic feet of gas daily, and two years later the British Petroleum Company struck oil, albeit modestly, at Eakring in Nottinghamshire.

By far the most important and dramatic event in Europe, however, was the discovery of a huge gas field at Slochteren in Holland in the summer of 1959. By 1970 Holland was not only meeting about 15 per cent of her own energy requirements but exporting valuable quantities of gas – possibly too much and too cheaply – to neighbouring countries. The Dutch discovery had important consequences for Britain, for it led to the finding of huge reserves of oil and gas beneath the North Sea: by an accident of geography, Britain – under the terms of the Convention of the Seas, agreed at Geneva in 1958 and ratified in 1964 – acquired the mineral rights of the whole of the western half of the North Sea, an area of some 100 000 square miles. Optimism about its value in terms of natural gas and oil was justified in 1965 with the discovery of the West Sole field, about 40 miles from the mouth of the Humber. In 1967 the first gas was brought ashore and only ten years later Britain had switched entirely from manufactured to natural gas. This was not, however, Britain's first experience of natural gas. As early as 1959 liquid natural gas had been imported experimentally from the USA and in 1963 regular shipments were bought in from Algeria, from which France, too, made large importations. The gas-liquefaction plant at Arzew could deal with 150 million cubic feet daily; it cost £31 million and was the largest industrial project in Africa except for the Aswan Dam. Liquid natural gas became a major commodity, shipped all over the world.

One comment only needs to be added to this necessarily brief summary of the history of the natural-gas industry. There are important chemical and physical differences between gas manufactured by traditional methods and natural gas; these are such that appliances that will burn one are quite unsuitable for the other. It follows, therefore, that if natural gas is to replace manufactured gas, within a given system, not only must the supply be established but a conversion programme must simultaneously be launched to modify or replace every existing gas appliance. In Britain, for example, 35 million appliances – containing in all about 200 million burners – had to be altered on $13\frac{1}{2}$ million premises. Moreover, the conversion had to be done in such a way that the supply to individual consumers was virtually uninterrupted. The British conversion programme, carried out between 1967 and 1977, cost £1000 million (including the value of gas-making plant made prematurely obsolete) and was a remarkable exercise in technological management. Similar conversion programmes were necessary in countries as far apart as Holland, Japan, and Hungary.

IV. PETROLEUM

Consideration of natural gas conveniently brings us to petroleum, for in nature the two often, but by no means always, occur together. This demands our particular consideration because not only is the petroleum industry essentially a twentieth-century phenomenon but it has also become one of the biggest and most powerful in the world. As we noted in the last chapter, the western world became dangerously dependent on it for the maintenance of its way of life. The political consequences of this dependence were enormous.

Not altogether surprisingly, for the two were established in very different circumstances, the shape of the oil industry differed in several important respects from that of coal. Generally speaking, the new industries of the Industrial Revolution established themselves at centres where most, and preferably all, of their principal raw materials were readily available. In Britain, for example, Northwest England became a major centre of chemical industry because of its deposits of salt, limestone, and coal. Of these determinative factors, coal was the most important, and the coalfields of Europe had an important influence on the location of industry. The natural occurrence of petroleum, however, is not normally closely linked to that of coal. The USA, it is true, became the world's largest producer as the twentieth century advanced, but we must remember that this is a vast country with distances comparable with those of the whole of Europe. Moreover, America did not retain her early lead: in 1913 she produced two-thirds of the world's crude oil but by 1950 her output had dwindled to barely half, and this was to be a continuing trend. Her principal rivals were Mexico, South America, and the Middle East, which together were producing one-third of the world's oil in 1950. Moreover, the nature of the principal product had changed: in 1900 it was largely illuminating oil (paraffin/kerosene) whereas in 1950 it was almost entirely fuel oil and petrol, in roughly equal proportions.

The new producing centres were not, however, centres of consumption; these lay in the countries which had grown rich on coal-based industries. There was, therefore, need to create a vast distribution system based on tank-vessels at sea and pipelines and road-tankers on land. These geographical factors, combined with the fact that the oil industry, unlike coal, is capital-intensive, were conducive to the growth of vast international companies such as Standard Oil in the USA, British Petroleum, and Royal Dutch-Shell. The growth of these powerful companies presented major social, economic, and political problems but as the demand grew and easily worked oilfields became exhausted, they alone could deploy the resources necessary for exploration and development. Of these resources, technological knowledge, experience, equipment, and capital were of major importance. Britain's North Sea programme for oil and gas, for example, could be carried out only with the close collaboration of the big international oil companies: without the necessary know-how, even government resources were insufficient.

BIBLIOGRAPHY

Beaton, Kendall. *Enterprise in oil: a history of Shell in the United States.* Appleton-Century-Crofts, New York (1957).

B. P. *Our Industry.* British Petroleum Company, London (1947).

—— *Gas making and natural gas.* British Petroleum Company, London (1972).

Brame, J. S. S. and King, J. G. *Fuel: solid, liquid and gaseous.* St. Martin's Press, New York (1967).

Eavenson, H. N. *The first century and a quarter of the American coal industry.* Pittsburg (1942).

Gerretson, F. C. *History of the Royal Dutch.* 4 vols. Brill, Leiden (1953).

Jensen, W. G. *Energy in Europe 1945-1980.* Foulis, London (1967).

Medici, M. *The natural gas industry.* Newnes-Butterworth, London (1974).

Political and Economic Planning. *The British fuel and power industries.* PEP, London (1947).

Report of a productivity team representing the Britih coal mining industry which visited the United States of America in 1951. Anglo-American Council on Productivity (UK section), London (1951).

Schurr, S. H. and Netschelt, B. C. *Energy in the American Economy, 1850–1975.* Johns Hopkins University Press, Baltimore (1960).

Tiratsoo, E. N. *Natural gas: a study.* Scientific Press, Beaconsfield (1972).

Webber, W. H. Y. *Gas and gas making.* Pitman, London (1919).

— *Town gas and its use.* Archibald Constable, London (1907).

Williams, Trevor I. *A history of the British gas industry.* Oxford University Press, Oxford (1981).

Williamson, H. F., Andreano, R. L., Daum, A. R., and Klose, G. C. *The American petroleum industry: the age of energy 1899–1959.* Northwestern University Press, Evanston (1963).

Wilson, D. Scott, *North Sea heritage: the story of Britain's natural gas.* British Gas, London (1974).

4

NATURAL POWER

In a sense all power is natural, in that it is obtained by harnessing the resources of nature. Coal and oil, for example, have lain buried in the earth's crust for many millions of years. They can be burned to produce heat which in turn can be converted to mechanical energy, as in steam-engines and internal combustion engines. Such engines can, in turn, be used to generate electricity in accordance with well-understood natural laws. Even uranium, the newest large-scale source of power, is natural in that it is just as much a mineral as oil or coal. Today, however, natural power is generally taken to comprise the power of the sun, the tides, rivers, wind, the heat of the earth's crust, and so on. Historically some of these were of major importance: in former times windmills and waterwheels were virtually the sole sources of mechanical power, except for man's own exertions and those of draught animals. Above all these, of course, the sun reigned supreme as a provider of energy to support life of every kind. It was the ultimate source, too, of all other forms of energy. Fossil fuels derive from the photosynthetic activities of ancient plants; the flow of rivers depends upon global water cycles powered by evaporation from the oceans; wind is the consequence of atmospheric disturbances largely dependent upon temperature changes. It has been estimated that, if it could be harnessed, the world's power requirements could be provided by about 1 per cent of the solar energy falling on the earth.

These natural sources of power were not of great significance in the first half of this century, except where local circumstances particularly favoured their exploitation. The main reason for this has already appeared. The abundance of cheap coal and oil, easily utilized anywhere in standard appliances and an easy route to electricity, gave little incentive to exploit less dependable sources. After the middle of the century, however, the situation changed radically. Latent fears about the exhaustion of fossil fuels came to the surface and the political and strategic perils of dependence on overseas oil became all too apparent.

I. SOLAR ENERGY

Because it is all-pervasive, solar energy may appropriately be considered first. In the present context we need not, of course, concern ourselves with the sun as a general provider of energy: our concern is with devices by which the radiant heat of the sun is trapped and concentrated to form the basis of power units. As the intensity of solar radiation is at best only about 1 kilowatt per square metre – falling to a small fraction of this in bad weather or high latitudes, and to zero at night – major installations would obviously have to be very large; the greater interest was therefore shown in small devices designed to provide a modest source of heat. Much development work has been undertaken since the Second World War at the French government's solar research laboratory at Mont Louis, in the Pyrenees; Israel, too, has been very active.

Many such devices were based on the familiar principle of the burning-glass: concave mirrors with a parabolic section were used to bring the sun's rays to a focus. Devices of this sort were widely used for cooking after the Second World War in poor countries such as India where fuel is scarce and expensive. In more temperate and cloudy climates, such as that of Britain, attention was directed to solar panels through which water – for example, that of a central heating system or a swimming-bath – is circulated by means of a pump which, incidentally, itself consumes power. Although the vagaries of the weather make such devices of only intermittent value, they might nevertheless make a worthwhile contribution to the heat requirements of an average household, especially in view of the increasing importance attached to house insulation. The latter reflected realization that conservation of power is as important as its production.

The special demands of space research – at first with high-altitude rockets and then with satellites – led to the development of solar batteries capable of converting light directly into electric energy. Early devices based on silicon developed up to 3 watts per square foot but more effective generation of low-voltage electricity was obtained with cadmium sulphide cells. Such devices could, of course, make no contribution to power production in the ordinary sense. What their future may be is a different matter.

II. TIDAL POWER

Tidal power has been harnessed for centuries, and tide mills antedate the Norman Conquest in Britain. The principle is very simple: the rising tide fills a basin, and, as the tide falls, the water is released to drive a water-wheel or, in more recent years, a turbine. That substantial quantities of power might be generated in this way is not open to question but in practice many difficulties have been encountered. Some of these are common to all sources of natural power, and will be considered at the end of this chapter, but others are peculiar to the use of tides. First and foremost, a substantial difference between high and lower water is essential: about 4 metres is regarded as the minimum. There are in fact rather few places in the world where this requirement is met. One is on the Rance in France south of St. Malo, where Électricité de France built a 750 metre barrage, completed in 1967, to provide 555 000 GW of energy. In this particular instance, the economic viability of the project was enhanced because the barrage carried a road which shortened the distance from St. Malo to Dinard by 30 kilometres. In this case, too, shipping facilities in the Rance estuary have been improved as a result, but in many situations the reverse might happen. The disturbance in the flow of water occasioned by the barrage itself may interfere seriously with seaways, fishing, and other existing rights which have to be protected. Among sites investigated, but not exploited, in this period were the Severn in Britain and the Bay of Fundy in the USA.

III. WATER POWER

The power of water flowing in rivers has been utilized almost since the dawn of civilization. In simple installations the natural flow of the river is used but, as this varies greatly according to weather and season, more sophisticated schemes include a reservoir to provide a steady head of water. At the end of the nineteenth century the advent of the electrical industry led to much interest in hydroelectric schemes: a massive installation was completed at Niagara Falls in the 1890s. Initially, it seemed that the great rivers of the world offered almost unlimited opportunities for electric power generation but it gradually became clear, as with tidal power, that in practice the requirements to be met severely limited the number of suitable sites.

Tidal power station on the Rance, France. This picture shows its appearance as it neared completion in 1966. The road across the barrage shortens the distance St Malo–Dinard by 30 km.

Although major hydroelectric schemes were completed in many parts of
the world in the first half of the twentieth century, their contribution failed
to keep pace with the growth of the electrical industry. In 1925, 40 per cent
of the world's electrical power was generated by water; by 1970 this had
fallen to 25 per cent. In 1950, 30 million horsepower of energy were
harnessed in hydroelectric schemes, and it was then estimated that this
represented around one-eighth of the water power potentially available. At
that time Switzerland and Italy were relatively the most advanced, having
harnessed one-third of their potential resources. It must be remembered,
that although high-voltage alternating current can be transmitted for quite
considerable distances without unacceptable loss, the development of
resources remote from centres of consumption presents many problems.

IV. WIND (AEOLIAN) POWER

The windmill is one of the oldest sources of mechanical power and in Europe
alone tens of thousands were built. They were, however, clumsy-
wooden devices and few generated more than 10 horsepower. Their
twentieth-century counterparts are made of steel lattice-work and were, and
are, widely used in country areas to pump water from underground sources
for domestic purposes or to water cattle, or drive dynamos to charge storage
batteries.

The number of such mills was enormous. It was estimated that in the

Wind motor for pumping water: American west.

1950s there were 300 000 pumping windmills and 100 000 electricity generators in the USA alone. Nevertheless, their total contribution was small, no more than 1 million horsepower/hours per day. This is not to say, of course, that they were unimportant to their individual owners, whose survival might depend on them.

The widespread use of windmills in former times encouraged the belief – especially during the Second World War, when fuel was short – that they might be developed to produce power on a scale appropriate to modern needs, but this hope was not realized. Scaling up from a few kilowatts' capacity – sufficient to meet the basic needs of a small farm – to the hundreds of kilowatts necessary for public supply proved insuperable in relation to the economics of the power industry generally. In the United States a 1250 kW wind-generator was erected at Grandpa's Knob, Vermont, and in Britain one of 100 kW capacity at Costa Hill in the Orkneys. Both had short lives; similar installations were experimented with in Germany, Russia, Denmark, and elsewhere, but with no greater success so far as a useful contribution to national power needs was concerned.

V. GEOTHERMAL POWER

Like more dramatic manifestations such as volcanoes, thermal springs provide clear evidence of hot spots in the earth's crust. They have been used since time immemorial, as in the Roman baths at Bath. However, there are few places in the world where this phenomenon is encountered on a scale and with a dependability that allows its exploitation for power generation. One of the best known is at Reykjavik, in Iceland, where water issues almost at boiling-point from the ground 3 kilometres from the city. In 1930 an experimental scheme of central heating was introduced to utilize this water. It was extended during the Second World War, exploiting springs at Reykjalaug, 17 kilometres from the capital. Despite the longer distance, and the cold climate, the use of insulated pipes limited the cooling in transit to $5°$C.

At Lardarello, in Italy, the water emerges, from depths of up to 600 metres, as high-pressure superheated steam at temperatures around $200°$ C. As early as 1905 the steam was used to drive a 25 kW generator, and in 1913/14 a 250 kW and three 1250 kW generators followed. By the outbreak of the Second World War capacity had risen to 135 MW. The installations were largely destroyed by the retreating Germans, but by 1950 capacity had risen again to 254 MW. Although steam can drive turbines direct, the presence of contaminants creates practical difficulties: in some of the installations, therefore, heat exchangers are used to provide clean steam. In a small installation in Ischia, where the temperature of the issuing water is only

55°C, ethyl chloride (boiling-point 12.5°C) was used as heat exchanger instead of water. In the USA an installation at the Geysers, California, came into operation in 1960.

The modesty of these developments reflects certain inherent difficulties. The temperature of geothermal water rarely exceeds 300°C, and is often much less. Even at this temperature the energy content of water is less than 0.1 per cent of that of an equivalent quantity of oil. The economic incentive to exploration and drilling is, therefore, correspondingly small unless local conditions are very favourable, as in the examples cited above. An important consideration is that heat loss in transit means that installations must be near the consumers. Another technical limitation arises from the fact that the rocks of the earth's crust are very poor conductors and this determines the rate at which heat can be extracted.

All these considerations have led to the direct use of geothermal energy being limited to applications where only low-grade heat is required – as in domestic heating systems, commercial glasshouses, and so on. This application was actively pursued in France from the 1960s. A prototype scheme was brought into operation in 1970 at Melun, near Paris, and was quickly followed by three others in the same area. These installations depend on water, at 60–75°C, drawn from a sandstone stratum lying at a depth of around 2000 metres. Each will cost about £300 000. The water is very saline and waste disposal is a serious problem: one scheme alone, at Creil, brings up 120 tonnes of salt daily. An important French innovation is reinjection of the water into its source after removal of its heat.

VI. LIMITATIONS OF NATURAL POWER SOURCES

The inevitable exhaustion of fossil fuels, rising costs, and the political and strategic considerations involved in their supply all contrived to direct more attention to the utilization of natural power sources. That they will be increasingly exploited in the future is beyond doubt but need not concern us in a historical work. We may, however, usefully conclude by considering some factors which are as relevant to their past development as to their future.

Modern civilization demands not only enormous quantities of energy but also a reliable supply. The almost immediate consequences of interruption of supply of a conventional fuel – as by cessation of shipments of Middle East oil, or a strike by tanker drivers, miners, or power-station workers – have repeatedly demonstrated the importance of continuity. But most natural power sources, by their very nature, are not reliable. Hydroelectric schemes, for example, are vulnerable to prolonged drought; wind power ceases in a calm; solar energy ceases at nightfall and is seriously curtailed by bad weather; tides run at different times each day and the extent of their rise and fall is seasonal.

On this basis, natural power sources can only supplement, and not replace, conventional ones whose output in normal circumstances is fully controllable. Power-stations fired by oil or coal, for example, could be proportionately shut down as wind power became available and brought into use again when the wind dropped. But this presupposes that conventional power sources have stand-by capacity sufficient to meet the full demand at all times. It is argued that if the stand-by capacity is there anyway, and can be brought into use at short notice, it is not only pointless but positively wasteful to spend money on installations based on capricious natural power. On capital expenditure there is no saving, but rather additional cost: the only saving is of fuel. On strictly economic grounds this is a convincing argument but, as we have already noted, the conservation of fuel has now become a major technological factor and may outweigh additional capital and running costs.

Another important factor is that the forces of nature can be very strong indeed, and equipment to harness them may be unavoidably exposed to their full fury. At the worst it may then be destroyed, and in any event it may have to be built with such a high margin of safety as fatally to affect the whole economics of its working. We can illustrate this general thesis by two specific examples. A wind generator must operate effectively at wind speeds as low as 5 metres per second and yet be able to withstand the sorts of gale that may be expected only once in ten years. Similarly, tidal devices must do more than operate under normal conditions: they must also be able to stand up to the pounding of waves which drive all but the largest ships to shelter. These basic requirements are not easily met within the limits of economic viability, as recent history has shown.

BIBLIOGRAPHY

Crabbe, D. and McBride, R. *The world energy book*. Kogan Page, London (1978).

Davey, N. *Studies in tidal power*. Constable, London (1923).

Garnish, J. D. Progress in geothermal energy, *Endeavour* (new series) **2**, 66, 1978.

Kruge, P. and Otte, C. (eds.). *Geothermal energy – resources, production, stimulation*. Stanford University Press, Stanford, Calif (1973).

Wahl, E. F. *Geothermal energy utilisation*. Wiley–Interscience, New York (1977).

5

NUCLEAR POWER

Part I. The scientific background

Strictly speaking, nuclear power should not figure at all in a work devoted primarily to the first half of the twentieth century, for the world's first commercial nuclear power station – at Calder Hall in Britain – did not become operational until 1956. Nevertheless, its technological roots stretch well back into the period of our present concern and it is most intimately connected with the Manhattan Project which produced the atom bomb that so abruptly terminated the Second World War.

Equally, were we to trace nuclear power back to its ultimate scientific origins we would be guilty not of going beyond the end of our present period but of anticipating its start. The key discovery in atomic physics was perhaps that of X-rays by W. K. Röntgen in 1895. This prompted the classic researches of such scientists as J. J. Thomson, Ernest Rutherford, Niels Bohr, A. H. Becquerel, Marie and Pierre Curie, Max Planck, William Crookes, and Frederick Soddy, to name only a few. Even to summarize their work would demand more space than is available here and we may conveniently begin our story with the situation that had been established around 1930.

By that time the atomic theory of matter, which had been mooted as a philosophical abstraction by the Greeks, had been put on a firm experimental basis. Atoms were far from the simple particles visualized by Newton as being 'so very hard as never to wear or break in pieces'. In fact, atoms had been shown to have complex structures, and the differing properties of the elements were seen to reflect differences in the structures of their atoms. Essentially, all atoms consisted of a small relatively heavy nucleus, carrying a positive charge, surrounded by negative electrons arranged in a succession of orbits, like planets round the sun. The atom had no over-all electrical charge: the positive charge on the nucleus was balanced by the sum of the charges of the orbital electrons. The charge on the nucleus increased in unitary steps from hydrogen, the lightest known element (atomic weight 1) to uranium, the heaviest (atomic weight 238). Contrary to earlier belief, the properties of an element had been proved to depend on the nuclear charge and not on the weight of the atom though the two often went hand in hand. Indeed, Soddy had made the important discovery that atoms with the same nuclear charge could have different atomic weights: such atoms were named isotopes. To take a specific example, atoms of magnesium all have a positive nuclear charge of 12, but may have weights of 24, 25, or 26. The nucleus was discovered to have a structure of its own, too, consisting of charged particles called protons and electrically neutral ones called neutrons.

Early atomic theories postulated that atoms, and hence the elements of which they were made up, were immutable: once an oxygen atom always an oxygen atom, as it were. This view was shown to be incorrect, at least in the case of certain elements, first noted by A. H. Becquerel in 1896, which spontaneously emit radiation and are called radioactive. Uranium, for example, spontaneously disintegrates to form radium, and the latter in turn decays to form lead. These changes are often relatively slow. The half-life of uranium (i.e. the time it takes for half of it to disintegrate) is 5000 million years, while that of radium is 1590 years. Other radioactive transformations are very swift, with half-lives measured in days or even seconds. With this discovery the dream of the alchemists was half realized. Spontaneous transmutation of the elements occurred in nature; it remained to be seen whether the same kind of change could be effected artificially.

Naturally radioactive substances proved to emit three kinds of radiation – alpha, beta, and gamma. The first two consist of particles, the third is a form of highly penetrating radiation akin to X-rays. Alpha particles proved to be positively charged helium atoms (atomic weight 4); beta particles to be electrons. When uranium transmutes itself into radium it emits three alpha particles and two beta particles. These particles are emitted at very high speed and in the micro-world which we are now exploring they are quite powerful projectiles.

Only a few natural elements display radioactivity, the great majority are quite stable. It seemed feasible, however, that if stable atoms could, as it were, be struck fair and square by one of these projectiles, they might disintegrate, at least partially. Alternatively, they might cohere after impact, like two snowballs. In that case, artificial transmutation would be effected. The alchemists' dream would then be fulfilled: one element could be transmuted into another at will. There were, however, evident difficulties. Atomic matter proved to be not compact, like marbles in a bag; it is more like a swarm of gnats with lots of space between them. An alpha particle might not score a hit at all, just as a rifle bullet could go through a swarm of gnats without hitting even one of them. To pursue the analogy further, a stream of particles would have a greater chance of registering a hit, just as a shotgun would be a better gnat-killer than a rifle. The chance of collision is still further limited by the facts that the atomic nucleus, like the alpha particle, is positively charged and that similar electrical charges repel each other. In this respect, a more promising projectile was the neutron: although possessing only one-quarter of the weight of an alpha particle, it has no charge and would not, therefore, be repelled by the nucleus as it approached. There seemed a fair chance, therefore, of neutrons and nuclei interacting.

In 1919 Rutherford reported a success he had predicted in 1914. By bombarding nitrogen atoms with alpha particles he was able to expel a hydrogen atom from them: it was rather like hurling a ball and chipping off a piece of the shell of a coconut at a fairground. In scale the achievement was almost infinitesimal: only about one nitrogen atom in every 300 000 in a very small sample was disintegrated. Nevertheless, it was an experiment that presaged great events.

Although alpha particles are the more massive, and thus seemingly the more destructive, theory indicated that the smaller protons – with unit positive charge and unit mass corresponding to a hydrogen nucleus – were more likely to penetrate the swarm of orbital electrons and interact with the nucleus. The chance of success would be increased if the proton could be speeded up, just as the bullet from a high-velocity rifle will wreak more havoc than one from a standard weapon. The fact that the proton was charged made it potentially easy to increase its velocity, because it could be accelerated in a high-voltage electric field.

Success along these lines was achieved by J. D. Cockcroft and E. T. S. Walton at Cambridge in 1932. In their famous 'atom-splitting' experiment they bombarded lithium atoms with protons accelerated in fields of up to 250 000 volts and obtained two helium atoms from every one of lithium. Although similar in principle, this splitting into two equal parts was far more dramatic than knocking a small hydrogen atom out of a relatively large nitrogen nucleus.

To appreciate the significance of these experiments in relation to the development of nuclear power we must, however, digress into a different field of science. At the end of the nineteenth century two fundamental scientific tenets seemed established beyond question: the conservation of mass and the conservation of energy. The one, advanced as an abstraction by some of the early Greek Philosophers, asserted that whatever changes occurred within a closed system no mass was destroyed, though its form might change considerably; a familiar example is combustion, in which a solid fuel disappears and gaseous products appear in its place. The other, formulated at the end of the eighteenth century, asserted likewise that energy was indestructible; if energy disappears in one form it reappears in another. Thus as an electric lamp consumes electrical energy, it produces heat and light. Moreover, the conversion of energy obeys precise rules, much as the various international currencies do (at least under stable economic conditions). A given amount of mechanical energy, for example, is equivalent to a specific amount of heat energy and so on.

At the beginning of this century, however, theoretical physicists, notably Lorentz and Einstein, undermined the very foundations of these two tenets. It appeared that mass and energy were not distinct and indestructible entities after all, but were interconvertible. In 1905 Einstein formulated his famous equation

$$E = mc^2$$

relating energy (E) and mass (m), in which the constant (c) represents the velocity of light. The velocity of light is so great that it appears to travel instantaneously; this was, indeed, believed to be the case up to the latter part of the seventeenth century, when a finite velocity of light had to be invoked to explain certain astronomical anomalies in the moons of Jupiter. In fact, the velocity of light is approximately 300 000 kilometres per second: it can girdle the earth in one-tenth of a second. Since energy equals mass multiplied by the square of the velocity of light, it follows that a very small mass represents an enormous quantity of energy. The precise figures, expressed in units familiar to scientists, are not very informative to most people. As a very rough approximation, however, one may say that if a lump of sugar were 'annihilated' and turned completely into energy, enough energy would be liberated to bring a million tons of water to the boil. By contrast, if the sugar were simply burnt as a fuel, it would boil no more than a pint of water. This possibility obviously brings us into completely new realms of energy production, but at the beginning of the century it was no more than a possibility, and a very controversial one at that.

Cockcroft and Walton's atom-splitting experiment was one of the earliest, if not the earliest, experimental verifications of Einstein's mass–energy equation. In this experiment a hydrogen and a lithium atom collided with production of two helium atoms. According to the theory of conservation of mass, the combined mass of the two original atoms should be precisely the same as the combined mass of the two new atoms produced. In fact, careful measurement showed a discrepancy. Some mass did in fact disappear in the collision process and – even more exciting – a quantity of energy was released which was precisely that predicted by Einstein's equation.

The excitement was muted, however, so far as power generation in a practical sense was concerned. The new generation of atomic physicists were not looking at matter in bulk; for the most part they were looking at collisions between single atoms, and atoms are incredibly small. Atoms of hydrogen, for example, are so small that a million million million million of them weigh little more than a gram; helium atoms are four times heavier and those of lithium seven times. Despite the enormous conversion factor relating mass and energy, the scale of operations would have to be vastly increased if the newly discovered phenomena were to be harnessed for power production. But this possibility did not loom large in the minds of most of the pioneers, whose primary interest was to discover the ultimate structure of matter. Rutherford, who had probably done more than any other single person to develop this new field of research, died in 1937 in the belief that the prospect of nuclear power was 'moonshine'. One exception was Soddy, who as early as 1908 saw the magnitude of the prospect, declaring that such energy (released in radioactive processes) could 'transform a desert continent, thaw the frozen poles, and make the whole world a smiling Garden of Eden'. Yet even he did not have sufficient faith in this glittering prospect to pursue it; in the years between the wars he sought to improve the lot of mankind not through nuclear power but through strange economic and political theories that gained no followers. Nor, indeed, would the faith of Soddy and a few like-minded contemporaries have availed much at that time: a clearer understanding of atomic processes was needed before power production came within sight of being a practical possibility.

So far as technology was concerned, at this stage interest lay mainly in machines by which particles could be accelerated in electrical fields, in order to increase their destructive power on impact. Among the earliest of these machines were linear accelerators, first constructed in the USA by D. H. Sloan and E. O. Lawrence, using a principle employed in Germany in 1929 by A. Wideröe. In these, particles were, in effect, given a succession of 'kicks' as they passed down a long tube from one field to another. Although using a source potential of only 42 000 volts, Sloan and Lawrence were able to obtain mercury ions of 1.26 MeV energy* by passing them through thirty successive voltage steps. In 1930 Lawrence conceived the idea of making accelerators more compact by using a magnetic field to force the particles to travel in a spiral instead of a straight line. Lawrence's first magnetic resonance accelerator – generally called a cyclotron after 1936 – gave protons of 80 000 eV energy, but by mid-century cyclotrons giving protons of 350 MeV energy were in use. By that time atomic physics was firmly established as a domain in which progress was to a considerable extent governed by the availability of very powerful and very expensive particle accelerators. The cost of these was eventually such as to demand government financing in many cases. In Europe, collaboration between

---

* One electron-volt (eV) is the energy acquired by a charged particle carrying unit charge when it falls through a potential difference of one volt.

governments was necessary to help to provide the necessary facilities: CERN was established at Geneva in the early 1950s to build a proton synchrotron intended to operate in the 10–20 GeV range.

In the years immediately before the war, however, the possibility of such machines being available within two decades was not even contemplated. In 1939 the most powerful machine of this kind in the world was Lawrence's 60-inch cyclotron at Berkeley, California, employing a magnet weighing 200 tons and capable of producing 40 MeV alpha particles. Lawrence had already embarked, however, on a far more ambitious machine with a 184-inch magnet containing 3700 tons of steel and 300 tons of copper, capable of producing 200 MeV alpha particles. In the event, however, the installation of this magnet at Berkeley was delayed four years, during which it was used for the electromagnetic separation of uranium isotopes at Oak Ridge, Tennessee, as part of the wartime atomic energy project. That, however, is a story to which we must turn later.

For the moment, we must give our attention to some fateful experiments made in Europe. Reference has already been made to the use of the neutron for bombardment of nuclei. In 1934 Enrico Fermi, in Rome, observed that if the aim was to cause the neutron to combine with the nucleus, rather than disrupt it, slow neutrons were more effective than fast ones. The necessary slowing-down could be effected by passing the neutrons through hydrogen-containing substances such as water or paraffin; such substances are called moderators.

Fermi discovered that if uranium – the heaviest of all the natural elements – was bombarded with neutrons it became radioactive. The Joliots in Paris had earlier achieved the similar results by bombarding light elements with alpha particles. Fermi believed that he had created a new element even heavier than uranium. The creation of such artificial elements was exciting and opened up the prospect of a trans-uranium series, which was in due course to be realized. For the moment, however, Fermi was at fault: others workers concluded that fission of the uranium nucleus had occurred. In 1938 Otto Hahn and F. Strassmann concluded that one of the products of fission was barium and not radium, as had been supposed. Hahn wrote about this conclusion to Lise Meitner, an Austrian Jewish physicist who had had to leave Germany in 1938 and eventually settled in Stockholm. It was there that she received Hahn's letter with his surprising news, and shortly afterwards she discussed it with her nephew, Otto Frisch, also a refugee from the Nazis. Early in 1939 they concluded that what had happened was that the uranium nucleus had divided into two smaller ones, of roughly equal size, and that this fission was accompanied by a great release of energy.

In April 1939 the Joliots, in Paris, showed that in the uranium fission process secondary neutrons were released. Very significantly, they noted that these secondary neutrons were more numerous than those used for the original bombardment. In principle, therefore, a chain reaction was possible, the secondary neutrons causing fission of more uranium nuclei. This is akin to the familiar domino effect: if dominoes are carefully stacked upright in line, the collapse of the one at the end will cause them all to fall. The fission would thus proceed throughout the mass of uranium, with colossal release of energy.

Later that same year Niels Bohr and J. A. Wheeler, in Copenhagen, showed that of the two natural isotopes – having atomic weights 235 and 238 respectively – the lighter, and very much rarer, ^{235}U was the most easily fissionable by slow neutrons. This crucially significant result was published on 1 September 1939, two days before the outbreak of the Second World War.

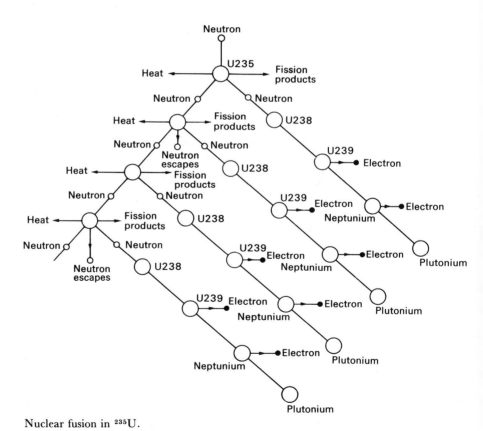

Nuclear fusion in ^{235}U.

This necessarily brief account does justice to neither the complexity of the basic research nor the contributions of individuals: for fuller accounts the bibliography must be consulted. It is hoped, however, that enough has been said to indicate that this is a tale of high drama and a remarkable example of the profound influence that science and technology can have on the course of history. The unleashing of atomic energy was unquestionably one of the most significant events in the long history of civilization.

Part II. Technological Achievement

At the outbreak of the Second World War it had been established that bombardment of uranium with slow neutrons could, in principle, initiate a self-perpetuating chain reaction liberating vast amounts of energy. The experimental and theoretical arguments had been published for all the world to read. The contestants therefore entered the conflict on precisely equal terms, yet in the event it was the United States – enlisting the aid of a large number of foreign scientists and engineers – who achieved total success: no other country, on either side, came within sight of it, though undeniably none tried very hard. The reasons for this are still somewhat obscure, though various explanations can be offered. Germany certainly had the capacity, but was handicapped by having driven out so many of the essential specialists, simply because they were Jewish. Moreover, she visualized a swift war on blitzkrieg principles and, by the time it was clear that this was not to be, the opportunity had passed. Russia might perhaps have successfully embarked on a project but may have supposed that the war would have ended before it could be completed. In addition it has been plausibly argued that Russian intelligence reported that there was nothing to fear from the Germans in this respect. Nor, as allies, had the Russians anything to fear – at least in the short term – from the Americans, of whose activities they must have been aware at least from 1942, when the spy Klaus Fuchs began to send them information. For the success of the USA we may perhaps need look no further than her commanding technological position: no other country at the time could deploy such industrial resources for the quick mastering of a technical problem of unique difficulty, complexity, and scale.

Although success was to be achieved in the USA, the initiative lay in, but not wholly with, Britain. Scientists there were of the opinion that, although atomic energy was a long-term possibility, the development of an atomic bomb within the duration of the war – uncertain though that was – was remote. Two refugee scientists who had made their way to Britain thought otherwise, however; they were R. E. Peierls and Otto Frisch. Early in 1940, during the so-called cold war, they drafted a paper persuasively arguing that an atomic bomb, equivalent to several thousand tons of TNT, could be made from as little as 5 kg of ^{235}U. In April 1940 the British government set up a committee of experts – under the chairmanship of G. P. Thomson and with the code-name MAUD – to consider the proposal. In July 1941 it reported that a bomb based on ^{235}U was practicable and might decisively affect the outcome of the war. It estimated the development time as three to four years, and that 10 kg of pure ^{235}U would be required. A bomb project was initiated under the code-name Tube Alloys. This MAUD report was disclosed to the US government – at that time interested in atomic energy primarily for power production in submarines – and led them to initiate a bomb project – later generally known as the Manhattan Project. Enthusiasm for this increased greatly after the Japanese attack on Pearl Harbor in December 1941. For a year the programme developed as an Anglo-American project but then, ironically, the British were excluded. Not until the late summer of 1943 were the British readmitted, but very much as junior partners.

In principle, the making of an atomic bomb was simple. All that was necessary was to extract ^{235}U from the natural element and then create a mass of critical size – that is, size such that the number of neutrons produced in the fission process just exceeds those lost by capture and escape. The process of capture is very rapid – perhaps a hundredth of a millionth (10^{-8}) of a second. If the number of neutrons doubles each time, it follows that the number would increase rapidly, in the series 1, 2, 4, 8, 16, 32, etc. After 80 generations there would be 10^{24} neutrons in the system – enough to cause fission throughout half a kilogram of uranium. As this juncture would be reached in less than one-million of a second, a vast amount of energy – roughly equivalent to that obtained from the explosion of 10 000 tons of TNT – would be generated almost instantaneously: the explosive force would be prodigious. In 1941 a new possibility occurred. In the ^{235}U chain reaction the ultimate product is a new element, plutonium, that is even heavier than uranium and is not found in nature at all. It appeared that plutonium, too, might be a suitable raw material for an atomic bomb.

In practice, these seemingly simple requirements presented immensely difficult problems. The first lay in the very nature of isotopes (p. 45): although these have different atomic weights they have identical chemical properties because the nuclear charge is the same. Yet in almost all ordinary metallurgical processes it is the difference in chemical properties that is utilized to prepare metals. Not only had recourse to be made to physical differences but to physical differences of a very minor order. The atoms of ^{235}U and ^{238}U differ in weight by only 0.13 per cent and natural uranium contains only 0.7 per cent of the ^{235}U isotope. In addition, there was the problem of critical size: the essence of an atomic bomb was that, above a critical size, a mass of pure ^{235}U would explode spontaneously as a result of the neutron chain reaction. It was necessary to calculate, on the basis of experiment, exactly what the critical size was. On the one hand, the utmost care had to be taken to see that this size was never attained in the course of manufacture. On the other, it was essential that when two subcritical masses were brought together in a bomb the critical size was actually attained: if it was not, no explosion would occur. The separation of plutonium, too, bristled with difficulties. Although chemical methods of separation could be used, the chemical properties of plutonium had to be discovered by experiment and only minute amounts were available. The final scale-up was probably unprecedented in industrial chemistry. The crucial chemical studies were made on a total of about half a milligram of plutonium – the size of the head of a very small pin – and on the strength of this a plant was built to produce successfully ten thousand million times this quantity.

Never before in history was so great a technological enterprise based upon such micro-scale experiments; only the unique circumstances of the Second World War made the risk, and the exceptionally high costs, acceptable. Above all this, was the problem of radioactivity. Many of the materials handled, both in the laboratory and in the factories, were dangerously radioactive; complex operations had to be carried out by remote control behind dense lead screens. Only the fact that a successful outcome might bring immediate victory made the gamble justifiable.

I. SEPARATION OF ^{235}U

Chemically ^{235}U and ^{238}U are indistinguishable and so chemical methods of separation were ruled out from the start. Physical methods of separation were therefore obligatory – they had already been used successfully for the separation of other isotopes. As early as 1931 it had been recognized that hydrogen existed in two isotopic forms. Ordinary hydrogen (^{1}H) contains about one part in 4500 of ^{2}H (deuterium, D) and by 1933 this had been prepared in the pure state by decomposing water by means of an electric current. Ordinary water (H_2O) has properties appreciably different from heavy water (D_2O): for example, heavy water freezes at 3.8 °C and boils at 101.4 °C (compared with 0 °C and 100 °C). Its specific gravity is 1.12 compared with 1.00. These differences are sufficient to make ordinary water decompose electrolytically appreciably faster than heavy water; consequently the residue left after a large volume of water has been decomposed is relatively rich in D_2O. The conditions here are unique however: hydrogen is the lightest of all the elements (atomic weight = 1) and deuterium is twice as heavy (atomic weight = 2). The ratio of the molecular weights of heavy and light water is 1.11. When we come to uranium, the heaviest of the natural elements, the situation is quite different: the ratio of atomic weights of the two isotopes is only 1.01. The disparity is similar when we consider uranium compounds. The one most commonly encountered in nature is pitchblende (U_3O_8). The ratio of the molecular weights of $^{238}U_3O_8$ and $^{235}U_3O_8$ is also a mere 1.01. A minor complication here is that oxygen itself exists in three isotopic forms: by far the most prevalent is ^{16}O, but ^{17}O and ^{18}O together constitute about 0.24 per cent of natural oxygen. Thus the molecular weight of $^{235}U_3{}^{17}O_8$ is 841, compared with 842 for $^{238}U_3{}^{16}O_8$. These marginal differences of weight correspond with marginal differences in other physical properties.

In practice, separation processes were based primarily on uranium hexafluoride (HEX, UF_6), partly because it is a liquid or gas at practicable operating temperatures (melting-point 65 °C) and partly because fluorine, unlike oxygen, exists in only one isotopic form and thus does not complicate the situation.

Of the various processes developed, the most important was that of gaseous diffusion, which had been recommended in the British MAUD Report. Briefly, this depends on the fact that light molecules pass through a suitable porous membrane faster than heavy ones, in proportion to the square roots of their molecular weights. In the case of HEX, this meant a theoretical enrichment of 1.0043 ($\sqrt{352/349}$) times for a single diffusion. In practice, because about half the total volume of gas must be allowed to diffuse through the membrane, the maximum enrichment possible is around 1.0014. On this basis something like 4000 diffusions would be necessary to raise the ^{235}U content from the original 0.7 per cent to the 99 per cent required. In ordinary industrial terms this was not worth serious consideration but, nevertheless, in December 1942 building of a diffusion plant was authorized and construction started at Oak Ridge in May 1943. It covered more than two million square feet.

It was a remarkable act of faith, for the foreseeable chemical engineering problems were not only formidable but unsolved. The design and construction of the compressors presented enormous difficulties, and the diffusion process had to be designed on a continuous basis, the very slightly enriched gas from one unit serving as the raw material for the next, and so on. But the real act of faith lay in confidence that suitable membranes – crucial for success – would be forthcoming: it is a remarkable fact that this vast enterprise was launched without any guarantee of this. Indeed, three months after work began, there was no satisfactory membrane, and as late as April 1944 only half the membranes produced in the laboratory met the required specification. Not until the end of that year – two years after the project was authorized – was a reliable and satisfactory product available, allowing the first ^{235}U to be produced on 12 March 1945.

Hopes had not, however, been pinned exclusively on the gaseous diffusion process. An alternative was thermal separation, originally proposed by Peierls and Frisch, depending on the fact that if a gas is passed through a narrow annular space between two tubes, one hot and one cold, light molecules will preferentially move towards the hot surface and heavier ones towards the cold one. Again, however, the degree of separation in one operation is minute and great batteries of thermal diffusion units were built at Oak Ridge. Work started early in 1942 and was completed within three months, a remarkable achievement. It fully achieved its objective, which was not in fact to produce pure ^{235}U, but to make enriched raw material for other processes. A main disadvantage was its enormous power requirement.

Finally, use was made of an electromagnetic process, the one used in the earliest laboratory studies of isotopes. This depends on the fact that if an electric discharge is passed through a gas the molecules acquire an electric charge. The resulting ions can be marshalled into a fast-moving beam by means of another electric field. If this beam is passed between the poles of a powerful magnet it is deflected and – most important – the heavier ions are deflected less than lighter ones. Thus the stream of particles is sorted out according to weight, much as white light is sorted out by a prism according to wavelength (colour). In the laboratory all this was done in an instrument called a mass spectrometer: the first of these was built at Cambridge as early as 1919 by F. W. Aston who, incidentally, had even before this effected some separation of neon isotopes by gaseous diffusion. The mass spectrometer was, however, essentially an analytical instrument; it was not intended for preparing pure isotopes in any quantity. Nevertheless, the possibility was there, and a large electromagnetic separation plant was built at Oak Ridge in 1942. It consisted of units with electromagnets about 4 metres in diameter, and the building of these stretched even America's technological resources to the limit. Sufficient copper was not available and silver bullion had to be withdrawn from the US Treasury to make wire for the windings. Lawrence's new cyclotron was pressed into service. Original designs proved unsatisfactory and modifications were necessary before successful operation was possible. By that time enriched material from the thermal and gaseous diffusion plants was becoming available and it was left to the electromagnetic separators to complete the purification and provide the ^{235}U used to make the first atomic bomb, dropped on Hiroshima.

Finally, mention must be made of technological defeat. Yet another method of separating particles of different weights is to use a centrifuge. If a gaseous or liquid mixture is spun at great speed – effectively increasing the gravitational field to which it is subjected – heavy particles will separate from light ones as cream separates from milk. An attempt was made to build a centrifugal separation plant at Oak Ridge in 1942 but had to be abandoned. It was calculated that some 25 000 machines would be required and these would have work for long periods at very high speeds: prototypes demonstrated endless and insuperable failures. It is interesting to note, in passing, that in the 1960s technology had caught up with science, and the centrifugal separation of isotopes became practicable. So far as the atomic bomb was concerned, however, centrifugal purification was a failure.

II. PLUTONIUM

Plutonium (Pu) does not occur naturally and has to be manufactured in nuclear reactors containing enough excess neutrons (n) to allow the reaction, via neptunium (Np):

$$^{238}U + n \rightarrow {}^{239}U \rightarrow {}^{239}Np \rightarrow {}^{239}Pu.$$

For technical reasons which need not concern us here, beyond remarking that they arise mainly from its four states of valency, the chemistry of plutonium is very complex. An additional complication is that it is radioactive, emitting alpha particles, so that chemical experiments cannot be carried out in the open laboratory: as a corollary to this, it was preferable to carry out microscale experiments even when larger amounts became available. Nevertheless, a very comprehensive and precise knowledge of the properties of plutonium was an essential prerequisite for its production.

The problems of producing plutonium were quite different from those for ^{235}U. Plutonium was a new element and not just an isotope of a familiar one; in consequence, chemical and not physical separation methods could be used. What the two problems did have in common was that the material to be extracted was present in only very small concentrations in the raw materials. For plutonium manufacture the raw material was irradiated uranium, in which plutonium was present to the extent of about 300 g per tonne. However, this in itself provoked no defeatist thoughts: as early as 1902 the Curies, working almost single-handed in Paris, had extracted 1 g of radium from 8 tonnes of pitchblende. The real problems were scale, speed, and purity.

Pitchblende was once again the starting-point. Up to 1939 the main interest in pitchblende was its radium content. The uranium, its main mineral constituent, was of very little interest even though it had been known since 1841; there was no industrial process for its manufacture, still less for manufacture to the exceedingly high standard of purity now required. Nor were its physical properties known with anything like the precision that was to be needed before it could be fed into the piles in which irradiation was to take place. Once again, a vast amount of basic chemistry and physics had to be carried out in an unbelievably short space of time.

Although uranium is widely distributed in the earth's crust, workable deposits are scarce, and at the outbreak of the Second World War the main source was the pitchblende mined in what was then the Belgian Congo. At the outbreak of war the considerable stocks held in Belgium were acquired jointly by Britain and the USA and shipped to safety in America. In the immediate post-war years stocks of uranium built up in Britain were to be powerful bargaining counters in negotiating an exchange of technical information with the USA. These supplies were augmented by pitchblende discovered in 1930 at Great Bear Lake in Canada. Unfortunately, this lies within the Arctic Circle and the ore had to be flown out 1500 miles and then sent by rail to a refinery at Port Hope, Ontario, some 4000 miles away. By 1958, however, Canada was the world's largest producer of uranium: ore was being milled at the rate of 42 000 tons per day. American output was comparable. It must be noted, however, that much of the ore contains no more than 2–5 lb of uranium per ton.

To return to the wartime project, the uranium metal was extracted by a series of chemical reactions, involving successively uranyl nitrate, uranium trioxide, and uranous oxide. The last had to be very highly refined, but by the summer of 1942 oxide of a purity seldom achieved outside the laboratory was being made at the rate of a ton a day. This was converted to the tetrafluoride (some of which went to make the hexafluoride, HEX) and then reduced to pure metal by reaction with calcium (later magnesium). This metal was extruded to form rods which were then enclosed in aluminium cans, at which stage irradiation was possible.

By the end of 1942 some 5 tons of metal were available and a large-scale experiment was possible to see whether the uranium–neptunium–pluto-nium chain reaction was in fact practicable. As late as 1940 only a few grams of uranium had ever been made, so this was indeed a remarkable achievement. Lumps of uranium and uranous oxide (the latter used in default of quite sufficient metal) were built into a complicated lattice with graphite blocks, which served as a moderator to slow down the neutrons. Strips of cadmium were inserted as a safety measure, to absorb neutrons if the chain reaction developed more quickly than expected. The manufac-ture of the graphite presented difficulties, for again very high purity was necessary: the presence, for example, of as little as two parts of boron in a

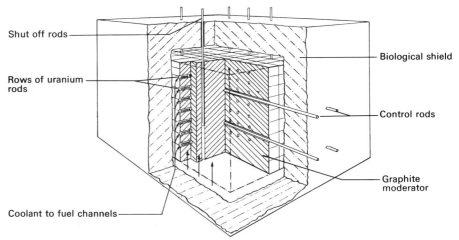

Schematic view of graphite-moderated reactor. The uranium rods penetrate right through the graphite structure.

million would have had a significant effect. Fortunately, very pure graphite was already being made industrially for electrodes, and the further degree of purification needed was achieved without too much difficulty. Great care was necessary, however, to avoid contamination in assembly.

This first nuclear reactor was referred to as a pile, for that indeed was what it was. Afterwards the name was retained for all such devices, partly for reasons of security: in itself, it gave no indication of its purpose. This first pile was built in a squash-court at the University of Chicago and, very appropriately, its first experimental run was initiated by Fermi. On the afternoon of 2 December 1942 the first self-sustaining nuclear chain reaction was demonstrated. On that first afternoon output was a mere 0.5 watt, but ten days later it was raised to 200 watts; safety considerations called a halt at this point, but when the pile was rebuilt the following year at the Argonne National Laboratory, output of up to 100 kW was recorded.

Within a month a contract had been placed for the construction of a 1800 kW pile at Oak Ridge. Work started in April 1943 and the pile was operative in November. By the end of February 1944 plutonium was being produced at the rate of several grams per month. One problem of all such piles was the need for effective and absolutely reliable cooling, without which the reaction might get out of hand. This first commercial pile was air-cooled but, when large-scale operation was undertaken, water cooling was chosen in order to achieve greater output. It entailed some possible extra risk and for safety reasons a remote site was chosen at Hanford, Washington, with abundant water from the Columbia River. The first reactor became operative in September 1944, and two more were completed within six months: each had a thermal output of 250 MW.

The speed and magnitude of the development of the Manhattan Project was impressive, and it was a triumph as much for management as for science and technology. At Hanford, alone, for example, 25 000 men were working on the site in the summer of 1943, and to accommodate them 175 barracks had to be built. The climax was reached on 16 July when the first atomic explosion took place on the testing site at Alamagordo, New Mexico, 150 miles south of another great scientific establishment, at Los Alamos, where the bombs were actually constructed. The explosion was equivalent to 20 000 tons of TNT. It was a dramatic moment, for until then nobody could be absolutely certain of success. As has been explained, the nature of the atomic bomb is such that it was impossible to proceed by way of small-scale models; it was literally all or nothing.

Ten days later President Truman, who had succeeded Roosevelt only three months earlier, issued a surrender ultimatum to Japan. It was refused and on 6 August 1945 an atom bomb based on ^{235}U was dropped on Hiroshima; three days later a plutonium-based bomb was dropped on Nagasaki; on 14 August Japan surrendered, and a war which had convulsed the world for six long years was over.

The cost was fearful – 120 000 dead in these two great cities – and the long-term consequences unforeseen and, indeed, unforeseeable. The decision to use the bomb against Japan remains highly controversial and it must be left to future historians to pass a dispassionate judgement. In the present narrow context we can do no more than note August 1945 as a turning-point: the whole world was suddenly aware that technology, whether harnessed for good or for evil, could dramatically alter the course of history.

III. MANUFACTURE OF THE BOMB

The Manhattan Project set up in 1942 was based on the simple but unproved theory that if a mass of ^{235}U or plutonium exceeded a certain size it would explode with devastating violence. All depended, therefore, on creating this critical mass from subcritical ones at the scene of action. It depended also on creating this critical mass almost instantaneously, for, as we have noted, the nuclear chain reaction takes no more than about a millionth of a second to complete: if creation of the critical mass is too slow only partial release of energy can be achieved.

Much of the fundamental scientific work necessary for the Manhattan Project was carried out at Los Alamos, at the research centre established in March 1943 with J. R. Oppenheimer as its director. At first his staff were American, but later, under the terms of the Quebec agreement, British scientists also participated. Initially, research on detonation was based on a gun technique; one mass of material could be literally fired at another to obtain a very high velocity of approach (around 1000 m per s), but by July 1944, when the first plutonium became available for tests, it was realized that the risk of pre-detonation was too great. This caused great dismay, for the whole vast project for producing ^{235}U and Pu would come to nothing if the conditions necessary for detonation could not be achieved.

Attention was, therefore, switched to an alternative implosion technique. Briefly, a spherical subcritical mass would be surrounded by an outer shell of conventional high explosive; when the latter was exploded powerful shock waves would travel inwards, as well as outwards, and momentarily compress the inner core so that it became critical. This was, in fact, the technique eventually used, but, although simple in concept it was exceedingly difficult to put into practice—again bearing in mind that no rehearsals were possible.

As we have noted, a successful explosion was effected in the New Mexican desert on 16 July 1945, but the preparations for this were made at leisure, with a large technical staff and facilities available, and with no risk of outside interference. Under combat conditions, however, the bomb had to be delivered by a B29 aircraft and to detonate over enemy territory at a predetermined time and height. There were, therefore, important constraints on size and weight. In the outcome, each bomb weighed about 4000–4500 kg and was 3 metres in length. The main difference was in diameter: that of the ^{235}U-bomb was only 70 cm, the plutonium bomb 152 cm.

IV. POST-WAR DEVELOPMENTS

The development of the atomic bomb had been a most closely guarded secret and it literally burst on a world quite unprepared for it and uncertain of its implications. Very soon, however, it was generally recognized that there must be two main and separate paths of development: a military path leading to more and more powerful weapons and a civilian one leading to nuclear power. In a history designed to terminate at approximately mid-century we cannot follow these paths far. This is fortunate, for they are very tortuous.

To most people, including, unfortunately, many in the American government and the American people, the atomic bomb savoured almost of witchcraft. There was a touching faith that there was some magical secret about its manufacture: if only this could be preserved by strict security measures, concealing the facts even from their closest allies, then the USA would have no military rivals. This misconception led to much political friction. The folly of it has already been shown: in 1939 the basic principles of the atomic bomb had been published for all the world to read. The difference in 1945 was that all the world knew that these principles could be translated into practice. American faith in inviolability was soon replaced by the belief that at least it would be a very long time before anybody else could possess the secret. Even this hope was shattered when, in 1949, Russia exploded her first nuclear bomb. Britain followed suit at Monte Bello in 1952, and in 1964 even China, starting from a relatively low scientific and technological base, entered the atomic race.

The American reply was to launch a new generation of atomic bombs, far more powerful than those used against Japan. They were based on a quite different kind of atomic reaction – essentially the fusion of atoms at temperatures around one hundred million °C: such temperatures could be reached at the centres of fission bomb explosions. In January 1950 President Truman authorized a massive programme of research to produce thermonuclear or H-bombs. Success was achieved in 1952, with an explosion several hundred times more powerful than that which shook the New Mexican desert in the summer of 1945. But in 1953 Russia too had a thermonuclear bomb and so did Britain by 1957. The world was once again engaged in a major arms race.

In Section II it was mentioned that one reason for locating the plutonium plant at Hanford was that the Columbia River provided abundant water for cooling the piles and dissipating the great heat generated. In nuclear power reactors this heat is not dissipated but is turned to practical use, principally for producing steam to drive turbines which in turn drive electrical generators. Once the war was over, greatly increased attention could be directed to this aspect of atomic energy, which was indeed the original sphere of interest of both the Americans and the French at its outset.

An Anglo-French team, working at Chalk River in Canada, completed in 1947 a reactor using heavy water as moderator and ordinary water as coolant. A later variant used heavy water for both purposes and gave rise to the CANDU reactors which were commercially successful, especially in developing countries. In Britain gas-cooled reactors were developed, and two reactors, primarily for plutonium production, were built at Windscale in Cumbria. In 1951 attention was directed to what were called fast breeder reactors. In these, the neutrons are not slowed down and are available to 'breed' more fissile atoms; such reactors are much more powerful than the earlier types. In 1956 the world's first nuclear power station was commissioned at Calder Hall and the electricity it produced was fed into the national grid.

In France, as in Britain, attention was at first directed to graphite-

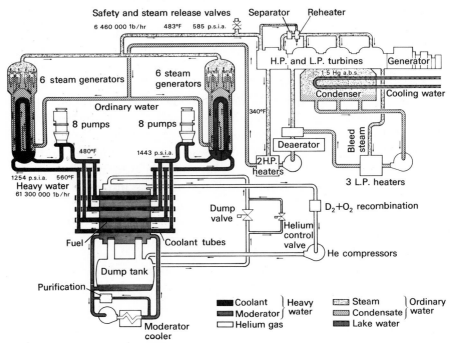

Schematic view of 500-MW CANDU power plant.

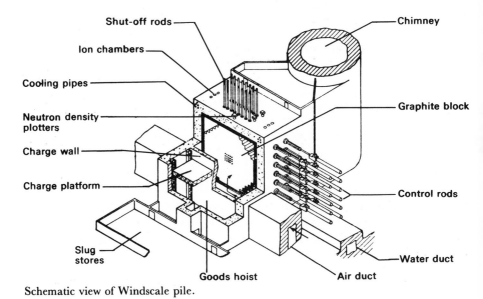

Schematic view of Windscale pile.

moderated gas-cooled piles, but subsequently cooling with ordinary water was favoured. In Russia, a power-producing reactor broadly similar to those at Hanford was operational by 1955, and in 1958 a reactor cooled with water under pressure was being used to power the icebreaker *Lenin*.

Surprisingly, American interest in nuclear power immediately after the war was limited. The Manhattan Project was wound up in 1946 and there was no great incentive to build commercial reactors because of the large reserves of oil. Only the Navy was really interested, and in 1955 the first of a series of atom-powered submarines, *Nautilus*, was launched, but a plan for an atom-powered aircraft-carrier was shelved. Not until 1957 was a commercial nuclear power station opened, at Shippingport; this was eighteen months after the opening of Calder Hall in Britain.

BIBLIOGRAPHY

Glasstone, Samuel. *Source book on atomic energy*. Macmillan, London (1952).
— *Effects of nuclear weapons*. USAEC (1964).
Gowing, Margaret. *Britain and atomic energy 1939–1945*. Macmillan, London (1964).
— *Independence and deterrence: Britain and atomic energy 1945–1952*. Vol. I: *Policy making* Vol. II: *Policy execution*. Macmillan, London (1974).
Hewlett, R. G. and Anderson, Oscar E. *History of the USAEC* Vol. I: *The New World 1939–1946*. Pennsylvania State University Press, University Park, Pa. (1962).
— and Duncan, Francis. *History of the USAEC* Vols. II and III: *Atomic shield*. Pennsylvania State University Press, University Park, Pa. (1969).
Hinton, Christopher. Axel Ax-son Johnson Lecture. Royal Swedish Academy of Science, Stockholm (1957).
Irvine, David. *The virus house*. William Kimber, London (1967).
Jay, K. E. B. *Britain's atomic factories*. HMSO, London (1954).
Kramish, A. *Atomic energy in the Soviet Union*. Stanford University Press, Stanford, Calif. (1959).
Modelski, George A. *Atomic energy in the Soviet bloc*. Melbourne University Press for Australian National University (1959).
Peck, M. J. and Sherer, F. M. *The weapons acquisition process: an economic analysis*. Harvard Business School (1962).
Pierre, A. J. *Nuclear politics*. Oxford University Press, London (1972).
Sherfield, Lord (ed.). *Economic and social consequences of atomic energy*. Clarendon Press, Oxford (1972).
Smyth, K. D. *Atomic energy for military purposes*. Princeton University Press (1945).
Spence, R. Chemical process development for the Windscale plutonium plant. *Journal of the Royal Institute of Chemistry*, May (1957).
York, H. *The advisors*. W. H. Freeman, San Francisco (1976).

6

THE ELECTRICAL INDUSTRY

The foundation of the electrical supply industry – an important addition to existing public utilities – had been firmly laid by 1900. S. Z. Ferranti in Britain and George Westinghouse in the USA had recognized that the future of the industry, like that of the gas industry some eighty years earlier, lay not with local generation but with generation at big central stations serving large areas. Implicit in this was transmission of high voltages, to avoid excessive power loss; this in turn implied the use of alternating rather than direct current because high-voltage direct-current generators are relatively difficult to construct; but, as we shall see, in certain circumstances direct current had important applications. Moreover, alternating current can easily be reduced from high voltage to low by means of the transformer (p. 73), a simple device involving no moving parts. Alternating supply did present a difficulty, however, if demand was such that generators had to be run in parallel: it was then difficult to keep the supply in phase. C. A. Parsons's steam turbine provided the high-speed engine desirable for driving the new generators: his first turbo-alternator – operating at 4800 revolutions a minute – was installed in the Forth Banks power-station in Britain as early as 1888. Water turbines, too, could be worked at high speeds, and hydroelectric schemes were developed where circumstances were favourable. Construction of the first major installation of this kind was begun at Niagara in 1886; it was designed to have an ultimate capacity of 200 000 h.p.

As we have remarked, the early electricity industry followed the same practice as the then established gas industry – distribution to many customers from a central supply. Electricity, however, was able to extend this principle further than gas: while it was economic to pump low-pressure coal gas for only a few miles, so that there were geographical obstacles to manufacturers pooling their supplies, electricity lent itself to distribution over large, even national, networks or grids. In the early years of the century, however, this technological possibility was sometimes frustrated by restrictive legislation.

By 1900 the main features of the modern electrical supply industry had appeared: central generation of high-voltage alternating current which could be stepped down for local use. There were, however, considerable variations in the frequency of supply and the choice of local voltages;

moreover, direct-current installations were by no means rare. One of the main tasks of the twentieth-century industry was to achieve a high degree of standardization in order to permit universal use of appliances. By the start of the century, too, most of the now familiar appliances had appeared: electricity was used for space-heating, cooking, and – above all – for lighting, an area in which it had become a powerful rival to the long-established gas lamp, which it had largely superseded by 1930. But the electricity industry of 1900 can fairly be compared to the motor-car industry of the same date: although the essential features had appeared the social impact was small. In most parts of the world both electricity and motor cars were novel enough to make people stop and stare; the measure of progress is that by 1950 it was the absence of these facilities rather than their availability that gave rise to comment. In both cases, the growth of the industry was spectacular. The growth of the electricity supply industry in Britain during the first half of this century exemplifies the speed of development. The year 1920 was something of a turning-point. In that year only 12 per cent of British households were wired for electricity, but by 1950 the proportion had risen to 86 per cent; a decade later electric wiring was almost universal (96 per cent). In the same period domestic consumption increased over fiftyfold, and in 1950 one-third of all electricity used in

Table 6.1

Year	Millions of units sold (U.K.)
1900	120
1910	1000
1920	3600
1930	8900
1940	2 3800
1950	4 6600

Britain was in the home. In the same year, however, industry – taking half the total supply – was still the highest single consumer, as it had been from the outset. Although their over-all consumption was relatively small, farmers were among those who most readily availed themselves of the advantages of electricity, resulting in a tenfold increase between 1930 and 1950.

As a final general comment on the electricity industry we must recall that it is, as it were, a secondary source of energy – a convenient way of utilizing other forms of energy. Electrical generators themselves require fuel of one kind or another – usually coal, oil, gas, or, more recently, nuclear fuels. Moreover, electricity is an intermediary, being converted back into other forms of energy – such as mechanical energy in motors or heat energy in electric fires – at the point of utilization. This point was not immediately taken. In 1874 the *Telegraphic Journal* thought that 'no one would be found so foolish as to employ a steam engine to produce electricity for the machinery. It would evidently be vastly more economical to drive the machinery by means of the engine itself without the intervention of any complicated apparatus.

The history of the industry can conveniently be considered under three main headings: the mode of generation; distribution to consumers; and utilization.

I. GENERATION

The early history of electricity generation mirrored that of the gas industry: an initial multiplicity of small undertakings – prompted, among other factors, by the potential difficulties of large-scale generation – gradually merged into a much smaller number of larger ones, though throughout the period of our concern small undertakings persisted in small communities too remote to warrant an extension from the large concerns. As with gas, there was from the beginning a mixture of private and public enterprise. In 1900 there were over 3000 electricity undertakings in the USA, of which 20 per cent were municipal. In Britain, by contrast, rather more than half of the 250 suppliers were municipally owned. In the USA, and in many other countries, the industry remained largely privately owned although, as a public utility, it was necessarily subject everywhere to a considerable degree of government control. In Britain, government intervention was more pervasive perhaps because, by international standards, the industry was in poor shape. In 1925 a Central Electricity Board was established to construct a national grid to connect a relatively small number of large stations: 58 operative stations, including 15 new ones, were selected and 432 recommended for closure. By the end of 1935 the national grid was virtually complete, with 4600 km of trunk lines and 1900 km of secondary lines. By a somewhat curious arrangement the Board purchased all the electricity generated and resold it to the distributors. This arrangement was perpetuated when the industry was nationalized in 1948; responsibility for generation and supply remained separate. Even then, slow progress had been made with the closure of small stations, 300 of which still survived.

In the inter-war years, Germany ranked third among producers of electricity with the USA in the forefront. In the 1920s generating capacity exceeded demand because industrial expansion was slower than expected; when expansion did begin there was a reluctance to invest in greater capacity. As a result, Germany entered the Second World War with barely sufficient capacity for immediate needs, and, as the war continued, lack of steel and other essential raw materials made expansion difficult. It has been argued that this was a contributing factor in Germany's defeat. In 1941 Germany had more than 2000 power-stations providing a public supply, but more than half the capacity lay in only 42 stations. By contrast, the national grid in Britain provided an invaluable means of meeting the changed pattern of demand resulting from the movement of a substantial part of the population and the relocation and expansion of industry. Generating capacity had been increased as part of the rearmament programme but in comparison with those of the USA the new units were small and worked at relatively low steam pressures.

From negligible proportions at the beginning of the century, world electricity production had risen by 1950 to no less than 800 billion Kwh per annum. By then the industry had, generally speaking, become mature and organized, a very different situation from that in the early years, when almost any entrepreneur with a taste for technical operations might set up a small electricity supply business. Such operators had no need to concern themselves with standardization and, as a result, a remarkable variety of supply was offered. There was, as we have already noted, the fundamental question of direct current (d.c.) or alternating current (a.c.) supply. It is probably true to say that by 1900 the question had been decided in favour of alternating current for general public supply, but fifty years later there were – even in great conurbations like London – pockets of consumers still receiving d.c. supply. Apart from choice between d.c. and a.c., the two other principal variables were voltage and (in the case of a.c.) frequency. High voltage was favoured for distribution in order to avoid excessive power losses: as early as 1889 Ferranti installed 10 000–volt alternators in Deptford power-station, some seven miles from central London, which was the forerunner of all modern central generating stations. But the supply voltage to consumers varied considerably from place to place: in Britain in 1935/6 it ranged from 100 to 480 volts. The supply frequency also varied considerably. By the mid-1930s it was commonly, though by no means universally, 50 cycles per second in Europe. About three-quarters of Britain's supplies then conformed to this standard but frequencies as low as 25 cycles per second were not unknown; at such very low frequencies electric lamps showed a perceptible flicker.

All major rationalization schemes had to take note of these differences, and eliminate them. Only then could manufacturers supply appliances capable of use without modification in any part of the system. In Britain, supply was standardized in 1945 as three-phase, 50-cycle, 240-volt a.c., but this was by no means universal. Elsewhere different combinations were adopted: in North America, for example, 120-volt systems were developed,

1000-kW turbo-generator supplied to Elberfeld, Germany (1900).

and frequencies of 60 and 25 cycles per second. Even today there is no universal standard. Legislation often favours low-voltage installation for safety reasons.

The central feature of a power-station is, of course, the generator; the most radical change in its design occurred early in the century. All the early generators had a rotating armature – driven at first by a reciprocating steam-engine but later increasingly by the much faster turbines – and a fixed electromagnet. About the turn of the century, however, the position was reversed and the field magnets rotated. This simplified both the wiring layout and the mechanical engineering problems associated with the high-speed rotation – around 3000 r.p.m. – necessary for satisfactory performance. The other main developments lay in the size of individual units and their efficiency.

In 1900 the largest turbo-alternators in the world were two 1500 kW machines supplied by Parsons to Elberfeld, Germany. Only four years later machines designed for the Carville power-station, Newcastle, operated at 6000 kW, well above the capacity for which they were designed. By 1933 Battersea power-station in London had a 105 MW installation, which was then the biggest in Europe. In the USA 208 MW had been attained in the 1930s, and in 1956 this threshold was raised to 260 MW. Not until 1963, however, was the million kW unit attained.

Turbine hall of Battersea A power station, London, 1933.

Greater efficiency went hand-in-hand with increased size. Parsons's 1500 kW generators for Elberfeld required 8.3 kg of coal per kilowatt hour, but the Carville units consumed only 2 kg per kwh. By 1964 levels below 0.5 kg per kwh had been attained. This improvement in efficiency was accompanied by higher operating temperatures for the steam. By the Second World War superheated steam at 370°C was being used; in the 1930s the temperature was raised to around 425°C; and by the 1950s the American industry was commissioning 150 MW units operating at 550°C.

It must be noted that various factors set a limit to the size of generating unit that could be assembled, and not the least of these were the size and weight of the indivisible units – alternator, rotor, and stator – that can be conveyed by road or rail from the factory to the power-station. In the USA this problem had been alleviated in the late 1940s by the introduction of hydrogen cooling for the largest installations; in Britain, however, this was not adopted until 1956. We may note, in passing, that this is a widely encountered industrial problem. High-pressure vessels for the chemical and petroleum refining industries, for example, could be satisfactorily assembled, and the welds tested, only at the manufacturers' own premises.

II. TRANSMISSION AND DISTRIBUTION

The distinction between transmission and distribution is not very sharp.

Generally speaking, however, we can regard transmission as consisting in the conveyance of electricity from the point of generation to the area of supply, and distribution as its further conveyance – through substations – to the consumer. In transmission, very considerable distances may be involved. Even in a compact country such as Great Britain, for example, the completion of the national grid in 1935 resulted in nearly 5000 km of primary transmission lines (66 000 V) and 2000 km of secondary lines (33 000 V).

One of the main problems of transmission is loss of power. In some cases this is desirable, as for example, when electricity is transmitted through the element of an electric fire or kettle: the object then is to transform electricity into heat as efficiently as possible. Unfortunately, energy is also lost, though less obviously, whenever electricity flows through a conductor: it is like supplying water through a leaky main. There are two chief ways in which power losses can be reduced: by using a good conductor and by using high voltages.

Generally speaking, the best conductors are metals, and the most commonly encountered are copper and silver. Silver is too expensive for large-scale use – though we have noted that the US Treasury had to be raided to provide silver wire for the electromagnetic separators of the Manhattan Project – so in practice the most commonly used conductor was copper. This too, was expensive and, after a substantial price rise in the 1920s, more attention was directed to aluminium. This is not such a good conductor as copper but by then it was cheaper. Nevertheless it had two serious defects: it was easily corroded, which made satisfactory electrical contact difficult, and it was not mechanically strong enough to support itself between widely spaced pylons. The first was overcome by alloying aluminium with a little magnesium and silicon to form Aldrey, a Swiss development; the lack of mechanical strength was overcome by wrapping Aldrey wire round a steel core.

More substantial reductions in power loss can be effected by increasing the voltage, since the power that can be supplied is proportional to the square of the voltage; but limits to what could be achieved were set by technological problems of generation and insulation. Nevertheless, as early as 1908, 110 000 volts was chosen for a 155-mile transmission line in California, and by the early 1930s 230 000 volts had been reached. Then the Hoover Dam project took the level to 287 000 volts for a 300-mile line. By mid-century 345 000 volts was looked on as standard and this had been substantially exceeded in a few instances. Sweden constructed a 600-mile line operating at 400 000 volts in 1952, and by 1961 Russia had broken the half-million volt barrier. Until the advent of the national grid in the 1930s,

Stranding aluminium electrical transmission cable for overhead use: a steel core is incorporated for strength.

Britain was beset – thanks in large measure to unimaginative restrictive legislation – by a multiplicity of small local suppliers, and long-distance transmission was relatively unimportant. It has been argued that elsewhere the development of long-distance transmission was encouraged by the availability of hydroelectric sources which, in the nature of things, are usually remote from the areas they supply.

For long-distance transmission there is a substantial saving in the cost of cable if direct rather than alternating current is used, but this has to be set against the cost of rectification and inversion at either end. With the advent of the mercury-arc valve in 1928, capable of handling very large outputs, this became a feasible proposition, and from mid-century a number of major d.c. transmission systems have been installed. These include links between Moscow and Kashira (1953); Volgagrad and the Donbas; the Swedish mainland and Gotland; England and France; North and South Island, New Zealand; Italy and Sardinia; and British Columbia and Vancouver Island.

For long-distance transmission across open country cables could be carried overhead on pylons, though with increasing voltages insulation at the points of suspension became a problem, especially in wet or icy conditions. Few visible features more fully epitomize twentieth-century technology than those tens of thousands of miles of overhead cable striding across country on long rows of pylons. In urban areas, however, and for ultimate distribution to individual consumers, underground cables were necessary though very much more expensive. For these, continuous waterproof insulation was necessary. For cables carrying light loads, rubber insulation was used, but at the turn of the century vulcanized bitumen was used for heavy-duty cable. The pioneer in its manufacture was W. O. Callender, who imported Trinidad asphalt into Britain for road-surfacing. Jute impregnated with resin or oil was also used. In the USA paper insulation was introduced in the 1880s by the Norwich Wire Company.

In the nature of things, all underground service installations tended to run close together; this was often the case with electric cables and telephone cables and it frequently led to interference with the telephone service as a result of inductance. To overcome this, concentric conductors, separated by insulating material, were introduced in the USA by R. S. Waring in 1886. The same principle was used by Ferranti for his 10 000-volt cable from Deptford to London. Rubber and jute proved unsatisfactory, and his tubular main was filled with paraffin wax.

After the First World War, higher voltages created new problems, largely owing to heat, which did not disperse freely as it did with overhead cables. It caused distortion in the insulation and a tendency to destructive internal sparking, leading to failure. One improvement was M. Hochstadter's screened cable, in which the electrical stresses were dispersed by a metal sheath, but the real solution was the oil-cooled cable developed in Italy in 1920 by L. Emanueli. Pressurized oil, which has better dielectric properties, was introduced by C. E. Bennett in 1931; its first major application was for a 135 kV cable for the Pennsylvania Railroad in 1935.

After the Second World War the plastics industry was firmly established and this had important repercussions for the electrical industry. Rubber insulation had a relatively short life because it became brittle and perished, but pvc and polythene had excellent insulating properties and a much longer life. Moreover, plastics lent themselves almost ideally to the cheap mass production of the multiplicity of electrical accessories essential for the

Oil-filled, paper-insulated, single-core cable (Sweden, 1952).

practical utilization of electricity. They rapidly replaced, for example, many of the wooden or ceramic parts traditionally used in domestic switches, lamp-holders, and so on.

III. THE UTILIZATION OF ELECTRICITY

The utilization of electricity is impossible without its effective control, and in this context we must consider two main devices – the transformer and the switch. The principle of the transformer is very simple: if primary and secondary circuits are wound on an iron ring an a.c. voltage in one circuit will induce an a.c. voltage in the other, and the two voltages will be related in proportion to the number of turns in each winding. In practice, things are very much more complicated, especially when large power inputs and outputs have to be handled. Problems then arise from the need to dissipate heat generated in the coils and the iron ring. For the latter it was necessary to develop new steels (4 per cent silicon) and adopt a laminated structure to reduce hysteresis. To effect a voltage change in a d.c. system, or to convert between a.c. and d.c., electromechanical devices were necessary until the advent of the mercury arc rectifier in 1928.

To cut off the electric supply when not required, some sort of switch is

10 000/2400 V transformer designed by S. Z. de Ferranti c.1891; it remained in use until 1924.

essential. For low-voltage systems carrying small loads this is no great problem. In the earliest days simple knife-blade types were used but these were unsatisfactory because of the sparks produced when they were opened or closed. Apart from their startling effect, the sparks caused mechanical damage to the contacts by pitting. While a brief spark is desirable, to prevent voltage surges, it should be extinguished as the a.c. potential passes through the zero point, which it does 100 times a second in an ordinary 50-cycle circuit. Quick-acting spring-loaded devices were introduced, as in the now universally used domestic tumbler switch. For heavy-duty switchgear, however, much more elaborate devices had to be developed. As early as 1900 the arc was deflected and extinguished by a magnetic field up to voltages of around 16 000, or it was struck in oil, which quickly cooled it. In the years between the wars, blasts of oil or air were used; these were effective for voltages up to 100 000 volts. The next development, in the 1940s, was a circuit-breaker operating in a sealed vessel containing the inert gas sulphur hexafluoride. However, as this liquefies at about 10°C, provision had to be made to keep the contact-breakers warm, and they were most commonly used indoors.

We shall consider in later chapters the development of prime movers such as the internal combustion engine and the various types of turbine, but the electric motor – which in effect is a dynamo working in reverse – is conveniently discussed at this point. Its far-reaching importance will become apparent in many contexts as we continue our survey of twentieth-century technology.

Electric motors are of two main types: the commonest is known as the induction motor, the other as the synchronous motor. Both were the

Experimental induction motor by N. Tesla, c.1888.

First British electric drill, 1914.

invention of the brilliant but eccentric Croatian-born inventor Nikola Tesla. For a time after his arrival in America in 1884 he worked with Edison, who favoured direct-current devices, but Tesla sensed the advantages of alternating current and by 1888 he had acquired a range of patents for a.c. dynamos, motors, and transformers which he sold to Westinghouse. This was a very important development in the construction of the world's first major hydroelectric installation at Niagara.

Although d.c. motors are favoured for electric traction, and some other applications demanding large horsepower and flexible control, because their speed is readily controlled and they can generate high torque at relatively low voltages, most motors are of the a.c. induction type. In this a current is induced in the rotor by the surrounding field windings. Speed is governed by the frequency of the a.c. supply and by the disposition of the windings. The first of these factors is fixed in all ordinary situations and the second is not easily varied. Induction motors, therefore, normally run at fixed speeds, but in 1957 the pole amplitude modulated (PAM) motor was introduced by G. H. Rawcliffe of Bristol. The synchronous motor, too, runs at a fixed speed; in this, the field is supplied either by a permanent magnet or by an electromagnet activated by direct current.

The electric motor made a powerful impact on industry, where the convenience of having separate engines for each machine – eliminating clumsy and noisy belting and capable of running for long periods without attention – outweighed the cost. The range of power is enormous, from massive machines developing several thousand horsepower capable of driving rolling-mills in steelworks, to tiny motors providing power for domestic appliances, electric razors, and clocks. Most of these small motors are now of the so-called universal type, capable of running on alternating or direct current.

This very brief survey of the growth of the electrical industry may appropriately conclude with its most dramatic application – the electric lamp. Again, the principle is very simple: if electricity is passed through a conductor the latter becomes hot, and if it becomes hot enough it will glow and emit light. In practice, certain problems are encountered. Firstly, the wire must have such a high melting-point that it does not fuse before the point is reached at which adequate light is emitted. For an acceptable white light very high temperatures are required: the red glow of lower temperatures is inefficient and unacceptable. Secondly, virtually all the suitable materials combine with oxygen at the very high temperatures required, and the filament had, therefore, to be enclosed in a high vacuum or (later) an inert gas.

By 1900 electric lighting was widely established in urban areas but a novelty elsewhere: it is estimated that there were then $2\frac{1}{2}$ million lamps in London alone. They were mostly of the carbon filament type, and had the disadvantage of blackening rapidly through evaporation of carbon from the hot filament and its condensation on the cold glass. Osmium filaments had been introduced as early as 1898, but this metal, like tantalum and tungsten, was difficult to work into the very fine wires required. The Dutch and Germans were pioneers of this development. In 1908 W. D. Coolidge devised a satisfactory method of making tungsten rod, which could be drawn into wire, by a powder metallurgy technique. These metals, too, tended to evaporate, like carbon though less readily, and in 1913 Irving Langmuir, of the General Electric Company, Schenectady, discovered that evaporation could be much reduced by using not a vacuum but an inert gas. These developments led to considerable increases in efficiency. Whereas the carbon filament lamp gave only 3.5 lumens per watt, the figure for the simple tungsten lamp was 8, and for the gas-filled tungsten lamp 12. By 1920 operating lives of 1000 hours were standard. A further improvement was made in 1934, with the introduction of the coiled-coil lamp, in which the filament is made not from a straight wire but from a very fine coil. Incandescent filament bulbs are commonly made in the 25–150-watt range but more powerful ones, up to 10 kW, are made for special purposes, such as cinema projectors, lighthouses, and so on. At the other end of the scale are very small lamps made for torches, bicycle lamps, and so on.

Although they are cheap and convenient, the efficiency of such lamps is low: only around 6 per cent of the electricity used appears as light, the remainder being dissipated as heat. From the 1930s attention was, therefore, directed at an alternative method of converting electricity to light, namely by discharge through a gas. This effect was well known – indeed, as lightning it was familiar to primitive man – but its first important practical

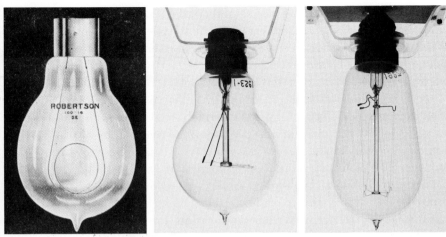

Early incandescent filament lamps (*left*) Carbon filament (1901); (*centre*) tungsten filament (*c.*1910); (*right*) coiled-coil, introduced in 1913.

application was the Cooper–Hewitt mercury discharge lamp (1901), which gave a bluish green light. The sodium vapour lamp – containing some neon to facilitate starting – gives a yellow light. Both became very popular for street lighting, where the colour of the light does not matter greatly, because although more expensive to install they were about three times as efficient as incandescent filament lamps. They were, however, quite unsuitable for domestic lighting. For this purpose the mercury vapour lamp was introduced in a modified form shortly after the Second World War.

The mercury lamp generates invisible ultraviolet as well as visible light, and when this falls on certain chemicals, known as phosphors, they fluoresce and emit visible light. By carefully blending the phosphors, and coating them on the inside of a discharge tube, a close approximation to daylight can be achieved. The efficiency of the fluorescent lamp is around four times that of the incandescent filament lamp, but even today the latter holds its own, especially in the home, because of its cheapness in both installation and replacement.

IV. BATTERIES

Hitherto, discussion has tacitly assumed the supply of electricity from some kind of mechanical generator, but in many applications a self-contained supply is necessary. This very important need is met by the battery, of which there are two important types: the primary cell, which can be used once only, and the secondary cell or accumulator, which can be repeatedly recharged. Both produce d.c. only. During the first half of this century the common primary battery was of the Leclanché type, with zinc and carbon electrodes. Storage batteries were mainly of the lead-acid type. The Edison battery, however, was of the nickel-iron alkali type, and during the Second World War the Germans developed the Jungner nickel-cadmium alkali cells. Other cells have been based on silver oxide-zinc. Dry primary cells were developed for use in torches and bicycle lamps, bell circuits, and, latterly, portable radios and hearing-aids. Early portable radios, however, used accumulators, because of the relatively heavy power demand of the valves. One of the most important uses of accumulators was in moving vehicles – motor cars, torpedoes, railway carriages, submarines, and aircraft. The ordinary accumulator, such as is used in motor cars, has a capacity of 70–100 ampere-hours and by mid-century could be recharged around 300 times before failure. The high weight : power ratio of accumulators has limited their application for traction. Nevertheless, electric delivery vehicles have had a continuous history of use since before 1900, and submarines relied on batteries when submerged.

The capacity of a storage battery is roughly in proportion to its size, and when heavy loads are required accumulators may run up to as much as a ton in weight. The voltage, however, is determined by the nature of the chemical reaction: for the lead-acid battery it is approximately 2 volts. Where, as is often the case, a higher voltage is required, a number of cells must be coupled in series, frequently within the same container: motor-car batteries, for example, are normally rated at 6 or 12 volts.

BIBLIOGRAPHY

Allen, N. L. The future of direct current transmission, *Endeavour*, **26**, 16 (1967).
British Electricity Authority. *Power and prosperity*. London (1954).
Byatt, I. C. R. *The British electrical industry 1875–1914*. Clarendon Press, Oxford (1979).
Dunsheath, Percy. *A History of electrical engineering*. Faber and Faber, London (1962).
Electrical World and Engineer, Staff of. *The electric power industry: past, present, and future*. McGraw-Hill, New York (1949).
Emanueli, L. *High voltage cables*. Chapman and Hall, London (1929).
Fleming, J. A. *Fifty years of electricity (1870–1920)*. London (1921).
General Electric Co Ltd. *Story of the lamp*. GEC, London (*c*. 1930).
Hannah, L. *Electricity before nationalisation*. Macmillan, London (1979).
Hennessey, R. A. S. *The electric revolution*. Oriel Press, Newcastle upon Tyne (1972).
Hinton, Lord. *Heavy current electricity in the United Kingdom: history and development*. Pergamon, Oxford (1979).
Hunter, P. V. and Hazell, J. T. *Development of power cables*. Newnes, London (1956).
Mellaway, John. *The history of electric wiring*. Macdonald, London (1957).
Salvage, B. Overhead lines or underground cables: the problem of electrical power transmission, *Endeavour*, **34**, 3 (1975).

Self, Henry and Watson, Elizabeth, M. *Electricity supply in Great Britain, its development and organisation*. Allen and Unwin, London (1952).

United Nations Economic Commission for Europe. *Organisation of electric power services in Europe*. Geneva (1956).

Walmsley, R. M. *Electricity in the service of man*. Vol. I. Cassell, London (1911).

7

THE CHANGING PATTERN OF
PRIMARY PRODUCTION

Primary production may be loosely defined as the large-scale production of basic raw materials for industry, but in the present context we need to be a little more explicit. On the one hand, we may regard primary products as those which exist in nature and are there for the taking; among such we may list air and water, coal and minerals, petroleum and salt, fish, and forest timber. But to be as rigorous as this would be to exclude other basic products commonly regarded as raw materials, such as steel, cement, and fuel oil, which are in fact the product of varying degrees – often considerable – of prior technological manipulation. These manipulations are not necessarily conducted at source: steelworks may be remote from iron-ore deposits and refineries from the oilfields. Very often there may be some sort of compromise: ore dressing may be effected at the mine to provide a richer product, reducing transport costs and simplifying the smelting process. In the present section we shall, therefore, take a broad view and consider the production of materials that are absorbed in bulk by industry and worked up into a diversity of finished products.

That the first half of the twentieth century saw profound changes in the pattern and scale of primary production scarcely needs stating, and we shall consider these changes in some detail in the chapters that follow. As a preliminary, however, it is instructive to consider some of the factors that have determined the change of pattern. While some useful general conclusions can be drawn, the role of individual factors is not easily definable, because it is of the nature of technology that developments in one field may have far-reaching repercussions in others. This interplay is well illustrated in transport: the growth of the transport system itself made it easier to bring to the manufacturing centres the raw materials necessary for greater production of the components of transport – ships, rails, locomotives and rolling-stock, and, later, motor cars and road-making materials. In this sense, the transport system was self-generating.

Transport was, indeed, a major factor. Except for air, raw materials – even water – are not universally available and occur sporadically and often in the less accessible parts of the world. We have already noted the political as well as the technological significance of the distribution of uranium, and many other examples come readily to mind. Chromium, essential for the manufacture of certain steels, is unevenly distributed: in 1958 world production of chromite ore was nearly 4 million tons, of which about a quarter was imported by the USA, which has only very limited indigenous supplies. Because it was an essential strategic material, the American need for chromium influenced political relations with the principal producers, which included Turkey, South Africa, and Rhodesia. No less significant, particularly in recent years, is the dominant position of the Middle East as a source of petroleum. Tungsten, too, is essential for certain steels and, as we noted in the context of the growing electrical industry, important for electric lamp filaments. At mid-century the largest single producer was China. The catalogue of essential minerals and the anomalies of their distribution could be almost indefinitely extended; we must note, too, that much the same can be said of plant and animal products. Until the latter part of the nineteenth century, for example, rubber had been obtained very largely from trees growing wild in the forests of South and Central America, but by 1900 a plantation industry had been established in Malaya, Ceylon, the East Indies, and Indo-China that was to meet the rapidly growing industrial demand, especially for motor-car tyres. The demand for rubber increased more than thirtyfold in the first half of the century. Whole new areas of grain production were opened up, notably in North America. The advent of refrigeration, and improvements in canning, led to the establishment of new meat-production areas as far apart as Australasia and South America. At sea, steam trawlers and refrigeration vessels made possible voyages to more distant fishing grounds. The reversal of seasons between the Northern and Southern hemispheres encouraged the import of out-of-season products. In short, basic food production began to be thought of in global rather than regional terms.

But these changes were possible only because reliable and economic transport facilities developed to move huge quantities of goods around the world. Rail and road transport – with emphasis on the former when large loads were in question – brought goods to the ports, to be carried onwards in steel ships. The opening of the Panama Canal in 1914 provided an important new link between the Atlantic and the Pacific: it carried 28 million tons of cargo in 1939. The opening of the Canal had two major effects: it shortened journeys between the east and west coasts of the USA and brought the west coast nearer to Europe. It also effectively brought the eastern Pacific – Japan, the eastern half of Australia, and New Zealand – nearer to New York than to London and the big European ports. Ships from Europe which used the Suez Canal to reach Australasia and the Orient returned home via Panama, collecting *en route* cargoes such as tin and rubber from Malaya. In the reverse direction cotton went from the USA westwards to Japan.

Special vessels were developed for special cargoes. The first oil-tanker appeared in 1886, and was of 3200 tonnes deadweight; by 1960 the 100 000 ton mark had been passed. Ore carriers, too, appeared around the turn of the century: the first ore was taken from Narvik in this type of vessel in 1903. In 1904 the largest vessel serving the Mesabi iron-ore mines was 11 000 tons. Similarly specialized vehicles appeared on the railways and roads to complement the traditional open wagons and trucks.

Global production statistics are not very reliable, at least until appropriate international agencies began to be established after the First World War. Many countries failed to keep proper records or were reluctant to disclose them, and products were not adequately defined. Some statistics for coal, for example, include all types of coal but others exclude lignite; none, apparently, make any allowance for inclusion of stone, clay, or water. Nevertheless, even with this reservation, the figures in Table 7.1, drawn from a variety of sources, give a fair indication of the magnitude of growth in production of a variety of basic raw materials.

Table 7.1.   (Weights are in tonnes)

Commodity	1900	1950
Iron and steel	70 M	320 M
Copper	525 000	3 M
Lead	850 000	1.6 M
Zinc	480 000	2 M
Tin	85 000	170 000
Aluminium	7 300	1.5 M
Rubber	50 000	1.6 M
Cotton	4 M	6 M
Petroleum		500 M
Coal	800 M	1388 M

The transport systems that served this worldwide trade also carried the workers necessary to provide labour in the new centres of production. The USA was the most powerful magnet. Its population rose from 75 million in 1900 to 131 million in 1940. A substantial part of the increase was due to immigration: 8 million in 1900–10, 6 million in 1910–20, and 4 million in 1920–30. By that time, however, the free land barrier had disappeared and immigration, once welcomed, began to be a problem. This first came to the fore with the need to assimilate veterans returning from the First World War, and in 1924 the Immigration Act came into effect. For some years afterwards immigration ran at around 300 000 per annum, but it was severely curtailed during the Depression. American immigrants were, of course, other countries' emigrants and the main source was Europe. The American census of 1930 showed that of 14 million foreign-born citizens all but about 2 million came from Europe: Britain, Scandinavia, Germany, Poland, Russia, and Italy each contributed more than a million. The majority of these immigrants settled in the industrial urban areas. Elsewhere, too, there were substantial movements of population; Australia, for example, received 280 000 immigrants in 1909–13.

Effectively, these vast flows of goods and people were carried by land and sea. Although the aeroplane appeared at the beginning of our period, and after two world wars had become a highly sophisticated machine by 1950, air transport made a relatively small over-all contribution until the second half of this century. Then, as we shall see later, it expanded with extraordinary speed, though mainly for passenger transport. For the carriage of basic raw materials it was generally too expensive, although we have already noted that urgently needed uranium ore was flown from mines in northern Canada.

In Europe and other centres of established population, the production of basic raw materials tended to continue along existing lines, with increasing recourse to mechanization. But much of the new production came from parts of the world which were sparsely populated and which therefore favoured large-scale operations. Examples are the new farming areas of Australasia and North and South America where producers thought in square miles rather than hectares or acres. The new rubber plantations of the Far East were also worked on a large scale; plantation crops as such, however, were nothing new, as exemplified by sugar, tea, and cotton. Mining operations, too, could be on a grand scale. Examples are the vast open-cast iron-ore workings in the Mesabi Ranges on the shores of Lake Superior, initiated by the United States Steel Corporation in 1901. By 1916 production had reached a peak of 42 million tons. The huge copper mines at Sudbury in Canada were worked on an open-cast basis until the 1920s, when the surface exposure was exhausted; large-scale copper mining was also established in Chile. In Malaya, large-scale dredging for alluvial tin was introduced in 1912.

Large-scale working demanded large-scale machinery, and the new demand for raw materials resulted in the development of much larger ploughs and harvesters, steam-shovels and dredgers. These developments were not always compatible with other social requirements, however. The finer washings from mines often polluted rivers and adversely affected farming downstream. Large agricultural holdings could lead to severe, sometimes disastrous, soil erosion.

As in so many other fields, this was a period of transition. Much of the world's raw materials continued to be produced in small units by traditional methods, using simple equipment and manual labour. At the same time, enterprises of quite a different kind were developing. Operating on a large scale in new parts of the world, capital-intensive rather than labour-intensive, and using the techniques and products of the new technologies which they were themselves feeding, they represented new sources of the increasing quantities of raw materials essential to modern industry, some of them – such as rubber, petroleum, aluminium, and uranium – scarcely needed at all at the beginning of the century.

BIBLIOGRAPHY

Bateman, A. M. *Economic mineral deposits*. Wiley, New York (1942).
Bunting, A. H. (ed.). *Change in agriculture*. Duckworth, London (1970).
Dixon, Colin J. *Atlas of economic mineral deposits*. Pergamon, Oxford (1979).
Jacks, G. V. and Whyte, R. O. *The rape of the earth: a world survey of soil erosion*. Faber and Faber, London (1939).
Lindgren, W. *Mineral deposits*. (2nd edn.) McGraw-Hill, New York (1933).
Oxford economic atlas of the world. Clarendon Press, London (1954).
Taylor, H. C. and Taylor, A. D. *World trade in agricultural products*. Macmillan, New York (1943).

8

AGRICULTURE: WITH A SECTION ON FISHING AND WHALING

Once again, in agriculture, we encounter a period of transition. While very considerable progress was made, traditional methods were not easily displaced: even in 1950 much agricultural practice might easily have been seen at least a century earlier. Nevertheless, by the middle of the century a variety of new techniques had been firmly established and were making a major contribution to the productivity of the agricultural industry as a whole: in the USA for example, agricultural production in 1953 showed a 77 per cent increase on that of 1910, even though 37 per cent fewer people worked on the land. We may conveniently consider these new techniques under two main headings: animal husbandry and food and industrial crops. Fishing and whaling, another important section of the food-producing industry, are also considered at this point.

I. ANIMAL HUSBANDRY

During this period there were no important developments in the basic materials produced for market. Beef, pork, lamb, and poultry were the staple meats, together with important complementary products such as milk, cheese, butter, and eggs. Here, in general, there were no very striking changes, though this is not to say that the industry did not emerge much strengthened at mid-century. Over-all production was much increased; productivity was improved; mechanization was greatly extended – though this was generally more important to the arable farmer; cross-breeding and the use of artificial insemination improved the quality of stock; management was better; and so on. On the whole, however, the very significant improvements were the consequence of improvements in existing methods rather than of major innovations.

As a matter of course, an expanding, and, in many areas, richer, world population demanded more meat, and the population of farm animals rose proportionately. To this there were, however, two exceptions – the horse and the buffalo or ox. For centuries, indeed millennia, these had been a major source of power, but quite suddenly, at around the turn of the century, they went into a sharp decline. This had, of course, already begun somewhat earlier, for the growth of the railway systems had driven the stage-coach off the road and steam ploughing had made some impression on the farm. Nevertheless, at the beginning of the century the horse was still supreme for local transport and as a draught animal on the land. It was the advent of the motor car and the lorry on the roads, and the tractor on the farms, that caused the rapid decline in the horse population. The USA, for example, had a horse population of around 40 million in 1913, but this had declined to 10 million in 1950. The position in Britain was similar; the horse population was just over 3 million in 1900, but had fallen to less than 2 million in 1924. At mid-century only 347 000 horses were employed on the land in Britain, and a decade later there were so few that they were no longer returned in the agricultural census. Nevertheless, these are examples from advanced western countries; it is estimated that there were still around 75 million horses in the world in 1950. The decline in buffaloes and oxen as draught animals was less dramatic, for they are associated with countries with less advanced agricultural systems, where change came more slowly. The world population of these two animals was about 70 million in 1950, rather more than 60 per cent of them in India and Pakistan and 12 per cent

Up to the First World War draught animals remained the main source of power on the land. Here they are being used to cut chaff for silage (Queensland, Australia, c.1910).

Farm mechanization was adopted much more slowly in Europe than in North America. In Britain, the need for greater home production of food encouraged the use of tractors during the Second World War: here drivers are being trained by a Ministry of Agriculture instructor.

in China. Indeed, India showed a slight increase on the pre-war total. Thus at mid-century there were still about 150 million draught animals in employment, though not all engaged in agriculture, compared with less than 6 million tractors, of which two-thirds were to be found in the USA.

The stock farmer has two variables to consider – quantity and quality – and these are not always compatible: high-yielding strains may yield inferior products. Moreover, the relative importance of each can vary from time to time. In wartime – and there were two world wars within our half-century – the emphasis was on quantity: in times of peace and prosperity quality was in demand. But even quality was a variable factor, for tastes changed. In Victorian England, for example, the demand was for large sheep with plenty of fat, around 90–100 pounds in weight, but by 1939 the preference was for lean lamb, at around 36 pounds weight. In the Second World War, however, this refinement of taste had again to give way to quantity.

These varying requirements taxed the resources of the breeders, in whose tactics there was a notable change. The breed societies had hitherto concentrated on the production of pure strains with distinctive physical features, but increasingly attention was directed to cross-breeding designed to improve the quality of the produce – creamier milk, more and larger eggs, heavier fleece, leaner meat, and so on. These were all easily measurable qualities, though the rate of improvement was slow, compared with what was being achieved in plant breeding, because it took so long to evaluate new strains. A major step forward was the introduction of artificial insemination in Russia in the 1920s. It was established first in Denmark and the USA, just before the Second World War, and was very widely used afterwards, especially after the discovery in 1949 of methods of freezing semen so that it could be preserved for long periods. By this means the qualities of the best sires can be perpetuated.

The physical features of an animal are determined by both nature and nurture. The animal breeder can help to improve a beast's genetic nature, but only a poor beast will result if it is not carefully reared. Advances in knowledge of the scientific principles of animal nutrition resulted in more systematic and efficient feeding, which was aided – from the 1920s – by an understanding that certain vitamins were as important to animals as to man. It was realized, too, that certain elements, such as copper, were essential dietary factors in trace amounts; their absence explained the

After the First World War increasing attention was paid to quantitative aspects of nutrition. This picture (Britain 1934) shows the daily ration (mangolds, silage, hay) for an Ayrshire cow.

diseases of stock reared on deficient pastures which were widespread in some parts of the world.

These important scientific discoveries percolated through to the farming community only slowly, and many continued – not unsuccessfully but without attaining maximum yields – with traditional methods. The large-scale farmers, with heavy capital investments, were of course the most receptive; they received the new knowledge through journals, lectures, and the veterinary services. The manufacturers of ready-mixed cattle foods, too, incorporated the necessary dietary ingredients in their products. Gradually, farmers began to think quantitatively rather than qualitatively, and to evaluate production of meat and milk in terms of feeding stuffs consumed.

Much progress was made, too, in the control of the more serious animal diseases, such as foot-and-mouth, swine vesicular disease, rinderpest, and tuberculosis. In the main, economic considerations demanded that control was exercised by the slaughter of infected beasts and beasts at risk, and by orders forbidding the movement of stock. Tuberculosis in cattle was particularly serious, because of the risk of communication to people through the milk. Many countries, therefore, adopted an official policy of regularly testing herds for tuberculosis, using tuberculin as the diagnostic agent, and building up attested herds free from the disease.

Sheep dipping; Queensland, Australia, c.1910.

Until the 1930s, the treatment of disease in farm animals had to be largely empirical, but then two major discoveries greatly improved the situation. Firstly, the advent of the first of the important chemotherapeutic agents, the sulphonamides, made practicable the successful treatment of a number of hitherto intractable diseases such as mastitis in cows and white scour in calves. Secondly, the discovery of the first of the important synthetic insecticides (DDT) by P. Müller in Switzerland (1939) opened up two very important possibilities: the improved control of many harmful external parasites and the eradication of insect vectors of disease, such as trypanosomiasis in Africa.

Mechanization had less direct effect on stock farming than on crop production, although we have already noted the very significant indirect effects resulting from revolutionary changes in the transport system, both long-distance and local. One of the most important mechanical innovations was in milking, a time-consuming daily chore on every dairy farm. Milking-machines began to appear about 1900 but made slow progress: in 1941 the great majority of cows were mechanically milked in New Zealand, but only 15 per cent in England.

Progress was made in farm managment, both to improve the health of the stock and to economize in labour. In general the size of holdings tended to increase; the average size of holdings in the USA increased from 220 to 336 acres between 1940 and 1954. As we have noted, in the great new food-producing areas ranches extending to several hundred square miles were familiar, though the density of animal population was relatively low – as in the less fertile parts of the Old World – and seasonal and unpredictable variations of rainfall were a serious hazard. At the other end of the scale, intensive methods of production were introduced, which we may compare, in the realm of crop production, with the market gardener and his glasshouses. The most familiar of these, and ones that were to evoke considerable public protest, were the rearing of hens in batteries, and the fattening of calves in close confinement.

II. CROP PRODUCTION

Although the techniques of growing crops are not essentially different whether the crops are destined for food or for use in industry, we may perhaps recall that there was in fact a generally steady expansion in industrial crops such as rubber, jute, cotton, and sisal, There were, nevertheless, areas of decline. Traditional vegetable dyes, such as indigo and madder, were replaced by synthetic products; in the latter part of our period synthetic antimalarial drugs largely replaced quinine from the cinchona plantations; man-made fibres challenged the supremacy of animal and

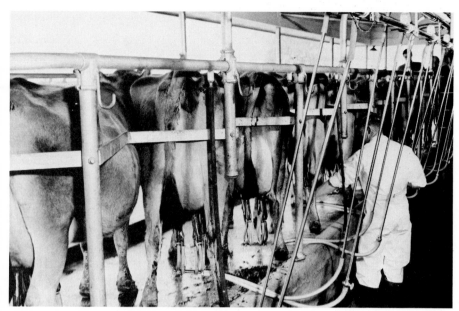

Machine milking; Victoria, Australia, 1959.

vegetable fibres used since the dawn of civilization. Some products reached the consumer only after chemical processing; oils, for example, were hardened by catalytic hydrogenation to meet the demands of the growing margarine industry. Margarine production rose from 400 000 tons in 1900, when it was based largely on tallow, to 2.1 million tons in 1950, more than half of it made from hardened vegetable oils.

The direct impact of mechanization was greater in crop production than in animal husbandry, mainly because of the wider range of tasks to be done. Most of the traditional operations – ploughing, harrowing, cutting, baling, reaping, binding, threshing, winnowing, and so on – lent themselves fairly readily to mechanization. There were, however, important areas where the old labour-intensive methods prevailed: such were, for example, the picking of fruit and tea. Towards mid-century, some success was achieved in mechanizing even such finicky tasks as potato harvesting and cotton picking. How far these new methods were applied naturally depended on local conditions: if labour was cheap and abundant, mechanization would show little advantage and might even be anti-productive. The USA was foremost in mechanization. In 1900, a remark made by Thomas Jefferson at the beginning of the nineteenth century remained true: 'In Europe the object is to make the most of their land, labour being abundant; here it is to make the most of our labour, land being abundant.' Availability of capital,

Ann Arbor pick-up bailer at work, England 1932. Note that the bales are tied by hand.

too, was an important factor, and it is not surprising that farm mechanization slowed up during the years of the depression between the wars.

We must, however, not neglect the indirect consequences of mechanization. For the urban dweller, it is easy to forget that a substantial proportion
of farm crops goes to feeding farm stock at times when grazing is inadequate.
The more efficient production of fodder crops resulting from mechanization
was therefore important in animal husbandry. We must remember too, that
unlike horses, tractors need no fodder at all, though their total dependence
on oil was to become a major problem in the 1970s as the energy crisis
developed. Energy consumption then became generally recognized as a
major factor in farming economics.

The basis of mechanization was the tractor, a very versatile machine. Not
only could it haul a variety of farm machines but it could also provide a
stationary source of power for such equipment as circular saws, chaff-
cutters, root choppers, and so on. By the Second World War, however, and
increasingly afterwards, machines such as combine harvesters were self-
propelled. Large machines of this sort were not suitable for working small
areas and contributed to the increase in the size of holdings. At the other end
of the scale, however, the needs of the smallholder practising intensive
cultivation were not forgotten and a range of small mechanical cultivators
was developed, particularly in Japan in the 1930s to suit the needs of small
rice growers.

The practice of farming large units of land contributed to one of the major problems of the new territories, namely soil erosion; this was particularly serious in Australia and the USA. The removal of hedges and other wind-breaks and the destruction of soil texture by repeated overcropping led to the creation of dust-bowl conditions in which the vital topsoil was literally blown off the land. In the USA the problem became of such magnitude as to warrant government intervention, and much attention was given to soil conservation techniques, originally as a matter of emergency and later as a normal part of farming practice.

Plant breeders had, as we have noted, greater success than animal breeders, primarily because life cycles are shorter and results could be evaluated correspondingly quickly. Cross-breeding led to some notable achievements and the selection of qualities that were particularly important. Yield and quality were by no means the only desiderata. The plant breeder was interested also in suitability for particular climatic conditions, resistance to disease, and ease of mechanical harvesting. Considerable progress was made, especially in cereals; important factors were the rediscovery by Hugo de Vries in Holland in 1900 of the basic laws of inheritance formulated on the basis of experiments with peas by Gregor Mendel in Czechoslovakia in 1865, and the enunciation of the chromosome theory of genetic inheritance by T. H. Morgan in the USA in 1911.

Some notable successes were achieved in the development of new hybrid corns, and contributed largely to increasing the average yield in the USA from about 26 bushels per acre in 1900 to nearly 40 bushels per acre in 1950. The new strains incorporated varieties from all over the world. Russia, for example, built up a collection of 30 000 varieties of wild wheats in the inter-war years at Leningrad. By 1918 the most important Canadian wheat was Marquis, resulting from a cross made in 1892 between Canadian Red Fife and an Indian strain, Hard Red Calcutta. Very early in the century Federation emerged as a wheat particularly suitable for Australian conditions. In Britain, Little Jos was introduced in 1910, as a result of cross-breeding experiments at Cambridge University, as a variety especially resistant to rust and suitable to East Anglian conditions. Similar successes were recorded over a very wide range of crops.

As with farm animals, however, the successes of plant breeders were frustrated unless crops were properly raised, and here there were two major lines of development. Firstly, better understanding of the principles of plant nutrition, coupled with technical developments in the chemical industry, led to widespread use of artificial fertilizers. Secondly, the attempts by plant breeders to provide disease-resistant strains were strongly supported by new chemical aids – insecticides, herbicides, fungicides, and so on – to protect the growing crops.

Harvesting Red Fife wheat with horse-drawn reapers and binders; Alberta, Canada, *c*.1908.

Spraying apple trees; Nova Scotia, Canada, *c*.1908.

By the beginning of the century it was well established that the principal mineral nutrients required by plants were nitrogen, phosphorus, and potash. Supplies of the last two of these were then adequate for the foreseeable future: phosphatic rocks were effective substitutes for bone meal as a source of phosphorus, and the vast salt deposits at Stassfurt provided the potash formerly applied as the ash of plants. Nitrogen, however, was a very different matter, for the world was largely dependent for this on caliche (mineral sodium nitrate) imported mainly from Chile. In 1900 total annual production was 1.3 million tons, most of which went to Europe; American imports amounted to 170 000 tons.

In 1898, in a memorable presidential address to the British Association for the Advancement of Science, Sir William Crookes stressed the risk of world starvation if reliance was placed solely on the rapidly dwindling deposits of caliche, and he suggested that an alternative might be found by chemically 'fixing' the universally abundant nitrogen of the atmosphere. An industrially viable fixation process was achieved in Germany in 1913 with the advent of the Haber–Bosch process (p. 134) for the synthesis of ammonia, though a decade earlier the Birkeland–Eyde process had had limited success in Norway because of the availability of cheap hydroelectricity. Ammonia could be used as a source of the nitric acid essential for the manufacture of military high explosives or of synthetic nitrogenous fertilizers. With the end of the First World War the first of these needs largely disappeared and a new system of arable farming developed which relied heavily on the availability of cheap synthetic nitrogenous fertilizers: cheapness was accentuated by world over-production and often by government subsidies to users. In 1938, 2.5 million tons of nitrogen was incorporated in synthetic fertilizers; by 1960 the figure had risen to more than 10 million tons. Two dangers implicit in this policy were not generally appreciated until after the Second World War. Firstly, while artificial fertilizers are in every way as nutritious for plants as those traditionally applied as manure, dried blood, bone meal, and so on, they do nothing to maintain the texture of the soil. This led to much controversy between the advocates of synthetic fertilizers – whose nutritional value can at least be precisely measured – and the so-called 'muck and mystery' school which advocated only 'natural' products. Secondly, changing technology in the chemical industry led, in the 1950s, to dependence on petroleum as a principal raw material, and after the OPEC increases of 1973 – by which time world consumption of nitrogen in fertilizers had risen to 32 million tons – this was no longer cheap. However, at mid-century these were still mainly problems for the future, and cheap nitrogen loomed large in farming economics.

Even in 1900, chemical warfare was no novelty in crop production. The

Experimental plots on Broadbalk wheat field, Rothamsted Experimental Station, England, 1925. This field had been used for fertilizer trials since 1834.

main targets were weeds, which smothered growth; insects which attacked growing crops; and micro-organisms – fungi and bacteria – which caused diseases. But at that time the weapons were relatively weak. Copper and arsenic salts and sulphur were effective against fungal diseases, and were widely used, for example, in vineyards; plant extracts such as nicotine and pyrethrum were used against insects; sodium chlorate and sulphuric acid were recognized as powerful, but non-selective, herbicides. From the 1930s, however, new synthetic products were introduced which were far more effective.

In weed control the need for selectivity is of prime importance: what the farmer needs is a product that will not affect his crops but will be fatal to weeds among them. This is to seek perfection, however, for what is a crop plant in one context may be a weed in another: there is no botanical distinction between them. Nevertheless, in the 1930s chemicals appeared that distinguished between the two major classes of plants – the monocotyledons and the dicotyledons. It so happens that all the cereal crops belong to the first class and many weeds to the second, so that selective weed-killers became feasible. The first of the important selective herbicides was dinitro-ortho-cresol (DNOC), patented in France in 1932, followed by 2,4-dichlorophenoxyacetic acid (2,4-D), developed in the USA and widely introduced during the Second World War. In 1945 nearly 1 million pounds of 2,4-D were produced in the USA; only five years later this had risen to 14 million pounds. When the exceedingly high biological activity of these products is considered, it will be appreciated that these are very large quantities.

In the insecticide field the most important of the new synthetic products was dichloro-diphenyl-trichloro-ethane (DDT), whose insecticidal properties were discovered in 1939. As we have noted, this proved highly effective not only against a variety of insects harmful to crops, but also against many insect vectors of disease (lice, fleas, mosquitoes). It had a conspicuous success in Naples in 1943/4 in controlling an outbreak of typhus. DDT was followed by a range of insecticides similar in that they were all chlorinated hydrocarbons, such as the gamma isomer of benzene hexachloride (Gammexane), but by the 1950s it was becoming apparent that these products were not as harmless to animal and human life as had been supposed and there was considerable public outcry about their use; moreover, the insect world had not quietly acquiesced in this sudden onslaught but had developed resistant strains. New insecticides were, therefore, sought among other classes of organic compounds, notably those containing phosphorus.

The reaction against synthetic insecticides encouraged the advocates of control by natural predators, which had already had some modest successes. At the end of the nineteenth century, for example, cottony cushion scale devastating citrus crops in California was controlled by means of ladybirds imported to destroy it. In the 1920s prickly pear, which had spread rampantly in Queensland, Australia, was largely eradicated by importing natural predators from Texas, including the moth *Cactoblastis cactorum*, the cochineal insect, and red spider (in fact, a mite). These were all highly specific to prickly pear, being adapted to feeding on no other host plant. This apparently attractive alternative to chemical agents is a panacea of only limited application and there are obvious hazards if the imported predator – finding itself in new surroundings – changes its way of life.

III. FISHING AND WHALING

Although commercial fish farming makes an important contribution to food production in the warmer countries of Africa and Asia – and with the introduction of new techniques has recently attracted greater attention in Europe and the USA – , fishing is essentially a survival of man's ancient role as a forager. So far as there were changes in the fishing industry in the first half of this century – and they were in fact quite considerable – they were concerned almost entirely with the adaptation of new technologies to the locating and catching of fish. This is in sharp contrast to agriculture where, as we have seen, the emphasis was on improved methods of production. In fishing the only developments on this side of the ledger were some international attempts to control fishing in the interests of long-term conservation, but these had very limited effect largely because of the difficulty of enforcement.

Of the new technologies, it was perhaps those related to transport that had the most far-reaching effects. On the one hand, larger vessels were able to travel further afield; on the other, their catches could be more widely and quickly distributed after landing. At the turn of the century steam trawlers were rapidly ousting sailing vessels. To take the British port of Grimsby as an example, we may note that, in 1902, 29 sailing trawlers remained there, compared with 686 only a decade earlier. Steam in turn began to be replaced by diesel engines, first in Japan and the USA. Engines, whether steam or diesel, were important not only for propulsion but also to provide power for hauling nets and otherwise alleviating the heavy physical toil of fishermen. In particular, this led to a great increase, after the Second World War, in the use of seining, in which a shoal of fish is surrounded by a circle of net which is then drawn close. Even before the war, Californian tunny fishers were using seine nets with a capacity of 100 tons of fish. After the Second World War there were important changes in the nets themselves, as such strong and rot-proof man-made fibres as nylon, Terylene, and polypropylene became available.

Such methods could be used, however, only when a shoal had been located, and here the echo-sounder, originally developed as an aid to navigation, became important in the later 1930s, and was much more widely used after the Second World War. Another navigational aid developed during the war was radar, and this quickly became standard equipment for fishing vessels seeking the more distant grounds.

Long before 1900, fishing vessels were carrying ice, produced in shore-based refrigeration plants, to assist in keeping the catch fresh until landed, and about 1910 experiments were made to install refrigerating plant on board ship. Not until the late 1920s, however, was this problem seriously tackled, when it was realized that the rich but distant new fishing grounds could be profitably exploited only by deep-freezing the catch on the spot either in the individual fishing vessels or in a factory vessel serving a fleet. By the mid-1930s, however, these early experiments were flagging, mainly for economic reasons, and deep-freezing at sea, which in recent years has transformed the industry, is essentially a post-war development.

For practical purposes whales may be classed with fish, although they are in fact mammals. Methods of catching them are determined largely by their size: the harpoon gun (c. 1870) has to replace the net. Unlike fish, it is the oil

The *Fairfree*, an experimental freezer trawler, was the first to use a stern ramp to haul nets aboard over the stern instead of the side. Leith, Scotland, 1947.

rather than the flesh that is prized. The oil is hardened for the margarine industry and the flesh and bones are turned into meal to be used in animal feed. Whales have to be sought in distant waters, even towards Antarctica. Traditionally, they were flensed while lashed alongside the catcher, but by the 1920s this task was being performed either on factory ships or at shore stations.

In the factory vessels, the whales were hauled aboard over a ramp in the stern, and after the Second World War this important device began to be used on fishing vessels to haul in their gear. One of the first vessels so equipped was the freezer trawler *Fairfree*, in 1947, owned by Salvesen of Leith.

Fishermen, operating unobserved in small groups in distant waters, cannot easily resist the temptation of a good catch, whatever international regulations may demand, and we have already remarked the comparative failure of international control conventions. The same proved true of whaling but with even more harmful results, and the industry suffered a serious decline after the middle of the century as a result of over-fishing.

BIBLIOGRAPHY

Bardach, J. *Harvest of the sea*. New York (1968).

Bunting, A. H. (ed.). *Changes in agriculture*. Duckworth, London (1970).

Burgess, G. H. O., Cutting, C. L., Lovern, J. A., and Waterman, J. J. *Fish handling and processing*. HMSO, Edinburgh (1965).

Callaghan, A. R. and Millington, A. J. *The wheat industry in Australia*. Angus and Robertson, Sydney (1956).

Carson, R. *Silent Spring*. Hamish Hamilton, London (1963).

Cuthbertson, D. P. (ed.). *Progress in nutrition and allied sciences*. Oliver and Boyd, Edinburgh (1963).

Evans, E. *Plant diseases and their chemical control*. Blackwell Scientific Publications, Oxford (1968).

Hall, D. (ed.). *Agriculture in the twentieth century*. Clarendon Press, Oxford (1939).

Hanson, S. G. *Argentine meat and the British market*. Stanford University Press, Stanford, Calif. (1938).

Harvey, N. *A history of farm buildings in England and Wales*. David and Charles, Newton Abbot (1970).

Held, R. B. and Clawson, M. *Soil conservation in perspective*. Johns Hopkins Press, Baltimore (1965).

Huxley, E. *Brave new victuals: an inquiry into modern food production*. London (1965).

Jacks, G. V. and Whyte, R. D. *The rape of the earth: a world survey of soil erosion*. Faber and Faber, London (1939).

Japan, FAO Association. *A century of technical development in Japanese agriculture*. Tokyo (1959).

Jenkins, J. T. *The herring and the herring fisheries*. London (1967).

Kristjonsson, H. (ed.). *Modern fishing gear of the world*. Fishing News (Books), London (1959).

Masefield, G. B. *A history of the Colonial Agricultural Service*. Clarendon Press, Oxford (1972).

Mellanby, Kenneth. *Pesticides and pollution*. London (1967).

Ministry of Agriculture, Fisheries and Food. *Animal Health: a centenary 1865–1965*. HMSO, London (1965).

—— *A century of agricultural statistics, Great Britain, 1866–1966*. HMSO, London (1966).

Ommaney, F. D. *Lost leviathan: Whales and whaling*. Hutchinson, London (1971).

Peacock, F. C. (ed.). *Jealott's Hill: fifty years of agricultural research 1928–1978*. Imperial Chemical Industries Plant Protection Division (1978).

Rasmussen, W. D. (ed.). *Readings in the history of American agriculture*. University of Illinois Press, Urbana (1960).

Russell, E. J. *A history of agricultural science in Great Britain*. George Allen and Unwin, London (1966).

Schlebecker, J. T. *Whereby we thrive: a history of American farming 1607–1972*. Iowa State University Press, Ames (1975).

Traung, J.-O. (ed.). *Fishing boats of the world*. Fishing News (Books) London (1955).

67084

United States Department of Agriculture. *After a hundred years. The Yearbook of Agriculture 1962*. United States Government Printing Office, Washington, DC (1962).

Von Brandt, A. *Fish catching methods of the world*. Fishing News (Books), London (1964).

Whetham, E. H. *British Farming 1939–49*. Thomas Nelson, London (1952).

White, E. W. *British fishing-boats and coastal craft*. Science Museum, London (1950).

9

MINING

Fishing and mining are not so disparate as might seem at first sight, for both are concerned with gathering the bounty that nature provides. The important difference is that, while the fish population can renew itself, the world's mineral resources do not, and this has important political and economic consequences.

Until roughly a century ago the miner was concerned primarily with solid minerals and metallic ores or coal – which might outcrop at or near the surface or have to be sought deep in the ground. Within the present century, however, liquids and gases have become of major importance. As we have noted, the petroleum industry, little more than nascent in 1900, grew by mid-century to be a principal source of energy. By that time, too, natural gas – with an already long history of exploitation in North America – began to be of major importance in the European economy. The very different physical properties of the minerals sought were inevitably reflected in different techniques of extracting them. In the present chapter we may conveniently consider the development of mining technology under three main headings: metallic ores, coal, and petroleum. Natural gas has already been considered (ch. 3) as a complement to manufactured gas.

I. METALLIC ORES

The techniques of extracting ores are largely independent of their chemical nature, but we must note that the twentieth century saw a demand for certain minerals previously of no great importance. One such, for example, was bauxite, the principal source of aluminium. In 1900 the aluminium industry was in its infancy; by 1950 world production of bauxite was nearing 10 million tons per annum. Chromium, tungsten, molybdenum, vanadium, and manganese became important as ingredients of special steels: in 1950 production of manganese ore (mainly pyrolusite and psilomelane) corresponded to more than 3 million tons of the metal. Old metals became increasingly in demand as new uses were found for them: lead, for example, one of the important metals of the ancient world, was required in large quantities for the manufacture of storage batteries and antiknock (lead tetraethyl) for petrol.

The first half of this century saw a big increase in open-cast mining as a consequence of the availability of powerful machinery to strip off the

Open-pit mining for iron; Kiruna, Sweden, 1936.

overburden of rock and then grab out the underlying ore. Initially steam-power predominated: steam-shovels loaded ore – if necessary previously loosened by blasting – into railway trucks to be hauled away by steam locomotives. With the availability of powerful diesel engines, however, more mobile machines, running on roads rather than rails, or on caterpillar tracks, were introduced. In the early 1930s drag-line excavators were introduced in the USA.

Some of these open-cast mines were enormous. When the Bingham copper mine opened in Utah in 1910 some 4000 tons of ore was milled daily: it was, indeed, only the huge scale of operations that made it economic to work these low-grade ores (2 per cent copper). Some pits, such as those opened up in 1901 to work the rich iron-ore deposits in the Mesabi Ranges on the shores of Lake Superior, eventually reached a depth of half a mile. With a non-renewable product, however, this process could not be continued indefinitely. Either the mine became exhausted or, as in the Sudbury copper mines in Canada or the Kiruna iron-ore mines in Sweden, the vein of ore ran steeply underground and could no longer be worked from the surface. Recourse to conventional methods of underground working was then necessary.

An alternative method of extracting valuable minerals lying at or near the surface was to bring them up by means of huge dredges floating on artificial lakes. This technique was used early in the century to extract gold-bearing gravel in the USA and tin ores in Malaya. An alternative method, hydraulic mining, was to wash out the ore by means of very powerful jets of

Gold dredger working in artificial lake, Colorado.

water; this, too, was used in tin-mining in Malaya, and it is estimated that in 1925 nearly half of the tin extracted there was obtained by this method. Both alluvial and hydraulic mining were open to the objection that they created vast quantities of waste that could be harmful to local agricultural and other interests. They were, therefore, often subjected to restrictive legislation. In California hydraulic mining was banned altogether in 1884, but allowed under licence during this century; in Malaya it was banned in 1933, but later permitted again in certain areas.

In mining underground there was little change in principle from traditional methods, but considerable changes in practice. Until 1900, transport underground was largely by horses in the main roadways, but by the 1930s small diesel locomotives had been introduced. The exhaust fumes presented a problem in the confined space but after the Second World War this was overcome by purifier attachments. Meanwhile, however, electric traction – using either batteries or mains supply – had been gaining ground; this, of course, presented no exhaust problem. The development of mechanical transport was accompanied by the development of mechanical loading, easing one of the heaviest of underground tasks.

In most mines considerable use was made of explosives to shatter the rock. In the latter years of the nineteenth century, Alfred Nobel's dynamite (1867) and blasting gelatine (1875) displaced the gunpowder which had been used for centuries. Whatever explosive was used, however, holes had to be drilled to receive it, and this was a slow and laborious job carried out by repeated percussion with a steel rod. Pneumatic drills had been introduced by G. Someiller for the construction of the Mont Cenis (1857–71) and St. Gothard (1872–81) tunnels, but their use in the much more constricted areas of most mines presented problems because of the clouds of dust created. The Leyner water drill of 1907 greatly reduced this hazard, and the introduction of drill bits tipped with tungsten carbide in the 1930s still further improved the performance of drills underground.

The quality of ore varies greatly and, in the absence of rich new discoveries, increasing use had to be made of lower-grade ores not worked in earlier times. Often, indeed, as with the lead mines of Mendip, yesterday's waste tips become today's raw material. Whatever the grade of ore, however, some concentration is necessary before smelting, and the twentieth century saw considerable improvements in this field. Concentration was commonly done at the mine, to avoid the cost of transporting worthless material. At its start, simple washing methods – based on the differing specific gravities of the mineral and the gangue – were used; these were similar in principle to the panning methods of traditional gold miners. From about 1910, however, flotation methods came increasingly into use. These are complicated in practice but simple in principle. Briefly, if the crushed ore is treated, under carefully controlled conditions, with a wetting agent,

Rock drilling with compressed air in a South African gold mine c.1920.

Hydraulic mining, using powerful jets of water to wash out gold-bearing gravel; California, early twentieth century.

and air is bubbled through the resulting slurry, the mineral particles can be concentrated in the froth on the surface. The method was first applied to sulphide ores and an early success was recorded in Chile by the vast Anaconda Company in 1915 when it increased copper recovery from 79 to 95 per cent. By 1950, some 200 million tons of ore were being concentrated annually by flotation methods.

II. COAL MINING

As in metal mining, the main development in coal mining was the increasing replacement of manual labour by mechanization. The form of the latter was to a considerable extent determined by the style of mining. In the older mines the method of bord-and-pillar working was used. In effect the seam was divided by roads (for which *bord* is a Saxon word) leaving pillars of coal to support the roof. The coal left as pillars was then extracted by paring them down and simultaneously supporting the roof with props. In deep mines, however, which had increasingly to be developed in Europe as more accessible seams became exhausted, the sheer weight of the overlying rock was liable to cause the pillars to collapse, and an alternative method of working, known as longwall, began to be widely used after about 1850. In this, as its name implies, extraction is effected in one operation from a long

Meco-Moore cutter-loader at work on longwall coal face, c.1943.

wall of coal – perhaps 100 yards long. If extraction was begun at the base of the wall, the weight of the overhanging coal helped to bring it down. This method lent itself more readily to mechanical cutting and loading than the more complex bord-and-pillar system.

Although mechanical cutters, driven by compressed air, had been introduced as early as 1863, their over-all contribution to coal production in Europe in 1900 was negligible: pick-and-shovel work on faces loosened by explosives was still the order of the day. From that time onwards, however, mechanical cutters – mainly of the chain type that cut a deep slot in the coal face – came increasingly into use to meet the heavy demand. Compressed air, which is inefficient when used over large distances, was increasingly replaced by electricity. If electricity was employed, however, the risk of explosive gas mixtures being present demanded great care in the design and use of equipment to avoid sparking hazards.

Mechanization at the coal-face demanded more efficient methods of removing the cut coal, and conveyor belts began to be introcuded around the turn of the century; one of the first – a Blackett steel-chain conveyor – was installed at Durham in 1902. The Meco-Moore cutter-loader, introduced in the 1930s and greatly improved during the Second World War, combined cutting and loading in one machine.

In the USA, where – in contrast to Europe – the seams are generally thick and relatively close to the surface, the ancient bord-and-pillar system continued to prevail and suitable machines for undercutting coal were introduced. As early as 1900, 25 per cent of American bituminous coal was cut in this way and at mid-century almost all was mechanically cut. Europe was not far behind; three-quarters of British coal was mechanically cut in 1947. Mechanical loading made slower progress, however. As late as 1923, 99 per cent of American coal was loaded by hand, and in 1950 the figure was still 30 per cent. A very important development in the USA was the continuous miner, a machine that claws its way along the seam and sends the cut coal back to a shuttle car or conveyor belt behind it. Thence it was automatically conveyed to the shaft, and the hoist to the surface was also automatically controlled. By 1964, 39 per cent of American coal was extracted by continuous mining. The effect on productivity was remarkable. In 1900 the American miner's average daily output was 3 tons: sixty years later continuous mining had raised this to 14 tons.

In 1936 Russia introduced hydraulic methods into coal mining and they were used also in the USA after the Second World War. In America, however, they were less attractive than in the generally more steeply pitched seams of Europe.

Other technical innovations underground were hydraulic roof supports in place of the traditional wooden pit-props; diesel or electric traction, as in metal mining; electric lamps which replaced the traditional Davy safety

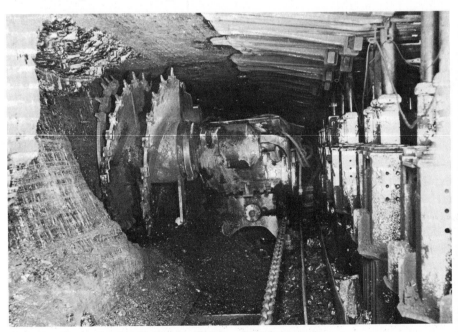

Anderton Shearer-loader; Kellingley Colliery, England, c.1955.

Trunk-belt coal conveyor *c*.1935.

As coal-mines were modernized after the First World War diesel traction began to replace the traditional pit ponies.

Electric lighting in coal-mines was an important innovation.

Hand sorting of coal; Virginia, USA, c.1910.

lamp; improved methods of ventilation; electric winding engines and improved lifting gear; and the replacement of beam pumps by powerful centrifugal pumps. This last development is a reminder of the constancy of certain factors in the history of technology. The first major use of steam-engines was to drain the Cornish tin mines, and two centuries later water remains a powerful enemy. In some coal-mines as much as 30 tons of water has to be pumped out for every ton of coal raised.

As has been noted, the coal deposits of the USA tended to be thick and to lie near the surface. This favoured open-cast or strip mining similar to that already described for minerals, although it did not make a significant (10 per cent) contribution to output until 1941. By 1964 one-third of American coal was gained by this method; it was highly productive, with yields of up to 30 tons per man per day. Enormous machines, with shovels capable of grabbing more than 100 cubic metres of coal at a time, were developed for this purpose. In Britain, economic and land-conservation considerations discouraged open-cast mining, but it was introduced during the Second World War to conserve manpower and has since been extended.

Machines were less discriminating than men, and one consequence of mechanization was dirtier coal, which had to be upgraded by cleaning. In the early years of this century grading had been done largely by hand picking, but even in the 1930s the amount of coal so treated in the USA was less than 10 per cent. After the Second World War, in the face of higher transport costs and more discriminating customers, it had risen to more than 60 per cent. By this time, however, washing techniques, similar to those used in metal mining, had been introduced.

III. PETROLEUM

Unlike minerals and coal, petroleum is a liquid and can therefore be extracted through a pipe, either forced out by natural pressure or by pumping. The basic technological problem is to locate a deposit and then to tap it through the intervening layers of rock, which may be some thousands of feet thick.

In 1900 drilling for oil was still almost entirely effected by percussion, a method used by the Chinese for centuries in their search for brine. It consisted in repeatedly dropping a heavy steel tool, using steam-power, so that its tip broke up the rock as it slowly descended: as the hole deepened, extensions were added to the drill, and the bore was lined with successive sections of iron or steel pipe riveted, or later screwed, together. To reduce the number of joints, the sections were kept as long as possible, and this was the reason for the tall derricks so characteristic of oilfields. This traditional method of drilling was slow but effective and in 1925 achieved a depth of nearly 8000 ft in Pennsylvania.

Meanwhile, however, an alternative method was being explored. This was the use of a rotary drill, well established in France in the 1860s to drill for water but making little appeal to American oilmen at the turn of the century, partly because the bits then available were suitable only for soft rocks. By 1910, however, rotary rigs began to be accepted and by the 1920s their use was fairly general, though even a decade later combination rigs were not unknown. This development was linked with the development of new alloy steels – and later tungsten carbide – capable of dealing with hard rocks.

At the same time, changes were made in the derrick. At the end of the First World War two-cylinder steam-engines replaced single-cylinder engines and these in turn were succeeded, especially after the Second World War, by diesel or diesel–electric units. In the early 1930s steel rigs began to replace wooden ones and the derricks became taller to enable larger sections of drill-pipes to be used. The wooden derrick extended to about 80 ft, but by the early 1930s steel derricks 136 ft high were in use. By that time wells some 10 000 ft deep were being drilled, and by mid-century this limit had been more than doubled; a depth of 20 521 ft was recorded in 1950.

These great depths – made necessary by the depletion of more accessible deposits – introduced new problems, notably loss of power through friction.

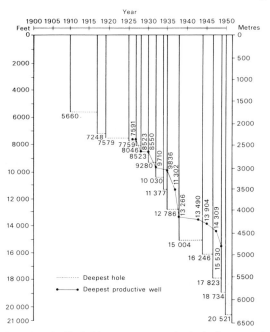

Cable-tool drilling rig with steel derrick, steam driven; Persia, 1909.

The first half of this century, particularly its second quarter, saw a rapid extension of depths achieved in the search for oil.

In the early days of rotary drilling it was noted that the work was easier when muddy conditions were encountered. This led to the practice of deliberately pouring mud – and subsequently much more sophisticated materials – down the well to lubricate the bit.

In normal circumstances the driller's aim is to bore a vertical hole, and, to prevent the bit wandering, it was customary to attach a heavy collar just above it. At times, however, an oblique approach was desired, as when the ground immediately above the oil was not available or suitable as a base. To achieve this deviation, wedges or whipstocks were driven in at intervals to deflect the drill; with deep wells, deflections of as much as a mile were possible at the bottom.

As drilling increased, mechanical problems arose owing to the sheer weight of the equipment; attention was therefore directed to an alternative method. In this, the power unit – actuated by the circulatory drilling liquid – is attached directly to the bit and lowered down the bore with it. Experiments with such turbo-drills were made in California at the beginning of the century, but much of the pioneer work was done in Russia. In 1925 the Russians introduced a single-stage turbine giving a drill speed of around 100 r.p.m., but this was not satisfactory and ten years later it was

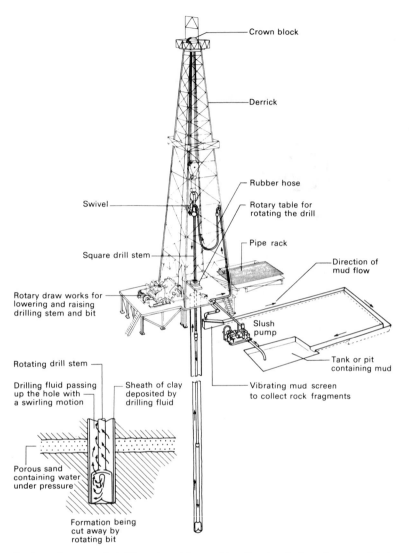

Crown block

Derrick

Rubber hose

Swivel

Rotary table for
rotating the drill

Pipe rack

Square drill stem

Direction of
mud flow

Rotary draw works for
lowering and raising
drilling stem and bit

Slush
pump

Rotating drill stem

Tank or pit
containing mud

Drilling fluid passing
up the hole with
a swirling motion

Sheath of clay
deposited by
drilling fluid

Vibrating mud screen
to collect rock fragments

Porous sand
containing water
under pressure

Formation being
cut away by
rotating bit

Schematic view of rotary oil-drilling rig, with provision for circulation of mud as lubricant;
early 1930s.

replaced by a multi-stage turbine. Difficulties arising from the cramped,
wet, and highly abrasive conditions in which such drills must work were
gradually overcome, and by 1950 some four-fifths of Russian deep drilling
was being done in this way. Further improvements were affected by
V. Tiraspolsky, and the maximum speed was increased to 500 r.p.m. In
Britain, a compact and relatively simple turbo-drill was developed by Sir
Frank Whittle.

In the 1920s the richest oil field in the world, in terms of yield per acre, was Signal Hill, California.

When first introduced into the USA, the Russian drill was not well received. Its main advantage – its relatively high speed of penetration – was offset by the fact that much of the time of a boring operation is occupied not in drilling but in setting up and dismantling the surface equipment. Moreover, the high speed of rotation led to rapid wear of the bit. Nevertheless, its flexibility was a major point in its favour; in particular it lent itself to the type of deflected bore-holes just described. This proved particularly valuable in the development of North Sea oil in the 1960s, enabling a large area of the sea-bed to be tapped from a single rig.

Although no previous project had equalled North Sea oil in complexity, offshore drilling was no novelty. In 1894 oil was discovered in California near Santa Barbara, and the Summerland oilfield was exploited from rigs mounted on piers extending out to sea. From this small beginning extensive underwater operations began. Rigs built on fixed platforms were established on inland lakes in Louisiana (1911) and Venezuela (1924). The first offshore operation seems to have been at Baku, on the Caspian Sea in 1925. In the 1920s and 1930s there was extensive offshore drilling in the Gulf of Mexico, partly from mobile rigs and partly from fixed platforms. After the Second World War these activities were greatly extended and very large self-contained mobile rigs were introduced. From these oil was taken ashore in tankers or by pipelines.

By this time, drilling had become very expensive and could be undertaken only by large international companies commanding immense resources of capital and equipment. The cost of a typical deep-sea rig – of which there were some 200 in 1965 – was around £2½ million; hire costs were around £10 000 per day. If one drilling in ten struck oil in the North Sea, performance was judged satisfactory, even though a single 10 000 ft well might cost £1 million in cash and three months in time.

In such circumstances increasing attention was paid to surveys designed to locate potential oil-bearing structures deep underground. The word potential is used advisedly, for in the last analysis only drilling can prove the existence of oil. Of various methods devised, those based on seismography, first used in Mexico in 1923, are the most important. In these, explosive charges are detonated in drill-holes 100 ft deep; the resulting shock waves are reflected back from the various underground strata and expert examination of the records can reveal much information about the underlying structures. At sea, things were easier and quicker, for the explosives could be fired at the surface while towed behind a boat. Whereas 10 shots a day was a good total for a team working on land, 200 shots was possible at sea in favourable weather.

A second, but less informative, method is magnetic surveying, introduced in the late 1920s. It reveals little except the variation in the depth of

Summerland Oilfield near Santa Barbara, California, as it appeared in 1903. This was the first offshore oilfield developed in the USA.

Self-contained drilling and production platform in the Gulf of Mexico about seven miles off Louisiana. At that time (1948) the oil was taken ashore in barges (right) but subsequently a submarine pipeline was built.

Depth-charge exploding during seismic tests to investigate deep structure of the sea bed, 1950.

sedimentary rocks but it can be conducted from the air, and is cheap and quick. Studies of local gravity variations are also informative but difficult to conduct at sea.

BIBLIOGRAPHY

Anderson, F. S. and Thorpe, R. H. A Century of coal-face mechanization. *Mining Engineer*, No. 83, 775–85 (1967).

Blainey, G. *The rush that never ended: a history of Australian mining*. Cambridge University Press, London (1963).

Brantly, J. E. *History of oil well drilling*. Gulf Publishing Co., Houston (1971).

Cooper, Bryan and Gaskell, T. F. *North Sea oil: the great gamble*. Heinemann, London (1966).

History of Petroleum engineering, American Petroleum Institute, New York (1961).

Howse, R. M. and Harley, F. H. *History of the Mining Engineering Company Limited, 1909–1959*. MECO, Worcester (1959).

Lupton, A., Parr, G. D. A., and Perkin, H. *Electricity applied to mining*. Crosby, Lockwood, London (1903).

MacConachie, H. Progress in gold mining over fifty years. *Optima*, September 1967.

Medici, M. *The natural gas industry: a review of world resources and industrial application*. Newnes-Butterworths, London (1974).

Pirson, S. J. *Elements of oil reservoir engineering*. McGraw-Hill, New York (1950).

Richardson, J. B. *Metal Mining*. Allen Lane, London (1974).

Stoces, B. *Introduction to mining*. Pergamon, London (1958).

Thompson, A., Beeb, Y. *Oil-field development and petroleum mining*. Crosby Lockwood, London (1916).

Webster Smith, B. *The World's great copper mines*. Hutchinson, London (1962).

Williamson, H. F., Andreano, R. L., Daum, A. R. and Klose, G. C. *The American petroleum industry. The age of energy 1899–1959*. Northwestern University Press, Evanston (1963).

10

METALS AND THEIR EXTRACTION

Twentieth-century technology created an almost insatiable demand for metals – familiar metals serving familiar needs, but on a far larger scale; familiar metals serving quite new needs; and new metals being introduced. To meet the huge demand in a competitive world, the efficiency of traditional processes was improved and, as technological development in other fields permitted, new processes were devised. The development of the electrical industry, for example, made possible the wide use of the electric furnace in steel making and the introduction of electrochemical methods on a large scale for metals such as aluminium. As in virtually all other industries, operations demanding heavy manual labour were increasingly mechanized. The working of metals will be dealt with in a later chapter, and at this point we will concern ourselves only with their extraction.

I. IRON AND STEEL

These metals remained of paramount importance throughout the period of our concern. Production of pig-iron rose from about 40 million tons in 1900 to 130 million tons in 1950; steel from 30 million tons to 190 million. Throughout the first half of this century the USA increased her lead as the dominant producer: 37 per cent of the world market in 1900 had risen to 52 per cent of a much greater market in 1948. Other countries, however, had greatly expanded their capacity. Russia, for example, produced only some 2 million tons of steel in 1900, but by 1948 this had increased eightfold to more than 16 million tons. Japan, with negligible production at the beginning of the century, was producing nearly 6 million tons in 1944, with a sharp but temporary drop immediately after the war.

If we take 1937 as a representative year, the biggest single consumer of steel in the industrialized countries (about 20 per cent) was the building and constructional engineering industry, followed by mechanical engineering (15 per cent), shipbuilding (10 per cent), and railways (7 per cent). Perhaps unexpectedly, another major use (9 per cent) was in quite small items – bolts, nuts, screws, and rivets – used to hold a huge variety of structures together.

Although a variety of iron ores were used, they fell largely into two classes which have to be processed rather differently. Firstly, basic ores rich in phosphorus and low in silicon; secondly haematite ores low in phosphorus

Newport Ironworks, Monmouth, England, as it appeared in 1890.

and high in silicon. Steel is made from pig-iron by reducing the carbon content and removing most of the phosphorus and sulphur. Increasingly, as the century advanced, special steels were made by alloying with small amounts of other metals such as chromium, tungsten, and molybdenum, and this demanded special techniques.

The first stage in production was still the blast-furnace, into which is packed a mixture of iron ore, coke, and lime or limestone to act as a flux. The charge is fired at the base and burns continuously as the furnace is reloaded at the top. As smelting proceeds, a slag forms on top of the liquid iron, and both are tapped off separately at intervals. The liquid iron may be immediately converted into steel or cast into pigs which must be re-melted later; the former process is, of course, the more economical of fuel. The slag can be used as road metal or for making cement, while the gas generated can be used to pre-heat the blast of air blown in through tuyères – about ten in number – supplying the bottom of the furnace.

This general description of a blast-furnace applied throughout our period, but there were important changes in practice. The most obvious was the change in size. In 1902 the largest furnace in the world, belonging to the United States Steel Company, stood 90 feet high with a 15-foot hearth: it produced 500 tons of pig-iron daily. Fifty years later the maximum height had not risen much (100 ft) but the hearth diameter had doubled and output had increased to 1400 tons per day. Hand-loading at the top of the furnace was steadily replaced by mechanical loading through a hopper. Efficiency was much increased. Although precise figures are meaningless – because they depend upon such variable factors as the quality and type of

Four blast-furnaces, named after British queens, at Appleby-Frodingham Works, England.
In the mid-1950s their capacity was 23,500 tons of iron for oxygen steelmaking.

Hand-charging a blast furnace: Carnegie Steel Company, Pittsburgh, 1900.

25-ton Bessemer
converter; United
Kingdom, 1950.

ore and coke and the care taken in operation – it would be fair to say that
whereas in 1900 some 22 cwt of coke was required for every ton of pig-iron
made, by 1950 this had fallen to around 13 cwt; by 1970 it had fallen below
10 cwt. The influence of the nature of the ore is appropriately illustrated by
reference to the vast Mesabi deposits, already described. Although of good
grade they were of very fine texture; this led to clogging of the charge in the
furnace and excessive dust in the escaping gas. The growth of the electrical
industry provided one solution to the second of these problems, namely
electrostatic dust precipitation.

    The biggest change in the industry was in steel making, for which two
processes were in use in 1900. The dominant one was the Bessemer (or
Kelly) process (1860) in which molten iron was converted into steel by
blowing air through it for 8–10 minutes in a converter carrying a charge of
perhaps 25 tons; the molten steel was then poured into a ladle and cast into
ingots. The alternative process was the open-hearth process (Siemens-
Martin), also introduced in the 1860s, which, among other advantages,
allowed the use of low-grade coal and scrap-iron. It was slower than the

Bessemer process and more easily controllable. The balance of advantage lay with the open-hearth process and as early as 1894 this outstripped the Bessemer process in Britain. In other countries the change-over to the open-hearth as the major process for steel production took place later – in the USA in 1908 and in Germany in 1925. For many years it was common for the two processes to be combined, a technique known as duplexing. In this, the first step towards steel making was taken in a Bessemer converter and the process was finished in an open-hearth furnace.

The original open-hearth furnaces were hand-loaded and fired by producer gas (p. 32), but in the early years of this century mechanical loading became general. About 1930 heavy oil was introduced as a fuel in the USA, where it was readily and cheaply available, but in Europe it was not extensively used until after the Second World War.

At mid-century both processes were challenged by a third in which oxygen was used in place of air to burn the carbon out of the molten iron. The idea was not new, for Bessemer had taken out patents as early as 1856, but it was impracticable until oxygen became cheaply available in tonnage quantities: this requirement was not met until after the Second World War although experimental plant had been operated in Germany as early as 1937, following the work of C. von Linde and M. Fränkl on the manufacture of liquid oxygen from air.

Another important development in steel making was the advent of the electric furnace (initially used for non-ferrous metals in 1886), another consequence of the development of the electrical industry. In this, coke or charcoal remained the reducing agent, but the heating was electrical. Although in use in both the USA and Germany in the first decade of the century, its contribution was relatively small until after the Second World War. American production was half a million tons in 1918 but had risen to 6 million tons in 1950, by which time furnace charges of up to 200 tons were in use. The electric furnace was particularly important for handling the new alloy steels, for many of the alloying metals – such as chromium, vanadium, and tungsten – were readily oxidized in the presence of air.

Throughout the period of our present concern steel making was essentially a batch process, the liquid metal being moulded into ingots for further processing. The advantages of continuous casting were recognized by Henry Bessemer as early as 1856, but a century was to elapse before the multiplicity of technical difficulties was solved. Experimental plant was operated in the USA and in Britain shortly after the Second World War, but the first major plant seems to have been built in Russia, where an output of 50 tons per hour was achieved in 1955.

Finally, mention must be made of furnace linings. We have already noted that some ores, in both Europe and the USA, contain a good deal of phosphorus; both the Bessemer and the Siemens-Martin process ran into difficulties when these were used, since they were incompatible with the siliceous refractories then in use. In 1875 S. G. Thomas discovered that phosphoric ores could be satisfactorily utilized if a basic refractory – such as magnesia – was used instead. Such refractories combined with the phosphorus – which if unremoved made the iron brittle – to form basic slag, which was a valuable agricultural fertilizer. Dolomite, a mixed calcium/magnesium carbonate, was widely used but the lime formed after calcining presented difficulties because of its liability to hydration. This difficulty was overcome by adding a little tar to prevent access of water. An important development, in Britain, was the introduction in 1937 of a process to make magnesia from dolomite and sea water.

II. NON-FERROUS METALS

The number of non-ferrous metals in common use during this century is too large to allow each to be considered in any detail, nor is it possible to range them in any meaningful order of importance. Only four metals – lead, copper, zinc and, latterly, aluminium – are produced in quantities at all comparable with those of iron and steel. But even in 1948 world production of all of these combined totalled barely 6 million tons, compared with some 230 million tons for iron, steel, and ferro-alloys. It is hard to say, however, that any of these were more important than, say, vanadium (1700 tons) or molybdenum (9000 tons), which are essential as alloying ingredients for special steels used, for example, in high-speed machine tools. How, again, can we rate these in comparison with platinum (essential for high-temperature electrical contacts and industrial chemical catalysts) or mercury (for mercury-arc rectifiers, thermometers, and barometers). We have already considered the suddenly unique importance of uranium and the methods developed for its extraction – and the extraction of the artificial element plutonium – during the Manhattan Project.

One consequence of developments in twentieth-century technology was that certain metals were required in an exceptionally pure state. At the beginning of the century, zinc, for example, was commonly prepared by reducing the ore with charcoal or powdered coal and collecting the volatile metal as it sublimed. This gave metal of high purity, around 99 per cent, though it usually contained some lead and iron, together with traces of cadmium, arsenic, and sulphur. In the 1930s, however, it was discovered that certain zinc alloys, commonly containing about 4 per cent of aluminium and traces of magnesium and copper, were readily shaped by pressure die-casting. These were known as Mazak alloys (magnesium, aluminium, zinc, and 'kopper'). For these, however, zinc of 99.99 per cent purity ('four nines' zinc) was necessary; it was obtained by fractional distillation of the metal. If less pure zinc is used, the castings fail mechanically. Copper was another metal for which, in certain applications, new standards of purity had to be set. The specification for electrolytic tough-pitch (ETP) copper, used as an electrical conductor, required a minimum copper content of 99.90 per cent. In 1913 the International Annealed Copper Standard was introduced to define copper of specific electrical conductivity.

In a necessarily selective account of developments in metal extraction processes we may perhaps take as examples four 'new' metals – aluminium, magnesium, titanium, and beryllium. The first of these was in fact no novelty in 1900, for it had been isolated in 1845 and the modern electrochemical extraction process had been independently invented in 1886 by C. M. Hall in the USA and P. L. T. Héroult in France. This depends on electrolysis of bauxite (aluminium oxide) dissolved in molten cryolite, another aluminium-containing mineral.

As it required about 20 000 kwh of electricity to make one ton of aluminium, the availability of cheap electricity was an important requirement; generally, this meant hydroelectricity, and the first commercial production (1887) was at Neuhausen in Switzerland, near the famous falls of the Rhine. For reliable operation of the cells the bauxite must be very pure, but a satisfactory purification process was developed by K. J. Bayer in 1889. Subsequently synthetic cryolite replaced the natural product, mined in Greenland.

By the end of the nineteenth century the technological and economic problems of aluminium production had been solved; today it ranks second only to iron and steel as a constructional metal. It is surprising, therefore, that aluminium got off to a very slow start: by 1900 world production amounted to only 7000 tons. By 1938 it had risen to rather more than 500 000 tons and during the Second World War reached a peak of about 2 million tons, thanks largely to the demands of the aircraft industry for a metal that is both light and strong. By 1950 it had dropped back to $1\frac{1}{2}$ million tons.

Magnesium is another important metal obtained by an electrochemical process. The raw material here is seawater which contains about 0.13 per cent of magnesium in the form of magnesium chloride. This is precipitated by treatment with lime and the precipitate is dissolved in hydrochloric acid to re-form magnesium chloride in concentrated solution. The magnesium chloride is decomposed electrolytically to form magnesium metal and chlorine; the latter is recovered to regenerate hydrochloric acid. During the

Plant for production of magnesium from sea water.

Second World War the alternative ferrosilicon process, which originated in Germany, was developed in Canada. For this the raw material dolomite (a mixed carbonate of calcium and magnesium, widely distributed in nature and especially identified with the Dolomite Alps) is used. This is calcined, pressed into small briquettes, and treated at a high temperature in a vacuum with ferrosilicon, an alloy of silicon and iron. Under these conditions the magnesia is reduced to magnesium metal, which condenses at the cool end of the retort.

Magnesium has limited value as a metal in its own right, and its fortunes are closely linked with those of aluminium. In 1909 Alfred Wilm in Germany accidentally discovered that an alloy of aluminium containing 3.5 per cent copper and 0.5 per cent magnesium slowly developed great strength after heating and quenching in water. This phenomenon is known as age-hardening. The rights in the invention were acquired by the Durennes Metalwerke at Duren and this type of alloy became famous as Duralumin. Its first major use was in the construction of Zeppelins; after the First World War it was widely used in the rapidly growing aeroplane industry.

The success of Duralumin prompted research to see whether age-hardening could occur in other aluminium alloys; and this led to the development of the so-called Y-alloy at the National Physical Laboratory in Britain. This contains 4 per cent copper, 2 per cent nickel, and $1\frac{3}{4}$ per cent magnesium; it has the merit of preserving its strength at high temperatures and is therefore suitable for making pistons and other parts for internal combustion engines. A related alloy, RR58, containing rather less magnesium, was widely used for aircraft cladding.

A disadvantage of these alloys as cladding materials compared with aluminium itself is that they are more readily corroded. To obviate this, a very thin sheet of aluminium was rolled on to each side of the alloy sheet, rather in the manner of the old Sheffield plate.

At the other end of the scale are alloys in which magnesium pre-dominates, with relatively small (up to 9 per cent) amounts of aluminium and traces of other metals such as manganese or zinc. Such alloys attracted much attention in Germany in the 1930s because of the risk of a shortage of aluminium in the event of war; magnesium could be made from indigenous raw materials and it is even lighter than aluminium. The alloys can be shaped by the ordinary metal-working processes. The magnesium-alum-inium-zinc alloys remain the most important, but in the late 1940s others based on the addition of zinc and zirconium were developed. Although the mechanical properties of such alloys are good, they lose their strength at quite low temperatures (150°C). This defect was much reduced by the addition of thorium and/or some of the so-called 'rare earth' metals (neodymium, praseodymium, etc.).

Like aluminium and magnesium, titanium was no novelty in the twentieth century for it had been isolated by the German chemist Martin Klaproth as early as 1794. Not until 1925, however, was it produced in a form pure enough to determine its physical properties. These proved very interesting, for they revealed an unusual combination of mechanical strength, ductility, and resistance to corrosion. These properties are dependent on high purity, and especially on the absence of oxygen. Deliberate addition of certain materials led to alloys with even better properties, of particular interest to aeronautical engineers. Typical alloying metals are aluminium, manganese, vanadium, and molybdenum.

The technology of titanium production is difficult, partly because of the need totally to exclude oxygen, and its manufacture had only just become feasible by the middle of this century. Production in 1948 was only 3 tons, but it had risen to 25 000 tons in 1957. The extraction process employed was rather similar to that used for uranium (p. 57). The ore (ilmenite, titanium dioxide) was converted into titanium tetrachloride which, after careful purification, was reduced to metal with magnesium (Kroll process) or sodium.

We may appropriately conclude this survey of twentieth-century metals with a brief mention of quite a different metal, beryllium. This, too, was of respectable antiquity, having been isolated by L. N. Vauquelin in 1798. It was prepared electrolytically in France by W. Bussy in 1898, and twenty years later A. Stock and H. Goldschmidt developed a commercial electrochemical process for its manufacture. In the 1920s and 1930s manufacture was undertaken in the USA, which became the major producer. The interest in the metal stemmed from the discovery by M. G. Corson in 1926 that beryllium would age-harden copper without destroying its electrical and thermal conductivity; from then until the Second World War interest in beryllium was largely in the context of copper alloys. The metal then assumed a totally different significance with the discovery that it has a very low absorption capacity for neutrons – the lowest of any structural metal. This made it of great interest to nuclear engineers. In 1956 world output of beryllium was 12 500 tons, about 80 per cent of it being manufactured in the USA.

The development of beryllium was clouded by the discovery in the late 1930s that it had unsuspected toxic properties which appeared only slowly but caused a high mortality rate. Many workers in the industry became affected, as did others who had been exposed to beryllium compounds used as phosphors to coat the insides of fluorescent lamps (p. 77). Recognition of this hazard led to the introduction of strict safety regulations for its handling.

BIBLIOGRAPHY

Aitchison, L. *A history of metals*. Macdonald and Evans, London, (1960).

Beck, A. *The technology of magnesium and its alloys*. F. G. Hughes, London (1941).

Benbow, W. E. *Steels in modern industry*. Iliffe, London (1951).

Burn, D. L. *The economic history of steelmaking*. Cambridge University Press (1940).

—— *The steel industry 1939–1959*. Cambridge University Press (1961).

Carnegie Steel Company. *The making, shaping and treating of steel*. Pittsburg (1925).

Carr, J. C. and Taplin, W. *History of the British steel industry*. Blackwell, Oxford (1962).

Dennis, W. H. *A hundred years of metallurgy*. Duckworth, London (1963).

Durrer, R. *History of iron and steelmaking in the United States*. American Institute of Mining, Metallurgical and Petroleum Engineers, New York (1961).

Edwards, J. D., Frary, F. C., and Jeffries, Z. *The aluminium industry*. McGraw-Hill, New York (1930).

Hogan, W. T. *Economic history of the iron and steel industry of the United States*. 5 vols. D. C. Heath, Lexington, Mass. (1971).

Inglis, N. P. and McQuillan, M. K. A progress report on titanium. *Endeavour*, **17**, 77, 1958.

McQuillan, M. K. and Farthing, T. W. Beryllium. *Endeavour*, **20**, 11, 1961.

Osborn, F. M. *The story of the Mushets*. Nelson, London (1952).

Street, A. C. and Alexander, W. O. *Metals in the service of man*. Penguin Books, Harmondsworth (1972).

Warren, K. *The American steel industry 1850–1970: a geographical interpretation*. Clarendon Press, Oxford (1973).

—— *The British iron and steel industry since 1840*. G. Bell, London (1970).

11

CHEMICALS

The chemical industry, although of great antiquity, first emerged in its present form only at the beginning of the nineteenth century with the establishment of large-scale manufacturing processes such as the Leblanc process for soda, the lead-chamber process for sulphuric acid, and Charles Tennant's manufacture of bleaching-powder. In mid-century William Perkin's discovery of the first synthetic dyestuff was to have significant repercussions, and a little later new processes for the manufacture of both soda and sulphuric acid appeared; these were respectively the Solvay (ammonia/soda) and the contact processes. By 1900, however, the industry was poised for a period of rapid and far-reaching change, which was to make it one of the largest and most sophisticated in the world. As a preliminary to reviewing the particular nature of these changes, it is useful to discuss them in general terms.

First and foremost we may note the over-all growth of the industry, but in the present context we need not consider this in detail. This is fortunate in that the chemical industry defies strict definition: in different countries, and at different times, its scope has been variously interpreted. Thus in Britain in 1948, the President of the Board of Trade requested the Association of British Chemical Manufacturers to make a comprehensive survey of the industry; in this, the Association stated the number of employees in the industry to be 160 000. In 1949, however, the Ministry of Labour put the number of employees at 429 000, nearly three times greater. The discrepancy was essentially due to a different basis of computation: the Ministry of Labour included many allied trades such as petroleum refining and the mixing and blending of chemicals to form such varied products as cosmetics, points, polishes, and proprietary medicines. Statistics relating to the industry therefore need to be very carefully interpreted, and in the present context we must restrict ourselves to particular products. One of the most important has always been sulphuric acid, and until quite recently its production was regarded as a measure of activity in industry as a whole. It is interesting to note, therefore, that world production of this acid rose from 4 million tons in 1900, to 9.5 million in 1920, and to 21 million in 1948; that is, roughly a fivefold increase in the course of the half-century.

It would not, however, be possible to produce comparable statistics for the whole industry, for during this period products were introduced which were virtually non-existent in 1900: such, for example, were plastics, synthetic rubber, man-made fibres, many new pharmaceutical products, and a range of synthetic insecticides, herbicides, and fungicides. Diversification, as well as massive expansion of total output, was an important feature of the industry during this period. Existing products were made by new processes from new raw materials; we have already mentioned the supplanting of the traditional Leblanc soda process by the ammonia/soda process developed in Belgium by the Solvay brothers. Another very important development was the Haber-Bosch process for the manufacture of synthetic ammonia, which demanded the development of a wholly new high-pressure technology, a technology which was later applied very effectively in other fields, such as the manufacture of polythene and of synthetic petrol from coal. The growing availability of electricity favoured the development of a range of electrochemical processes, as for the manufacture of caustic soda; for the fixation of atmospheric nitrogen in the short-lived Birkeland-Eyde process; and for the electric-arc manufacture of calcium carbide.

These highly sophisticated developments were made possible by advances not only in engineering but in pure chemistry. Looking back to the early days of the industry it is astonishing how much was achieved by men with no formal chemical training and, indeed, at a time when chemical science was still very primitive even among its most skilled practitioners. By the middle of the nineteenth century this situation was changing rapidly, but it is interesting to recall that as late as 1856 William Perkin's synthesis of the dye mauvein was based on what we now recognize as a totally misconceived attempt to synthesize quinine. As the twentieth century advanced, and the basic principles of chemistry were firmly established, progress became increasingly dependent on the recruitment to the industry of highly qualified professional chemists from the universities. Chemical engineering began to emerge as a profession in its own right: the American Institution of Chemical Engineers was founded in 1908. In Britain the Institution of Chemical Engineers was founded in 1922 as an offshoot of the much older Society of Chemical Industry (1881), which was not able to offer formal qualifications because of the terms of its charter.

In 1890 the appointment of a Swiss chemist, Ferdinand Hurter, as head of an industrial chemical laboratory in Britain (United Alkali Company) was something of an event. Half a century later every major chemical company had its own research and development organization, which was a major charge on its income. In 1950 a survey showed that chemical companies allocated an average of about 5 per cent of their annual turnover to research, though this obscured major variations. In the progressive pharmaceutical branch of the industry, for example, it was often twice as much; in explosives manufacture, a good deal less. When ICI was formed in 1926, its annual research and development budget was roughly £350 000: by 1939 it was £750 000, and in 1950 nearly £5 million and rising very rapidly. By that time development, and the final cost of plant, had become the greater financial problem. While research was becoming expensive, and much of it unavoidably unproductive, the cost of turning a useful invention into an economically viable product rose even more alarmingly. A commitment to a research programme was relatively small, and could be broken off at short notice; a commitment to development, involving contractual obligations, was a very different matter. This was one of the main reasons for the emergence of the great international chemical companies: increasingly, only very large concerns could deploy the money and resources necessary to establish a major new product.

This period saw a major change in the centre of gravity of the chemical industry. On the eve of the First World War, Germany had a commanding lead. The value of her chemical production was £120 million compared with about £77 million for the USA, which had begun to forge ahead about 1880, and £25 million for Britain on the comparable basis. Her technological superiority was exemplified by the Haber-Bosch process, highly advanced by the standards of the day, which began to be worked in September 1913. Her versatility was demonstrated by the range and complexity of her products. By the eve of the Second World War the balance had decisively changed: the USA had established a commanding lead in the world as a whole. Though Germany was still the largest producer in Europe, her output of chemicals was already matched by Japan, a comparative newcomer. The American lead had been established in the 1920s, and by mid-century, after the tremendous stimulus to demand given by the Second World War, it seemed unassailable. These are all, however, subjects of relatively recent importance, and for the moment we must turn our attention to developments at the beginning of the century. In doing so, it is convenient to follow two separate but related paths – the inorganic and the organic. Traditionally, inorganic chemistry concerned itself with the inanimate, mineral world; organic chemistry with substances derived from living organisms – such as proteins, fats, and sugars. For many years, however, organic chemistry has been taken as comprising the compounds of the extraordinarily versatile element carbon, which are numbered literally in millions compared with hundreds or thousands for other elements. Relatively few occur in living matter, and a number that do can now be made synthetically. Organic compounds include solvents, dyes, drugs, and plastics. Nevertheless, natural organic compounds have not been wholly eclipsed; ones of continuing importance during our period of interest include insulin, penicillin, animal and vegetable fats and oils, rubber, and quinine.

These two major fields of industrial chemistry are not altogether separate, however. Inorganic sulphuric and nitric acids are very important intermediaries in the making of many organic compounds; organic petroleum is a source of hydrogen for synthesizing important inorganic substances such as ammonia. In two world wars another link was of permanent importance: high explosives are made by the action of nitric acid on a variety of organic substances such as glycerine, toluene, and phenol. They are also, of course, essential in mining and civil engineering.

I. HEAVY INORGANIC CHEMICALS

This section may appropriately begin with an account of sulphuric acid, the so-called 'barometer of industry'. Up to the 1930s the title was probably appropriate, in that, because it is so widely used, the demand for the acid was a fair measure of national industrial activity. Throughout the first half of this century the traditional lead-chamber process was worked as well as the more recent contact process, with the latter steadily gaining ground. In 1950 about 70 per cent of sulphuric acid in the USA was made by the contact process, but in Britain the two processes made roughly equal contributions. A major factor was that the contact process produces the strongest form of acid (oleum), which was increasingly required as an intermediary for the manufacture of many organic chemicals – dyes, drugs, explosives, etc. – as well as for refining petroleum. For this purpose, lead-chamber acid was not suitable. An important change was in the source of sulphur, traditionally derived mainly from pyrites (iron sulphide). From the beginning of the century, however, increasing use was made of the vast deposits of almost pure native sulphur in Louisiana and Texas. This was made possible by a process invented by Herman Frasch, and effectively worked from about 1900, in which the sulphur was melted underground with superheated water and forced to the surface by compressed air. The process had economic disadvantages, notably the large volume of super-heated water required, but was crucially important in that it provided very effective means of penetrating the overlaying quicksands which resisted traditional mining methods. By mid-century half the world's sulphur was being obtained by this novel process. In addition, larger amounts of sulphur were obtained as a by-product of the purification of petroleum and natural gas: in 1954 France made $1\frac{1}{2}$ million tons of sulphur in this way. Apart from sulphuric acid, sulphur was also required in increasing quantity for other purposes, most notably for vulcanizing the huge quantities of rubber absorbed by the motor-tyre industry.

The basis of the contact process is the oxidation of sulphur dioxide to sulphur trioxide in the presence of a catalyst (a substance which promotes a chemical reaction but is not consumed in it). Success depends on careful purification of the gases – to avoid 'poisoning' the catalyst – and careful choice of the catalyst and the way in which it is prepared. At the beginning of the century platinum-based catalysts were widely used, but during the First World War the German chemical industry (Badische Anilin-und Soda-Fabrik, BASF) began to use vanadium-based catalysts, first patented by E. Haen in 1900. By the 1930s vanadium was in general use.

A second very important mineral acid was nitric acid, also required in the preliminary stages of the manufacture of many organic compounds. It was required, too, for the manufacture of nitrogenous fertilizers and in two world wars was vital to all combatants for the manufacture of high explosives such as TNT (trinitrotoluene). Until the end of the nineteenth century nitric acid was made by treating sodium nitrate, imported from Chile as caliche (p. 95), with sulphuric acid, but early in the present century new methods were developed.

The first of these depended on the fact, discovered by Henry Cavendish as long ago as 1784, that at high temperatures – as in an electric spark – nitrogen and oxygen – the principal constituents of air – will combine to form nitric oxide, which can be converted into nitric acid. In the process developed by K. O. B. Birkeland and Samuel Eyde in Norway in 1901–4 a mixture of nitrogen and oxygen was passed through an electric arc fanned out with electromagnets into a disc some 3 metres in diameter. The efficiency of conversion was low and the consumption of electricity correspondingly high; hence its development in Norway, where hydroelectricity was cheap. The Birkeland–Eyde process was worked, after various vicissitudes, until the 1920s but it was only moderately successful: production in 1919 was around 100 000 tons.

Meanwhile in Germany, O. Schönherr, working with BASF, had developed a similar process and the community of interest led eventually to the formation of a joint company, Norsk Hydro. However, BASF soon withdrew as they became interested in a much more promising process invented by Fritz Haber, for which he was awarded a Nobel Prize for Chemistry in 1918. In this, direct combination of nitrogen and hydrogen took place at high pressure and high temperature in the presence of a catalyst; the resulting ammonia could then be oxidized to form nitric acid, neutralized with sulphuric acid to form ammonia sulphate for use as a fertilizer, or used in other ways. The process presented great engineering difficulties, in that working pressures of around 200 atmospheres were required at about 600 °C, originally using uranium carbide or osmium as catalyst; later, an activated iron compound, which was cheaper, was found satisfactory. Nevertheless, the practical problems were steadily solved and a strong patent position built up. The first works was opened at Oppau in 1913 and had a capacity of 8700 tons of ammonia a year; by 1915, with a then rapidly rising demand for nitric acid for manufacturing explosives, capacity had been stepped up to 60 000 tons of ammonia a year. In 1916 a second plant was opened at Leuna, and the combined output was almost 100 000 tons annually. Not until the 1920s was the Haber process worked outside Germany – for its technology was not easily mastered – but by 1930 roughly half the world's production of fixed nitrogen was made by this process or variations of it. By 1950 the proportion had risen to more than 75 per cent. The advent of huge quantities of cheap nitrogenous fertilizers had important consequences for the world agricultural industry (ch. 8).

Finally, we must mention one other way in which atmospheric nitrogen was brought into a useful form of chemical combination. This was the cyanamide process; it starts with a mixture of lime and coke, which is fused in an electric furnace to form calcium carbide. In the 1890s this was used almost solely to generate acetylene gas for lighting – from bicycle lamps to lighthouses – simply by letting water drip on it; later, the intense heat of the oxy-acetylene flame was widely used in welding and metal cutting. At the turn of the century, however, it was discovered that if calcium carbide is heated in an atmosphere of nitrogen it is converted to calcium cyanamide $(CaCN_2)$, from which ammonia can be generated by treatment with superheated steam. There were early difficulties in working the process, but

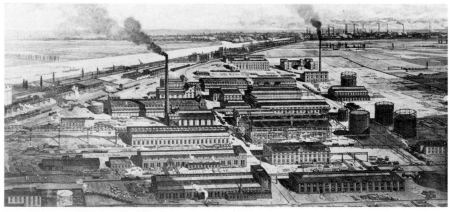

First large-scale Haber-Bosch plant; Oppau, 1914.

by 1910 world production totalled about 20 000 tons with plants in Canada, France, Germany, Italy, and Japan. Wartime demands had raised this total to 600 000 tons by 1918. In 1947 there were nearly fifty plants producing cyanamide, much of which was used as an agricultural fertilizer.

Calcium carbide had meanwhile found a new role as a chemical intermediary, as acetylene was a convenient starting material for many organic syntheses. During the Second World War, German production of carbide, which could be made from indigenous raw materials, rose to nearly 3 million tons a year.

In the chemical sense, the antithesis of acid is alkali, exemplified by soda. For the manufacture of ordinary washing-soda (sodium carbonate) the old Leblanc process was almost totally eclipsed by the Solvay process – in which ammoniated brine is treated with carbon dioxide – by the beginning of the century. The Solvay process is significant in being the first major industrial chemical process to be operated continuously – which is basically the most economic method – rather than as a batch process. Of 1.8 million tons of soda made in 1902, more than 90 per cent was made by the Solvay process; nevertheless, the Leblanc process – with its accumulating piles of noxious waste – was not wholly obsolete until the end of the First World War. For many purposes – such as soap making – a stronger form of soda (caustic soda, sodium hydroxide) is required. For this an electrochemical process was developed by an American chemist, H. Y. Castner, and first worked in 1897 by the Castner-Kellner Alkali Company in Cheshire. It was based on the electrolysis of brine in a specially designed cell incorporating a mercury cathode; hydrogen and chlorine were produced as by-products. From 1900 to 1925 the Castner cell was modified in various ways, and in the USA a number of diaphragm cells were developed over the same period. Where electricity was not cheaply available, however, soda continued to be causticized by treatment with lime. Electrolytic processes were also developed for the manufacture of sodium chlorate, used mainly as a weed-killer; at mid-century production was about 20 000 tons annually.

II. ORGANIC CHEMICALS

As Lucretius sagely remarked two thousand years ago, *nil posse creare de nilo* (nothing can be created out of nothing). For the making of organic chemicals an essential requirement is a sufficient and suitable source of the carbon atoms of which they are basically composed, and in our present context the biggest change in the organic chemical industry took place in its raw materials. In the industry's infancy, when its main products were synthetic dyestuffs, reliance was placed wholly on coal tar; indeed, these products were commonly referred to as coal-tar dyes. This tar was abundantly and cheaply available from the rapidly expanding coal-gas industry. In 1900 it was roughly £1 per ton in Britain, and ten years later the price had fallen to little more than half this. Distillation of the tar yielded a range of simple organic chemicals that were convenient starting-points for industrial chemicals. They included, for example, benzene, toluene, naphthalene, phenol, pyridine, and anthracene. On these, with the use of sulphuric and nitric acid as auxiliaries, extraordinarily varied chemical changes could be rung; as we have noted, the number of organic chemical substances now known runs literally to millions. From about 1900 further supplies of tar became available from the steel industry, which partly explains the fall in price. Up to that time most metallurgical coke was produced in so-called beehive ovens, from which the volatile by-products were simply discharged into the atmosphere, creating environmental problems. These were gradually replaced by by-product ovens, in which the tar and gas were recovered; as we have noted (p. 31), some of the latter could be utilized by gas companies. World production of coal tar was about $2\frac{1}{2}$ million tons in 1901. Ten years later, Germany alone was distilling a million tons of tar a year; by-products then included 30 000 tons of naphthalene, 4000 tons of anthracene, 20 000 tons of benzene, and 3000 tons of toluene.

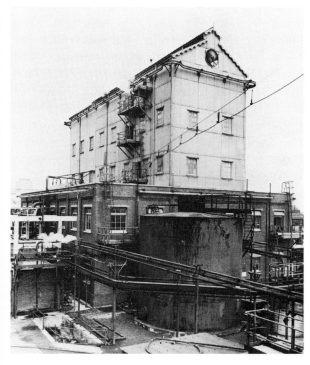

Coal-tar distillation works of the Gas Light and Coke Company at Beckton Works, near London.

While coal tar was a convenient and concentrated source of carbon in a variety of useful constructional units, it was not the only one; petroleum was an obvious alternative. Some four-fifths of petroleum consists of carbon and by distillation it can be fractionated – like coal tar – into a variety of building units. These are rather different from the units derived from coal tar, the carbon atoms being mainly arranged in short chains (aliphatic compounds) rather than rings (aromatic compounds). Nevertheless, these units can be pressed into service, and by the end of the First World War the petrochemical industry was established in the USA. Gaseous propylene was turned into liquid isopropyl alcohol by Standard Oil in 1920; from this could be made acetone and other solvents. In 1927 ethylene glycol, an increasingly important antifreeze for motor cars, began to be made from ethylene. At the same time, better methods of fractionating petroleum made it possible to extract some of the relatively small amount of aromatic substances present in it. Some crudes, such as those from Borneo, are relatively rich in aromatics and had in fact been used as a source of benzene and toluene in Britain during the First World War. However, the petrochemical industry was essentially an American development, and by 1940 more than a million tons of chemicals a year were being made in the USA from petroleum. In Europe, for the most part with no indigenous oil, there was no parallel to these developments: until well after the Second World War the basic raw material remained coal tar. Thereafter, there was a rapid and virtually complete change, but even in 1950 western Europe produced no more than 100 000 tons of chemicals from petroleum. The change to an oil-based chemical industry was accelerated by the phenomenal growth of the polymer industries, because these in fact require the aliphatic straight-chain units of petroleum rather than the aromatic rings predominating in coal tar. We have already noted a half-way position in Germany during the Second World War, when there was considerable dependence on acetylene, closely allied to ethylene. Another important factor was the replacement of the traditional method of making town gas by carbonization of coal in retorts; this dried up one source of tar.

The synthetically important ingredients of coal tar – benzene, toluene, etc. – are mostly not present in coal itself but are created in the carbonization process used to make gas and coke. Nevertheless, coal itself contains around 90 per cent carbon and there was, therefore, much interest in the direct use of coal as a raw material. Although no general success was achieved, some important processes emerged. We have already noted the use of calcium carbide, made from coke and lime, to generate acetylene as a basic synthetic organic intermediary. In the 1920s BASF began to make synthetic methanol from the products of passing steam over red-hot coke, which produces water-gas. At the same time Franz Fischer and Hans Tropsch found that, if hot water-gas is passed over a suitable catalyst, low-boiling hydrocarbons akin to petrol are formed. In 1913 F. Bergius developed a process for hydrogenating coal at high pressure to form petrol. This was developed in Germany in the 1920s for strategic reasons and it was an important source of aviation fuel during the Second World War for both Germany and the Allies. It was not basically economic, however, except in the special circumstances of wartime shortage or, as in South Africa, an abundance of cheap coal.

Finally, we must briefly mention fermentation as a source of organic chemicals. The production of alcohol by yeasts from sugary or starchy materials has been known from antiquity, as also the fact that the fermentation can proceed further to produce vinegar (acetic acid). Other fermentations became of industrial importance. During the First World War acetone, essential for making explosives, was made in Britain by a process devised by Chaim Weizmann, later President of Israel. This was based on a bacterial fermentation of grain, resulting in production of a mixture of acetone and butyl alcohol. In the years between the wars fermentation processes were widely used to produce other products such as citric and lactic acids and glycerol. During and after the Second World War it was the sole route to the production of penicillin, the most important of all chemotherapeutic agents. With the growth of the petrochemical industry, however, synthetic methods became generally more attractive.

The organic chemical industry was originally based firmly on the manufacture of dyes; Germany quickly became dominant in what was originally a British development. This dominance was forcibly brought to Britain's notice on the outbreak of the First World War: she could not even provide the dyes for army uniforms. Britain, like other countries, improvised as best she could – even reviving cultivation of the indigo plant in India, despite the fact that the synthetic dye had been available since 1897 – and emerged from the war far stronger than she had entered it. In the inter-war years, dyestuffs manufacture increased enormously and the range of colours and their fastness was greatly improved. Important developments were Caledon Jade Green (1920) and Monastral Blue (1935), an example of an entirely new class of dye. The great expansion of man-made fibres (which we will consider later, in the context of textiles) after the Second World War gave a powerful stimulus to the development of new dyes, for these new products had dyeing characteristics quite different from the natural fibres with which the textile industry had hitherto been almost exclusively concerned. This problem was overcome: partly by modifying existing dyes and techniques of using them; partly by modifying the fibres to make them

As in many industries, modernization did not proceed evenly; here casks are being laboriously washed by hand at ICI Dyestuffs Works, Blackley, 1943.

more receptive to the available dyes; and partly by developing new products. For example, the disperse dyes, invented by A. Clevel in 1922, were of a novel type particularly suitable for cellulose-based fibres.

III. POLYMERS

Man-made fibres are important examples of a totally new kind of synthetic (or semi-synthetic) chemical product which developed during the twentieth century. To the chemist, all these products are polymers, but for practical convenience we can consider them under two headings – plastics and man-made fibres. Broadly speaking, plastics cover the multitude of moulded items now commonplace throughout the world – electrical fittings, telephones, kitchen-ware, ball-point pens, bottles and jars, refuse bins, water tanks, and hundreds of others. In addition, plastics appear for constructional purposes in the form of tube, sheet, and rod. Man-made fibres include cellulose acetate (Celanese), nylon, Terylene (Dacron), polypropylene, polystyrene, and acrylics. These two classes are not mutually exclusive: for example, nylon, cellulose acetate, and polypropylene are used to a limited extent for moulding.

Whatever their nature, and the uses to which they are put, polymers have one feature in common – they consist of very large molecules made up of tens of thousands of small ones (monomers), usually identical but sometimes of two different kinds, rather as beads are threaded to make a necklace. Many important natural products are polymers – rubber, cotton, and wool, for example – but in 1900 artificial polymers had scarcely appeared. Those that there were, were largely modifications of natural polymers. Such, for example, was Alexander Parkes's nitrated cellulose (1862) which disappointed his hopes that it would be useful as an electrical insulator. In the USA J. W. Hyatt developed the same material as celluloid, primarily as a substitute for the ivory no longer available in sufficient quantities for making billiard balls and also for a variety of other purposes such as shirt-cuffs. Later it became essential as the base for photographic and cinematographic film, and in the 1920s British production alone rose to 40 000 tons annually. Subsequently, however, it was replaced increasingly by cellulose acetate, which is much less flammable and therefore safer.

In the First World War cellulose acetate was an important dope for the aircraft fabrics then in general use; light alloy cladding was essentially a post-war development. The end of the war, therefore, saw a surplus capacity for its manufacture and led to its production in the form of fibre, acetate rayon. This had the advantage over viscose rayon, invented by C. F. Cross and E. J. Bevan in 1892, that it was soluble in acetone and was suitable for injection moulding. The manufacture of viscose rayon had been established by Courtaulds at Coventry in 1907.

First knitted sample of Nylon (Du Pont).

Around the beginning of the century, another semi-synthetic material began to be manufactured from milk protein by treating it with formaldehyde. Discovered in 1897 by A. Spitteler, it was first marketed in Germany, under the name Galalith, and subsequently elsewhere. It became widely used for small items, especially buttons.

Celluloid and Galalith exemplify two important classes of plastic—thermoplastic and thermosetting. Celluloid is thermoplastic: on heating it softens and can be moulded to any desired shape, the process being repeatable. Casein plastics are thermosetting: the reaction involved in their formation is irreversible and once formed in their mould they cannot be reshaped.

The first important thermosetting plastic was based on phenol–formaldehyde. In the 1880s it was well known that these substances combine to form a resinous solid, but the organic chemists of that time were primarily interested in pure crystal substances; such resins they contemptuously dismissed as mere *Schmiere*. But two men – L. H. Baekeland in the USA and James Swinburne in England – thought differently and carefully studied the reaction and the processing of the resin with a view to obtaining a useful new product. Baekeland – who had already made a fortune by selling the rights in his Velox gaslight photographic papers to George Eastman (Kodak) –was interested in mouldings; Swinburne, an electrical engineer, was looking for a new insulator. Both achieved success almost simultaneously in 1907, but Baekeland lodged his patent for Bakelite one day before Swinburne. He established manufacture in the USA and Germany, and eventually the American and British interests were merged. When Baekeland died in 1944, world production of phenolic resins was 175 000 tons per annum. A major customer had been the motor-car industry, since Charles Kettering's introduction in 1910 of a new system of electrical ignition. Bakelite proved an ideal material for making distributor heads, rotor arms, switches, and so on. It was also very suitable for the multiplicity of small items necessary in domestic wiring circuits. In this context the dark dull colour of phenol–formaldehyde did not matter, but for other purposes a colourless resin–capable of being coloured by addition of pigments – was desirable. Such were the urea–formaldehyde (Beetle) resins that appeared in Britain in 1928 and the melamine–formaldehyde resins developed in Germany in 1935.

In the 1930s Germany was greatly concerned with the strategic importance of rubber and was anxious to find substitutes for it; American industry was also interested although at that time American interest lay almost entirely in synthetic rubber, having special properties. One important substitute in the field of electrical insulation was polyvinylchloride

Early high-pressure reaction vessel for preparation of polythene (1937/8); mercury sealed gas compressor (1937/8); and very early (pre-war) sample of polythene.

(PVC), the manufacture of which was established in both countries before the Second World War. Even more important in this context was polythene (polymerized ethylene), which had greatly superior electrical qualities: it has been said by Sir Robert Watson-Watt that, without it, radar – one of the most decisive of all wartime developments – would have been virtually impossible. Polythene was a British invention, made in the laboratories of ICI, and production began on the day the Second World War broke out. Initial expectations that relatively small quantities – a few tons annually – would be required for specialized electrical equipment were confounded: it proved a most versatile thermoplastic moulding material, and millions of tons have been used for making household and other equipment and as a wrapping material – the polythene bag is so ubiquitous that its disposal has become a major environmental problem. Production in the USA alone rose from 25 000 tons in 1950 to one million tons in 1963. Originally, polythene was made by a high temperature/high pressure process giving what is now called low-density polyethylene (LDPE). Since 1953 a high-density form (HDPE) has also been made according to a low temperature/low pressure process invented by Karl Ziegler and Giulio Natta, working in Germany and Italy respectively, who shared the Nobel Prize for Chemistry in 1963. Closely related to polythene is polytetrafluoroethylene (PTFE) in which the presence of fluorine confers high resistance to heat and chemical agents. As

By mid-century the chemical industry was rapidly becoming capital intensive rather than labour intensive. This picture shows the control panel of ICI's No. 3 olefines plant at Wilton (1959).

it is expensive and difficult to make and work, however, it has found only restricted application. Natta went on to discover (1954) that propylene, closely allied to ethylene, can also be polymerized by Ziegler-type catalysts, and Montecatini in Italy began commercial production of polypropylene in 1957. Now produced throughout the world, polypropylene is used not only in the form of mouldings and film but also as a cheap fibre.

The 1930s saw the birth of other important plastic materials. One of these was polystyrene, a hard transparent material. Originally it was rather brittle, but improved high-impact polystyrene was introduced in the USA after the Second World War; this was a valuable alternative to cellulose acetate, which between the wars was the most important injection moulding material. In 1960, for example, British production of cellulose acetate plastics was only 11 000 tons as against 40 000 tons of polystyrene. Another range of transparent plastics, based on esters of acrylic acid, was also developed in the 1930s. In 1927 Rohm and Haas in Germany produced poly(methyl acrylate), but an improved version of this was discovered by Rowland Hill of ICI in 1931. It was poly(methylmethacrylate), a glassy substance with good optical properties; it transmits 90 per cent of daylight. Its commercial exploitation was much improved in 1933 with the development, also by ICI, of an economic route to the monomer. In Britain this material was developed as Perspex, and during the Second World War it was widely used in aircraft for windscreens and observation domes, where its combination of toughness and lightness was very important. Later it found wider use, particularly as corrugated transparent sheeting for the building industry.

IV SYNTHETIC RUBBER

This brief survey of the history of the plastics industry cannot omit discussion of synthetic rubber. Rubber, as has been noted earlier, is a natural polymer, consisting of thousands of simple hydrocarbon units. The units were fairly easily identified, and it was inevitable that attention should be diverted to synthesizing these simple building bricks and building them up into giant molecules equivalent to those of natural rubber. The incentive to invention was, again, particularly strong in Germany because of the risk of such a vital commodity being cut off in the event of war. In the First World War Germany produced some 2500 tons of a rather poor 'methyl' rubber (from dimethylbutadiene). In the 1930s, I. G. Farben produced two much better, but expensive, types, Buna–S (butadiene/styrene) and Buna–N (butadiene/acrylonitride); by 1938 output was 5000 tons per annum. The USA supposedly had no need for such substitutes for natural rubber, but the over-running of the rubber plantations by the Japanese in 1941 transformed the situation almost overnight: by a tremendous effort, US output of synthetic rubber was raised to nearly one million tons in 1945. One advantage of synthetic rubber is that by modifying the manufacturing process the product can be tailormade for the particular purpose it is to serve.

The impact of the dramatic rise of the new technology of synthetic polymers was enormous. In 1900 it was perfectly possible to imagine life going on much as usual in a world in which the infant plastic and synthetic fibre industry had been extinguished. Half a century later the collapse of the new industry would have been catastrophic, for it had invaded virtually every branch of industry and was a commonplace in every home. This revolutionary development would have been impossible without a corresponding development in our understanding of the basic science involved. The resins which the older organic chemists had refused had become the headstone of the corner. Polymer chemistry became a highly reputable field of academic study and those knowledgeable in the art were eagerly recruited by industry for its rapidly expanding research and development departments. Worldwide, polymer chemists were numbered in thousands. It would be invidious to select a few for special mention, but it would probably be widely agreed that this new and highly productive field of inquiry was epitomized by the original and creative work of the German chemist Hermann Staudinger who was born in 1881 and died in 1965. From 1926 to 1951 he was professor in the University of Freiburg; he was awarded the Nobel Prize for Chemistry in 1953.

BIBLIOGRAPHY

Baud, P. *L'Industrie chimique en France*. Masson, Paris (1932).

Beer, J. J. *The emergence of the German dye industry* . University of Illinois Press, Illinois (1959).

British Dyestuffs Corporation. *The British dyestuffs industry 1856–1924*. Manchester (*c*.1924).

Brunner, Mond and Co. *The first fifty years of Brunner, Mond and Co*. (1923).

Campbell, W. A. *The chemical industry*. Longman, London (1971).

CIBA Ltd. *The story of chemical industry in Basle*. Urs Graf Publishers, Lausanne (1959).

Cook, J. Gordon. *Your guide to plastics*. Merrow Publishing Co., Guildford (1964).

Findlay, Alexander. *Chemistry in the service of man* (7th edn.) Longmans, Green and Co., London (1947).

— and Williams, Trevor I. *A hundred years of chemistry*. Duckworth, London (1965).

Goldstein, R. F. History of the petroleum chemicals industry. In *Literature resources for chemical process industries*. American Chemical Society (1954).

Haber, L. F. *The chemical industry during the nineteenth century*. Clarendon Press, Oxford (1958).

— *The chemical industry 1900–1930*. Clarendon Press, Oxford (1971).

Hardie, D. W. F. *A history of the chemical industry in Widnes*. Imperial Chemical Industries, London (1951).

— and Davidson Pratt, J. *A history of the modern British chemical industry*. Pergamon Press, Oxford (1966).

Haynes, W. *American chemical industry*, 6 vols. Van Nostrand, New York (1945-54).

ICI Mond Division. *A hundred years of alkali in Cheshire*. Kynoch Press, Birmingham (1973).

ICI Plastics Division. *Landmarks of the plastics industry*. Kynoch Press, Birmingham (1962).

Imperial Chemical Industries. *Fifty years of progress : the story of the Castner-Kellner Alkali Company 1895–1945*. Kynoch Press, Birmingham (1947).

Kaufman, M. *The first century of plastics – celluloid and its sequel*. Plastics Institute (*c*.1962).

Kirk, R. E. and Othmer, D. F. *Encyclopaedia of chemical technology*. 15 vols. Interscience, New York (1947–56).

Lunge, G. *Coal-tar and ammonia*. (5th edn.) Gurney and Jackson, London (1916).

Marshall, A. *Explosives*. (2nd edn.) Churchill, London (1917).

Metzner, A. *Die chemische Industrie der Welt*, 2 vols. Econ-Verlag, Dusseldorf (1955).

Miall, S. *A history of the British chemical industry*. Ernest Benn, London (1931).

Morgan, G. T. and Pratt, D. D. *British chemical industry : its rise and development*. Edward Arnold, London (1938).

Reader, W. J. *Imperial Chemical Industries : a history.* 2 vols. Oxford University Press, London (1970, 1975).

Reuben, B. G. and Burstall, M. L. *The chemical economy.* Longman, London (1973).

Sherwood Taylor, F. *A history of industrial chemistry.* Heinemann, London (1957).

Thomas, R. W. and Farago, Peter. *Industrial chemistry.* Heinemann, London (1973).

Warrington, C. J. S. and Nicholls, R. V. V. *A history of chemistry in Canada.* Pitman (Canada) (1949).

Williams, T. I. *The chemical industry, past and present.* Penguin Books, Harmondsworth (1953).

Wilson P. J. and Wells, J. H. *Coal, coke and coal chemicals.* McGraw-Hill, New York (1950).

Winnacker, Karl. *Challenging years: my life in chemistry.* Sidgwick and Jackson, London (1972).

12

TEXTILES

I. THE PATTERN OF THE INDUSTRY

The outstanding development in the textile industry in the twentieth century was the appearance of artificial fibres, first those based on natural cellulose and later wholly synthetic ones based mainly, but not exclusively, on polyamides and polyesters. We have alluded to these briefly in the previous chapter in the context of the chemical industry, for it was this that manufactured the new products in bulk and converted them into fibre form. The textile industry, however, retained the task of converting the fibres into fabric.

It is convenient to consider the history of natural fibres separately from that of synthetic ones. This is logical in view of the nature of the raw materials, for one is a traditional agricultural product and the other a novel product of the chemical industry. Their fates, however, were similar during the first half of this century simply because the chemical manufacturers were obliged to supply a product that could be processed on conventional textile machinery; without this, their hopes of breaking into a suspicious and highly conservative industry were slim to vanishing-point. Initially, too, chemical manufacturers played a defensive role, pleading that their products would complement rather than supplant the natural fibres, to the profit of both. In 1950 combined world production of wool and cotton was around 7 million tons, of which about one-quarter was wool. By contrast rayon production was about 1.5 million tons – compared with 14 000 tons just before the First World War – but the production of truly synthetic fibres, exemplified by nylon (polyamide) and orlon (acrylic), totalled only 76 000 tons. Terylene (polyester) was then only just appearing above the industrial horizon, though it had been discovered in 1941. Twenty years later, however, as the chemical industry lost its inhibitions and, as was no less important, offered better and more varied products, world production of wholly synthetic fibres had risen to nearly five million tons, 70 per cent of the market being accounted for by polyamides and polyesters. While much of this was blended with natural fibres, some manufacturers had begun to take synthetic fibres neat: an era of direct competition had opened.

II. NATURAL FIBRES

Apart from semi-synthetic fibres based on cellulose, which appeared before the First World War but really took off only in the 1930s, natural fibres had the field virtually to themselves up to the Second World War. The main contestants were cotton and wool, while linen was of minor importance and silk retained a place in the luxury market. Total production did not vary much, but the centres of manufacture show important changes. Thus, according to the New York Cotton Exchange statistics, world production of cotton was about 5 million tons in 1911/12 and about 6 million in 1938/9, with little year-to-year variation in between; by mid-century it had not greatly increased, at around $6\frac{1}{2}$ million tons. The picture for wool is not dissimilar: world production in the late 1920s averaged $1\frac{1}{2}$ million tons and stayed pretty constant up to the Second World War; in 1950 it was about $1\frac{1}{4}$ millions tons.

If we follow common historical practice and use the number of spindles in operation as in index of the cotton industry, Britain continued up to 1939 to maintain the lead she had established in the Industrial Revolution, but she was steadily losing ground. Her 56 billion spindles in 1913 represented 40 per cent of the world total; 39 billion spindles in 1939 only 27 per cent. She had lost ground not so much to the United States, where activity also declined, or to other European countries which had managed to remain fairly constant, but to Japan and India. Between 1913 and 1939 the number of Japanese spindles increased sixfold, from 2 billion to 12 billion, attaining one-third of the British total. The Indian increase was less spectacular but started at a higher level: in 1913 it was already 6 billion, and in 1939 it was half as big again at 9 billion. The reasons for the change are arguable, but the importance of the availability of cheap labour in Japan and India has probably been over-emphasized. By using better machinery and improved organization – and the creation of larger manufacturing units – the USA and other European countries at least managed to hold their own. Britain, on the other hand, was undeniably put at a disadvantage by the unavoidable loss of valuable export markets during the First World War, which were not regained afterwards.

The position in woollen goods was similar. In the late 1930s Europe was the dominant producer with rather more than half the total market, but the USA supplied nearly a quarter of it, establishing a slight lead over Britain. Japan, however, with a negligible woollen manufacturing industry at the beginning of the century, had managed, thanks to tariff protection, to gain 6 per cent of the world market. In 1950, 60 per cent of the world's wool was produced in the Southern hemisphere but it was almost entirely spun in the Northern hemisphere.

Quantitatively, the production of silk has never been more than a minor industry in the global sense, but one by no means unimportant in certain areas. Peak production was in 1934, with a total (excluding China) of around 6000 tons, of which 80 per cent came from Japan. This represented the life's work of 300 000 million silkworms. In that year mulberry trees, on which the silkworm feeds, covered one-tenth of the cultivated land of Japan, and more than 2 million peasants were occupied wholly or partly in sericulture: a quarter of a million workers were engaged in weaving silk. By far the largest market (70 per cent) for raw silk was the USA. The most important European producer was Italy, but her production was only 6 per cent of Japan's.

The peak production in 1934 was the sequel to a fairly steady increase since the beginning of the century. In its early years a change in women's fashion, with shortening skirts, led to a demand for more elegant stockings, and for these silk was ideal. The advent of rayon brought a cheap but not very strong competitor; nylon, however, was a very different matter. When nylon stockings first appeared in 1939, 64 million pairs were sold in the first year. Fashion was to influence the market again after the Second World War: the miniskirt promoted nylon tights in place of stockings and tights needed more material for their manufacture. Moreover, tights were thrown away in pairs, whereas stockings might be discarded individually: on such vagaries do the fortunes of industry depend.

As a generalization, it is fair to say that during the first half of this century advances in the technology of natural fibres lay more in the improvement of existing processes than in innovation. The purification of raw cotton continued to be entirely mechanical but it had to be more thorough because the introduction of harvesting machinery resulted in a dirtier crop. In the preliminary treatment of wool, washing in warm soapy water continued to be the usual practice but during the Second World War synthetic detergents became an acceptable and lasting alternative. The purpose of this treatment is primarily to remove natural grease, and in a relatively few mills this was done by extraction with hydrocarbon solvents.

Some important developments took place in spinning and weaving. In the USA shortage of labour led to the widespread introduction of ring spinning – improved by Casablanca's invention of high-draft spinning in 1904 – and by 1913 this was almost universal there and in Japan. By that time, roughly half the world's cotton was being spun in this way. In Britain, however, this new development made very slow headway, partly because of the conservatism of the industry but also because the kinds of cotton predominantly made there were more suitable for the old mule spindles, which in 1913 still accounted for four-fifths of British production.

Another major development, also American, was the Northrop automatic loom. This was developed in the 1890s but not widely adopted until after 1900. It is often regarded as representing the watershed between labour and capital expenditure in weaving. In 1909, 200 000 were in use in the USA, compared with only 8000 in Britain, where it first appeared in 1902. To an extent its adoption was related to ring spinning, in that ring-spun yarns were particularly suited to automatic looms. As in many industries, the 1920s saw the widespread replacement of overhead belt drives by electric motors serving individual looms.

Ring spinning and the automatic loom were the major developments up to the 1930s, combining to increase productivity and lower labour costs. In the 1950s an important new improvement was centrifugal, pot, or open-ended spinning, which eliminated spindles, rings, and bobbins and greatly speeded up the process. In this, airborne fibre is collected inside a cup-shaped rotor spinning at about 50 000 revolutions per minute. The fibres are collected in a groove and peeled off continuously. Although this method seems to have been thought of as early as 1901, the principles of modern machines – to which contributions were made in Czechoslovakia, Holland, and Australia – were embodied in a patent lodged in 1937 by the Danish engineer S. Berthelsen. The 1950s also saw the appearance of the first Sulzer shuttleless loom and of looms in which the shuttle is propelled either pneumatically or hydraulically.

Two important developments in textile finishing deserve mention here. One was the introduction of crease resistance by, in effect, forming urea-formaldehyde resin within the fibres of a cloth: this was developed by Tootal Broadhurst Ltd. in Manchester in the 1920s. In the 1930s wool began to be made shrink-resistant by exposing it to chlorine gas and in other ways.

III. MAN-MADE FIBRES

As has been noted, the differences between natural and man-made fibres are to some extent obscured by the fact that the latter had to be produced in a form that the textile industry could handle with its existing machinery. Nevertheless, the differences are important. Firstly, the mode of manufacture of artificial fibres results in their appearing as long continuous fibres (monofil); they thus resemble silk but differ markedly from cotton and wool, which consist of short fibres that have to be spun to form a thread. Secondly, they are thermoplastic and melt at quite low temperatures. Thirdly, their dyeing characteristics are different.

One feature of textiles is that, unlike many industrial products, they are judged not only on their measurable physical properties but on less tangible ones such as texture, drape, capacity to retain shape, and visual attractiveness – they must lend themselves to comfort and style. The importance of these subjective qualities was not immediately recognized nor, once recognized, were the qualities easily attained. Rayon, for example, was optimistically first marketed as artificial silk – partly, no doubt, because Courtaulds had long been established in the silk industry – but the resemblance was far too tenuous to deceive the public: it was at first demonstrably an inferior imitation. In particular, the monofil was a single shiny thread, whereas natural silk consisted of a number of exceedingly fine threads from individual cocoons twisted together; this process trapped a little air and helped to give the softness, warmth, and moisture-absorbing capacity for which real silk has long been prized. Furthermore, the early rayons lacked the elasticity of silk, and stockings made from them – an important outlet – were apt to bag inelegantly. Some of these disadvantages were obviated by cutting fine monofil into short lengths and then spinning these like wool and cotton. With increasing confidence, manufacturers began to sell rayon on its merits – which were considerable – rather than as a substitute for silk. The merits of rayon did not limit its use to clothing; with development of a stronger fibre, substantial industrial outlets were found, in industrial fabrics of various kinds and for reinforcing motor-car tyres.

In the clothing industry, nylon and Terylene presented similar problems. For example, shirts made from nylon monofil are uncomfortable in hot weather because they fail to absorb sweat: they feel cold and clammy. This was overcome partly by using staple fibre, as with rayon, to impart some absorptive porosity and partly by utilizing their special physical properties. They could, for example, be bulked by crimping the slightly softened fibre and making this permanent as it cooled; another alternative was to draw the heated yarn over a knife-edge (edge crimping), making it curl up, like the ribbons used by Continental shopkeepers to wrap fancy parcels.

Resistance to shrinking is as important in man-made fibres as in natural ones, and is achieved by heat-setting them under tension at about 200 °C. It is particularly important where an artificial fibre is blended – as is very widely done – with a natural one. If the two shrank to a different extent on washing the effect on the garment would be disastrous. Initially, it was not realized that this consideration applied equally to make-up: nylon garments sewn with cotton thread became distorted on washing through differential shrinking.

Machine knitting has been practised since the late sixteenth century and its recent developments might, therefore, have appropriately been considered under the heading of natural fibres. Their consideration has been deferred to this point simply because the knitting industry was greatly stimulated by the advent of artificial fibres. An important development was the fly needle frame (FNF), using a hollow needle, devised by Courtaulds in the 1930s, but not introduced until after the Second World War. It was

F.N.F. (fly needle frame) warp knitting machine.

particularly successful for making rayon underwear. The setting properties of nylon made it possible to knit fully-fashioned stockings on simple circular knitting machines, finishing the work on a heated frame shaped like a leg.

In Europe, the machine-knitting industry concentrated, though not exclusively, on made-to-measure items, but in the USA large machines became popular for producing jersey-type fabric for the garment make-up trade. After the Second World War knitted fabrics became generally used for a wide range of purposes.

Man-made fibres presented special difficulties in dyeing because their affinity for traditional dyes was different from that of natural fibres. The development of suitable new dyes has already been discussed in the context of the dyestuffs industry. Generally speaking, these new dyes were designed to be used in the normal way, that is to say by dipping the fibre or fabric in a liquid dye-bath. An alternative method, known as dope dyeing, was to incorporate the dye or a pigment in the molten polymer, in which it would be permanently imprisoned on cooling in fibre form. This was simple, and conferred great fastness to washing and light.

IV. THE GARMENT INDUSTRY

Until the invention of the sewing-machine in the middle of the nineteenth century, all sewing had of necessity to be done by hand, and garments were made individually. The speed of the sewing-machine, however, which had risen to several thousand stitches per minute by 1900, made it desirable to speed up pattern cutting and other processes and paved the way for the rapid growth of the ready-made clothing industry. Not surprisingly, in view of its readiness to accept labour-saving devices, the lead was taken by the USA, but Britain, too, with its long tradition of textile manufacture, was an important centre.

Where factory production was established, an elaborate subdivision of labour was organized on the assembly-line principle: a single garment might pass through fifty different hands before completion by a combination of hand and machine workers. In the 1930s factories employing several thousand workers were established, but to a considerable extent the manufacture continued to be organized as a cottage industry, the work being sent to out-workers working in their own homes. In the big urban centres such as New York and London this system was favoured by the presence of large numbers of poor immigrants, often Russian Jews. The exposure, by Charles Kingsley and other social reformers of their exploitation led to some statutory improvement in their lot, chiefly by laying down minimum rates of pay, but the regulations were not easily enforced.

However the work was organized, the first requirement was to cut the fabric quickly to pattern. For this a band-saw was originally used, but in the 1880s G. P. Eastman developed an electric knife with a reciprocating blade, later made self-sharpening. Depending on the fabric, up to 50 thicknesses could be cut at once. The patterns used, and the number of each size of garment made, depended on a statistical analysis of the physical dimensions of the potential customers, and these analyses became increasingly sophisticated. Much skill was devoted to laying out the patterns to achieve the greatest economy in the use of cloth.

Although the principal machine was the sewing-machine, many other basic operations were mechanized at an early date. The first button-holing machine was patented in 1881, and in 1908 the 'Hand Hole' machine appeared for men's suits, so called because its finish closely resembled the hand-sewn product. Button-sewing machines also appeared before the turn of the century.

These developments created something of a social revolution. In 1900 every gentleman had his tailor and every lady her dressmaker; ready-made clothes were for the working classes. By mid-century the situation had greatly changed. Ready-made clothes were of better quality, fitted better, and available in a much wider range of fabrics and styles; no less to the point, they were relatively cheap in a world in which labour had become expensive. Twenty years later still, the quiet revolution was virtually

Singer oscillating-hook machine (1908), partly dismantled to show mechanism.

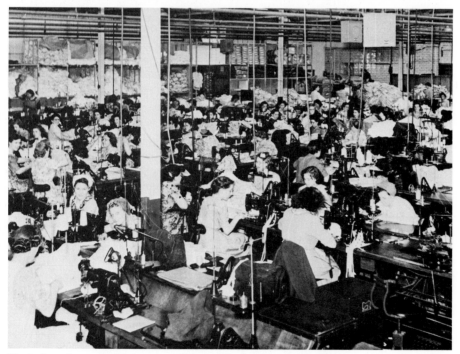

Typical sewing room in the 1920s showing line and bench shafting.

complete. The vast majority of the population of the western world bought ready-made clothes as a matter of course; only a tiny minority, the rich and fastidious, had their clothes made specially for them.

BIBLIOGRAPHY

Board of Trade (Kemsley, W.F.F.) *Women's measurements and sizes.* HMSO, London(1957).
— *Board of Trade Working Party Reports: Heavy clothing.* HMSO, London (1947).
— *Light clothing.* HMSO, London (1947).
— *Rubber proofed clothing.* HMSO, London (1947).
Coleman, D. C. *Courtaulds: an economic and social history.* Clarendon Press, Oxford (1969).
Dobbs, S. P. *The clothing workers of Great Britain.* Routledge, London (1928).
Duxbury, V. and Wray, G. R. (eds.). *Modern developments in weaving machinery.*Columbine Press, Manchester (1962).
Gilbert, K. R. *Sewing-machines.* HMSO, London (1970).
Hague, D. C. *The economics of man made fibres.* Duckworth, London (1957).
Lyons., Allen, T. W., and Vincent, W. D. F. *The sewing-machine: an historical and practical exposition of the sewing-machine from its inception to the present time.* John Williamson, London (c.1930).
O'Brien, R. and Shelton, W. C. *Women's measurements for garment and pattern construction.* US Department of Agriculture, Washington, DC (1941).
Poole, B. W. *The clothing trades industry.* Pitman, London (1947).
Popkin, M. E. *Organisation, management and technology in the manufacture of men's clothing.* Pitman, London (1929).
Pransky, A. I. History of the cloth cutting machine. *Bobbin,* **7**, 23 (1965).
Press, J. J. (ed.). *Man-made fibres encyclopaedia.* Interscience, New York (1950).
Rainnie, G. F. (ed.). *The woollen and worsted industry: an economic analysis.* Clarendon Press, Oxford (1965).
Riches, W. *Lock-stitch and chain-stitch sewing-machines.* Longmans, London (1938).
Robson. R. *The cotton industry in Britain.* Macmillan, London (1957).
— *The man-made fibre industry.* Macmillan, London (1958).
Simons, H. *The science of human proportions.* Clothing Designer Co., New York (1933).
Wray, G. R. (ed.). *Modern yarn production from man-made fibres.* Columbine Press, Manchester (1960).

13

THE INTERNAL COMBUSTION ENGINE

The history of the internal combustion engine in the twentieth century follows three different paths, of which only one led to a complete innovation. These paths were represented by the Otto engine, the Diesel engine, and the gas turbine; it was the last of these that was wholly new. All paths led eventually to a heavy dependence on oil products as a source of motive power, particularly for transport, the consequences of which became painfully apparent in the 1970s.

I. THE OTTO ENGINE

The development of this engine was conservative, in that a young engineer familiar with it in 1900 would, in his old age, see no radical changes in the latest models of 1950. This is not to say, of course, that there were not very great improvements in efficiency and reliability: over the half-century the critical weight: power ratio, for example, was improved eightfold and engines could, unlike their early ancestors, run for long periods without a falter.

In 1900 the engine devised by N. A. Otto, originally fuelled by coal gas, was already rather more than twenty years old, and Gottlieb Daimler had adapted it to run on petrol; by 1900 about 9000 petrol-driven motor cars had been built. Powered flight had not then been achieved, but with growing understanding of aerodynamic principles it was becoming apparent that the internal combustion engine, unlike the steam-engine, had potentially a low enough weight: power ratio to make it feasible.

To understand the development of the Otto engine, and the incentives that prompted it, we must first briefly consider its basic principles. In a steam-engine the elastic fluid that drives it is generated by burning fuel to heat water and convert it into steam in a separate boiler; in the Otto engine the fuel is burned inside the engine and the products of combustion provide the driving force. A mixture of air and petrol vapour is drawn into the cylinder, where it is compressed by the piston. Roughly at the point of maximum compression the mixture is exploded, driving the piston back and providing the power stroke. Finally, the returning piston expels the products of combustion and the four-stroke cycle starts again. In the early engines ignition was effected by briefly drawing a flame into the cylinder or, a little later, by inserting a hot tube into it. For the light, fast-running engines

Pair of Allis-Chalmers double-acting four-stroke gas engines. Fuelled by natural gas, each drove a 750-kW generator (*c*.1915).

necessary for road vehicles, and later aircraft, the operation of the engine – but not its basic principles – had to be changed.

For satisfactory running, accurate control of the petrol : air ratio is necessary, but this had been achieved by Wilhelm Maybach's float-feed carburettor of 1893. At roughly the same time Karl Benz introduced electrical ignition, using a sparking-plug, very like those in use today, to give more precise control of the moment of combustion in the cylinder. Faster running, say 1000 revolutions per minute, presented a problem in the dissipation of heat, necessary to stop the engine over-heating; this, too, was accomplished by 1900 with the introduction of water cooling using a honey-comb-pattern radiator to give a large heat-transfer surface. The earliest engines had to be started by swinging them by hand – and the starting-handle is still not extinct – but by 1910 C. F. Kettering had devised a starting motor driven by the same battery that supplied the ignition. This innovation sprang from his realizatiy that an electric motor can be heaviiy overloaded for a short time, and thus its size and weight can be kept within bounds when it is used for only a few seconds to turn an engine over.

For many purposes a single-cylinder engine suffices. Ransom E. Olds, for example, the first mass-producer of motor cars (18 000 in 1901–5), based his early models on a single-cylinder 7 h.p. engine. Today, such engines are used in millions for driving light motor cycles, lawn-mowers, chain-saws, and a wide range of small machines. For motor cars, however, four cylinders were usual, as in the elegant Mercedes and the famous Model T Ford which was to make motoring universally popular – 16 million were built between 1908 and 1927. The larger number of cylinders made for smoother running, and for more expensive models the principle was extended. Six-, eight-, or even twelve- and sixteen-cylinder engines were introduced; up to eight cylinders could be arranged in line without making an excessively long engine and putting excessive strain on the crankshaft, but for many multi-cylinder cars a V-shaped arrangement of cylinders was favoured. In 1923 the Lancia Lambda incorporated a V-4 engine. The multi-cylinder engine complicated the ignition system in that a specific order of firing had to be organised.

The efficiency of an Otto engine can be improved by increasing the compression ratio, which by 1920 was commonly about 4:1. From the strictly engineering point of view there was no great problem in improving on this; the difficulty was the phenomenon known as 'knock', with which all motorists are familiar. In this, the fuel in the cylinder explodes rather than burns and hammers rather than pushes the piston. Avoidance of knock proved a major problem, and it has in fact never been wholly solved. Numerous attempts were made, including varying the design of the combustion chamber and altering the composition of the fuel by introducing a higher proportion of aromatic (benzene) fractions. This raises the

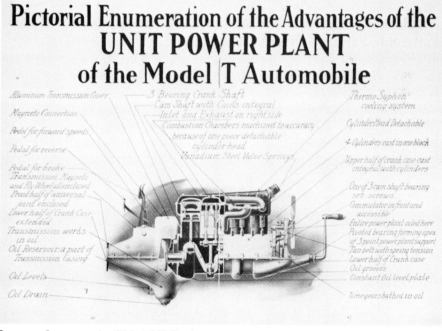

Layout of power unit of Model T Ford.

'octane number', a measure of the amount of compression that can be achieved without knocking. The major advance, however, came in 1921 with the discovery, in the laboratories of General Motors in the USA, that knocking was much reduced by addition of small quantities of a chemical known as lead tetraethyl. By this means compression ratios were raised to 7:1 by 1950 and octane numbers from 55 to 85; roughly speaking, this meant that a gallon of petrol went twice as far as it would have done thirty years earlier. This was judged very satisfactory until environmentalists raised objections to the use of leaded petrol on the grounds of dangerous pollution of the atmosphere.

For small, cheap engines – such as are required for lawn-mowers – a modified Otto cycle is used. They work on a two-stroke principle in which an auxiliary air pump introduces new fuel mixture as it blows out the exhaust gases. As a result, there is one power stroke for every rotation of the crankshaft. This system is, however, less efficient than the four-stroke system and the engines tend to be less easy to start, especially when warm. The two-stroke cycle was originally used for gas engines and was invented by Dugald Clerk in 1879.

Initially, engines for aircraft paralleled those for motor cars. The Wright brothers' first powered flight in 1903 was achieved with a four-cylinder, water-cooled engine that they built themselves. Its performance was poor even by the standards of the day – about 6kg per h.p.; S. P. Langley had already achieved 2kg per h.p. with an engine built for an unmanned model. This contrast stresses what was to be the main target of designers: to achieve the maximum power for the minimum weight. The limit was just about reached with the Wright Cyclone R3350 of 1944, used for the famous

Wright Cyclone R3350 air-cooled 18-cylinder radial engine (1944).

Gnome aircraft engine. This
French air-cooled rotary engine
was widely used by the Allies
in the First World War.

American B29 bombers: this was rated at 0.5 kg per h.p. On theoretical
grounds not much further improvement could be looked for, but in any case
the days of the piston-engined aircraft engine were then numbered, except
for light machines.

In the early years of this century the main centre of design and
manufacture of aero-engines was France, possibly because of the interest
aroused by the Wright brothers' exhibition flights there in 1908/9 and Louis
Blériot's cross-Channel flight in 1909.

Two new types of aero-engine began to emerge in addition to those of the
design used in motor cars: radial and rotary. The radial engine followed in
principle that developed for motor cars, the pistons driving a crankshaft that
in turn rotated the propeller; the difference was that the cylinders were
arranged radially instead of in line or in V-array. The rotary engine was
similar in appearance but the whole engine, driving the propeller, rotated
round a stationary crankshaft. The rotary engine had a slight advantage in
weight and was widely used for the relatively light fighter aircraft of the First
World War. One of the most famous was the French Gnome, rated at 1.5 kg
per h.p. After the war, however, the rotary engine fell into disfavour because
of the engineering difficulties arising from the larger sizes then required.

The USA was well behind Europe at the outbreak of the First World
War, but emerged with the famous V-12 Liberty engine, of which 20 000
were made for the American and British air forces. It was water-cooled and

Liberty engine. This American water-cooled V–12 engine was widely used in the American and British air forces in the last year of the First World War.

rated at 1 kg per h.p. A product of the automobile industry, and designed like a motor-car engine, it went from drawing-board to production in six months, between May and November 1917. Liberty engines were used extensively in the 1920s, many of them being war surplus.

As aircraft developed, higher altitudes were attained and this introduced a new problem: the rarefied air of the upper atmosphere caused loss of power, as well as aerodynamic difficulties. The problem was overcome with the supercharger, which in effect was a pump that drew in the thin air and compressed it to sea-level equivalent, or sometimes higher.

The Second World War gave the final stimulus to the piston engine for aircraft. In the USA the Liberty engine was succeeded by the Allison which had roughly the same capacity but three times the power. For bombers, the Wright Cyclone – already mentioned – emerged as a 28-cylinder 2200-h.p. radial engine, the cylinders being arranged in four circles of seven. In Britain, the famous Hurricanes and Spitfires were powered by 1000-h.p. V-12 Merlin engines. Germany, too, entered the war dependent mainly on V-12 liquid-cooled engines.

II. DIESEL ENGINES

It has been said that science owed more to the steam-engine than the steam-engine owed to science, and it is also true that the early development of the

Otto engine was effected largely by practical mechanics rather than trained engineers. The Diesel engine was different, however, in that it was designed on strict thermodynamic principles by Rudolf Diesel, who had trained under Carl von Linde – founder of the liquid-gas industry – at the Technische Hochschule in Munich.

Like the Otto engine, Diesel's worked on a four-stroke cycle. The first stroke fills the cylinder with air, and the second compresses it. The compression is so high (around 15:1) and so rapid that the air is heated above the ignition point of the fuel, which is injected at the start of the next stroke. The last stroke, as in the Otto engine, expels the burnt gases. In passing, we may remark the resemblance to the fire-gun of certain primitive tribes. In this, tinder is ignited by driving a wooden piston into a hollow tube, such as bamboo.

The main advantages of the Diesel engine are that the high compression makes for high efficiency (30–40 per cent); a relatively cheap, and safe, fuel can be used; and no electrical ignition system is required. In practice, however, it presented great difficulties; Diesel patented his engine in 1892 but it was five years before he produced an engine that even ran, and five

First successful diesel engine, built by Maschinenfabrik Augsburg 1898.

more before he made one that was reliable enough to be a commercial proposition.

The main problem from the start was fuel injection. For maximum efficiency this had to be injected at such a rate that it burned in step with the admission of air to the cylinder and the temperature remained constant. This condition is more easily met in the Otto engine, in which the fuel and air are intimately mixed in the carburettor before entering the cylinder. In the Diesel engine a jet of fuel has to be injected at each stroke, and mixing has to be effected in a very brief space of time.

The fuel must be injected at high pressure, if only to overcome the back-pressure of the compressed air in the cylinder. Originally this was effected by a compressed-air tank served by a separate pump, external to the engine and driven by it, but this absorbed an undue amount of the energy generated. Later, the compressor was incorporated as an integral part of the engine.

By 1908, about a thousand Diesel engines were in use as stationary engines; most were in the 50–100 h.p. range. It had become recognized as a reliable alternative to the steam-engine, and new opportunities for application began to be explored, especially for fairly large units. A major obstacle was the weight of the Diesel engine relative to power developed: in this respect it did not begin to comptete with the petrol engine. Attention was, therefore, directed to applications in which this was not an insuperable obstacle, as it was in flying, and where the Diesel had advantages over the steam-engine. One such field was in the submarine, then developing as a new naval weapon. Here diesel fuel was safer than petrol and it offered an extended range in terms of fuel load. By 1914 all the leading navies had Diesel-powered submarines, based on units of around 500 h.p. They provided power for cruising on the surface and for recharging the accumulators which were essential under water. However the weight: power ratio was still high compared with the contemporary petrol engines we considered earlier: an average figure was around 30 kg per h.p.

Such ratios were acceptable also for ships, though much more powerful engines were required. Oil-firing for marine steam-engines was already in use and the resulting convenience and economy in manpower and space was appreciated. The Diesel engine promised to extend these advantages still further. Engine-room staff could be reduced, more space made available for cargo, and refuelling simplified.

Although several hundred Diesel-powered ships were in service before the First World War, the day of the giant marine diesel engine did not come until the 1920s. By 1930 Maschinenfabrik Augsburg-Nürnberg were building engines of around 6000 h.p. and half the new ships being laid down were designed for Diesel propulsion. By mid-century the marine Diesel

Typical marine diesel engine of the 1930s. It was an 8-cylinder double-acting, 2-stroke engine with direct fuel injection, built by Maschinenfabrik Augsburg. It developed 6350 hp at 140 rev/min.

reigned almost supreme: its main rival, the steam turbine, was to be found only in large naval vessels and liners (Ch. 21).

Such large engines were relatively slow turning – about 100–200 revolutions per minute – and quite unsuitable for road vehicles. In the 1920s the growth of the motor-car industry, and the clear indication that the growth would continue, led to a fresh look at problems which had seemed daunting in the face of the limited market prospects of pre-war days. The biggest problem – apart from the old weight : power ratio, which was much improved by elimination of the separate air compressor – was still that of fuel injection. In slow-turning marine Diesels the fuel-injection stroke might take one-tenth of a second, and this was initially regarded as little enough time in which to achieve uniform mixing of the injected spray of oil. For the fast engines essential for road vehicles, the allowable time was nearer to a few thousandths of a second. Nevertheless, a solution was found by designing combustion chambers to produce violent turbulence and pumps which very accurately controlled the time and quantity of fuel injection. Pioneers in this field were H. R. Ricardo in England and P. L'Orange in Germany. By the 1930s satisfactory units of a size suitable for lorries – say 80–100 h.p. began to appear, but this was by no means the limit. By mid-century Diesel engines small enough for motor cars, taxis, and other similar vehicles were widely used, though not seriously challenging petrol-driven vehicles of comparable size. Their undoubted economy in fuel was offset by rather noisier and rougher running; the attention necessary to keep the fuel-injection system working satisfactorily; and the fact that fuel and service stations were not so readily available.

While Diesel engines did not compete very effectively with petrol for light road vehicles, they did so very effectively indeed for a wide range of heavy vehicles designed for special purposes. These included military tanks, earth-shifting equipment, excavators, combine harvesters and other heavy agricultural machinery, and so on.

In the USA a dominating motor-car manufacturer was General Motors, founded by William Durant in the year that Ford launched his Model T. After the First World War they directed a great deal of attention to developing Diesel engines for railway traction. Here, as in ships, weight was not a major obstacle – the Union Pacific 'Big Boy' steam locomotives reached a maximum of 346 tonnes in the 1940s – and there were some notable advantages to be gained over steam, then in its heyday. Basically Diesel traction was more efficient; it consumed fuel only when it was running; servicing was easier and less frequent; it required no stoker. The main problem, as we shall see in our discussion of railway transport (Ch. 20), was the transmission of power from the engine to the wheels. In Poland in the 1920s, a 1200-h.p. Diesel locomotive was designed with electrical transmission, but generally speaking European engineers considered this unsatisfactory; Germany, for example, favoured direct geared drive or pneumatic transmission. General Motors, however, uncompromisingly favoured electrical transmission from the outset. In 1937 they restricted themselves further by concentrating on a single 1800-h.p. unit, which was far below the capacity of the largest contemporary steam locomotives. This limitation was overcome by a modular system, by which these units could be coupled until the desired power – say 5400 h.p. – was reached. This device ensured that each unit was working at near maximum capacity, at which the Diesel engine is at its most efficient.

III. GAS TURBINES

In this form of internal combustion engine the burning fuel drives not a set of reciprocating pistons but a continuously rotating turbine. The latter has a double duty to perform: it drives a compressor to compress the air entering the system and it provides power for whatever purpose is required – the propeller of an aeroplane, an electrical generator, or the wheels of a locomotive. In theory, continuous action is preferable to reciprocating , but, in practice, the amount of additional work to be done inside the engine very considerably limits the useful power generated; nevertheless there are circumstances in which this is discounted by other characteristics, notably the turbine's capacity to generate high thrust quickly.

A major technical problem of the gas turbine is its high operating temperature. Combustion of the fuel is continuous, as opposed to intermittent in an Otto or a Diesel engine, and this makes heavy demands on the turbine blades: the suitable constructional materials available are limited. A number of gas turbines were built between the wars – indeed, Holzworth built one in Germany as early as 1908 – and used for a variety of purposes including marine propulsion, railway locomotives, and electrical generators. Notable pioneers were Brown Boveri in Switzerland. It was, however, military demands – where economic considerations became of secondary importance – that stimulated large-scale development. As early as 1930, Frank Whittle, a Royal Air Force officer, pointed out the potential advantages of the gas turbine for high-altitude flight: he recognized that the gas turbine could not only drive a conventional propeller but could be adapted for jet propulsion. His views attracted little attention at the time, but quite soon, with rearmament under way and war in prospect, the situation was very different. Whittle meanwhile had been sent to Cambridge University and had established in Cambridge a small firm, Power Jets, to develop his ideas. Their first jet engine ran in 1937. They won a small government contract and then – quite suddenly – massive support.

Meanwhile events were taking a rather similar course in Germany. By 1937 Hans von Ohain had a jet engine running, and the world's first jet flight, in a Heinkel He 178 aircraft, took place on 27 August 1939, literally the eve of war. At another aircraft manufacturer's, Junkers, Herbert Wagner had been working on conventional lines, using a gas turbine to drive a propeller, but just before the war the research programme was

Stationary gas turbine built by Brown Boveri in 1937.

switched to a turbo-jet system. From this emerged the Junkers 004 jet engine, of which more than a thousand were built during the war and used for the twin-engined Messerschmidt 262. In Britain, events moved rather more slowly but perhaps more surely. The first jet flight, in a Gloster Meteor powered by a Whittle jet engine, took place on 15 May 1941.

The USA was relatively slow off the mark in this field, as in the First World War, and the emphasis in research was on turbo-prop systems rather than turbo-jet. In the event, reliance was placed on development of the Whittle engine, in which American improvements were effected by the development of better materials for the turbine blades.

In the post-war world jet-propulsion and turbo-prop systems developed side by side in both military and civilian use, with the former gradually gaining the ascendancy. The first jet transport plane was the de Havilland Comet of 1952, though it had to be temporarily withdrawn owing to serious structural weaknesses in the original fuselage. Piston-engined aircraft also remained in use, and dominated the light aeroplane field. The great advantage of the jet engine is its very high power : weight ratio – which has been a crucial factor in the aircraft industry from its very beginning – and its ability to give a pretty constant thrust at all speeds.

On land, the gas turbine found many applications after the Second World War. An important use was to drive centrifugal compressors on gas pipelines, the units of which ranged from 2000 to 14 000 h.p. They are also used in the petroleum industry to drive compressors to maintain pressure in oil wells and to compress gas in refineries. They are used in the steel industry to provide hot compressed gas for blast-furnaces. At sea they have found a variety of applications in units of up to 6000 h.p., especially for craft required to operate at high speeds over short ranges. The first warship powered with a gas turbine (MGB 2001) went to sea in 1947; nine years later, the converted Liberty ship *John Sergeant* was the first merchant vessel so equipped.

For all such applications some form of petroleum-derived fuel is normal. After the Second World War there was some interest in the use of powdered coal as a fuel for gas turbines, especially for railway locomotives, but great difficulties were encountered as a result of damage to the turbine blades by fly ash.

BIBLIOGRAPHY

Beaumont, W. W. *Motor vehicles and motors*. 2 vols. Constable, London (1900, 1906).

Bryant, L. 'Rudolf Diesel and his rational engine'. *Scientific American*, **221**, 108 (1969).

Clerk, Dugald. *The gas, petrol and oil engine*, 2 vols. Longmans, London (1916).

Dickey, P. S. *The Liberty engine 1918–1942*. Smithsonian Institution, Washington, DC (1968).

Donkin, B. *Gas, oil and air engines*. Charles Griffin, London (1911).

Ewing, J. A. *The steam engine and other heat engines*. (3rd edn.) Cambridge University Press, London (1920).

French, J. W. *Modern power generators*. Gresham Publishing Company, London (1908).

Hubbs, L. S. *The Wright brothers' engines and their design*. Smithsonian Institution, Washington, DC (1971).

Kilon, I. *The evolution of the heat engine*. Longman, London (1972).

Palmer, S. *The impact of the gas turbine on the design of major surface warships*. Royal Institution of Naval Architects, London (1973).

Ricardo, H. R. 'The development and progress of the aero engine'. *Journal of the Royal Aeronautical Society* **34**, 1000 (1930).

Smith, G. G. *The modern Diesel*. Iliffe, London (1944).

Stodola, A. (trans. Loewenstein, L. S.) *Steam and gas turbines. 2 vols. McGraw-Hill, London (1927)*.

Tatton-Brown, P. *Main propulsion gas turbines in the R. N.* North East Coast Institution of Engineers and Shipbuilders (1970).

Schlaifer, R. and Heron, S. D. *Development of aircraft engines and fuels*. Harvard University Press, Boston, Mass. (1950).

Walshaw, A. C. *Heat engines*. (4th edn.) Longmans Green, London (1956).

Whittle, F. The early history of the Whittle jet propulsion gas turbine. *Minutes of Proceedings of the Institution of Mechanical Engineers*, **152**, 419 (1945).

14

TURBINES

Gas turbines were discussed in the last chapter, as they are effectively part of the internal combustion engine family. As we saw, their practical impact was not important until the advent of the jet engine during the Second World War; their other main development, as the turbo-prop engine, occurred at the very end of our period. With other turbines the situation was different – both water and steam turbines were well established by the beginning of the twentieth century. Indeed, Benoît Fourneyron's first water turbine, generally regarded as the direct ancestor of modern engines of this kind, appeared in 1827. It was an outward-flow engine, and long before 1900 other important variants had appeared, based on axial (Jonval, 1837) or inward flow (Francis, 1849). The Pelton wheel, crude beginning of another line of development, was invented in the 1870s. The measure of progress is given by the fact that by the end of the century the Niagara hydroelectric project embodied water turbines capable of generating 10 000 h.p.

Relatively speaking, the steam turbine was a newcomer, for Charles Parsons's first successful steam turbine was not built until 1884, primarily in response to the need for a high-speed engine to drive the dynamos required for the rapidly expanding electrical industry. Even before this, in 1883, the Swedish engineer, Gustav de Laval had produced a workable, but not very efficient, turbine. In 1894 Parsons turned his attention to the use of steam turbines for marine propulsion and in 1897 his *Turbinia*, the first vessel ever to be propelled by a turbine, made a dramatically successful appearance at the great Spithead naval review. This marked the start of a new line of development in ships' propulsion.

I. WATER TURBINES

At the beginning of this century only two types of water turbine were of major importance – the Francis inward-flow reaction turbine and the Pelton wheel or impulse turbine.

In the Francis turbine, which runs fully submerged, water from a series of peripheral nozzles is directed on to vanes fixed to the vertical rotating shaft. The power of the jet is proportional to the head of water above the installation. In 1903 the Canadian Niagara Power Company had a series of double Francis turbines operating under a water head of 133 ft; they generated 10 000 h.p. at 250 revolutions per minute. By 1923, 70 000-h.p. Francis wheels had been installed at Niagara. The Grand Coulee Dam project, on the Columbia river, the first stage of which became operative in 1941, included 115 000-h.p. units working under a 485-ft head.

A variation of the Francis turbine is that developed by Victor Kaplan, a Czech engineer, between 1910 and 1924. In this, the horizontal rotor resembles a ship's propeller and the pitch of the blades can be varied while the machine is running, thus allowing high efficiency at varying loads. It works best under relatively low heads of water, up to say 100 ft. Kaplan turbines are relatively expensive to build and are not economic in all situations. The Deriaz turbine, developed by English Electric, combines features of both the Francis and the Kaplan turbines.

In the impulse turbine a jet of water impinges on curved vanes or buckets of ellipsoidal shape attached to the edge of a rotor. Ideally, these completely reverse the flow of water. In the early machines, the buckets were bolted on to the rotor, but from 1920 the rotor began to be cast in one piece, an innovation introduced by the firm of Gilkes in Britain. About the same time, this firm introduced the turgo impulse turbine. In this, a jet of water strikes one face of the rotor and is discharged through curved vanes attached to the other; this arrangement improved efficiency for a given wheel size. At mid-century the largest Pelton-type installations in the world were those near

Water turbine based on Pelton wheel.

Sao Paulo in Brazil. They generated 78 000 h.p. and operated at 360 revolutions per minute under a head of 2200 ft of water.

The development of water turbines presented formidable engineering problems, many of which arose from the weight of the rotating parts. Even at the beginning of the century this might exceed 100 tons. Another problem was caused by the phenomenon known as cavitation. Briefly, this arises from the formation of bubbles of water vapour which collapse when they strike a solid surface. This causes something like a hammer blow, insignificant in itself but capable over a period of time of causing pitting of the metal sufficient to cause mechanical failure. Unfortunately for turbine designers, cavitation is most serious at the high speeds they seek to achieve. It is difficult to eliminate completely but can be greatly reduced by careful design. It is, as we shall see later, a phenomenon also very familiar to the designers of ships' propellers.

II. THE STEAM TURBINE

Like the water turbine, steam turbines are of two types – the reaction type and the impulse type. It was the first type that Parsons developed; the second is particularly identified with C. E. A. Rateau in France and C. G. Curtis in the USA. In Parsons's design the steam expands on passing through a succession of rotating blades; in the other type there is no expansion in the blades but it takes place instead in fixed nozzles. Early in this century Curtis turbines tended to be favoured by the world's navies – in the USA, Britain, Japan, and Germany, for example – but as the years went by the difference became more apparent than real, the designers of impulse engines having to introduce some degree of reaction. Curtis-type turbines are, however, universally used to provide power for going astern, essential for manœuvring ships in restricted water. All these types are axial-flow turbines. In the Ljungström variant, developed in Sweden before the First World War, the steam flows radially outward through bladed discs rotating in opposite directions, each disc being directly mounted on the shaft of one of two overhung dynamos; it is made in sizes up to around 40 000 kW, beyond which stress on the blades becomes excessive, and it must be coupled with an axial-flow unit.

To improve the efficiency of turbines the temperature and pressure of the steam must be varied, and this creates considerable engineering problems. In particular, the arduous working conditions demand careful choice of materials of construction. The casing is made of steel alloyed with molybdenum, vanadium, chromium, etc. The rotors are also of steel, either forged in one piece or consisting of a series of wheels shrunk on to the shaft. The common material for the very large number of blades is a low-carbon steel containing 12–13 per cent chromium. Nevertheless, during the first half of this century great progress was made in both respects: pressure and temperature were raised from about 140 lb per square inch and 200°C in 1900 to 1400 lb per square inch and 565°C in 1950.

The above figures should be regarded as representing general practice; in some cases they were substantially exceeded. Thus in 1951 two 125 000-kW turbo-generators installed at Sewaren, New Jersey, operated – using re-heat steam – at 1500 lb per square inch and 565°C. At Twin Branch, in the USA, much higher pressure (2500 lb per square inch) was used but at a lower temperature (505°C). By mid-century thermal generation of electricity depended almost wholly on turbines, and over-all thermal efficiency was in the region of 30–5 per cent, representing something like a doubling in fifty years. After the First World War regenerative feed heating was introduced, exhaust steam being used to heat the feed water.

Turbo-alternators have to run at a constant fixed speed to maintain steady voltage output. This is 3000 and 3600 revolutions per minute to suit the standardized voltages of Europe and North America respectively. For ships, requirements were different, as propeller-shaft speeds of 200–500 revolutions per minute were needed; another consideration was that higher speeds caused cavitation damage to the propeller similar to that encountered in water turbines. The need for high torque at low speeds

Large marine turbine, fitted with reduction gearing, built for the *Maheno*.

presented a problem, for the turbine is not at its most efficient under such conditions. In the earliest battleships to use turbines and in the great transatlantic liners *Lusitania* (1907) and *Mauretania* (1907) direct drive was used, but in 1909 the cargo vessel *Vespasian* was fitted with reduction gearing and this became general practice. A common arrangement, devised by de Laval, was to locate the turbines in such a way that helical gears on their shafts meshed with a large central gear on the propeller shaft. In some American battleships, and the pre-war French liner *Normandie*, an electrical system of power transmission was used, turbo-generators supplying low-speed electric motors directly driving the propeller shafts.

While the improved steam turbine was quickly adopted for large liners and warships, it was rivalled, as we have noted, by the Diesel engine which was quickly developed to produce very large power units. The steam turbine did not, therefore, attain in marine propulsion the almost total ascendancy it had acquired by 1950 in thermal electricity generation. After the Second World War, however, its scope increased with the appearance of new tankers and container vessels requiring exceedingly powerful engines. Both these had a challenger in the gas turbine.

BIBLIOGRAPHY

Althin, T. *Life of de Laval*, Stockholm, (1943).

Appleyard, R. *Sir Charles Parsons*. Constable, London (1933).

Baumann, K. Recent steam turbine practice. *Journal of the Institution of Electrical Engineers*, **48**, 768 (1912).

—— Some recent developments in steam turbine practice. *Journal of the Institution of Electrical Engineers*, **59**, 565 (1921).

Parsons, C. A. Steam turbines. *Transactions of the First World Power Conference 1924*, vol. 2, page 1477.

Parsons, G. L. (ed.) *Scientific papers of Sir Charles Parsons*. Cambridge University Press, London (1936).

Parsons, R. H. *The development of the Parsons steam turbine*. Constable, London (1936).

Stodola, A. *Steam and gas turbines*. McGraw-Hill, New York (1927).

Stoney, G. Steam turbines (Cantor Lecture). *Journal of the Royal Society of Arts*, **57**, 954 (1909).

15

THE WORKING OF METALS

Earlier (Ch. 10), we considered developments in the extraction of metals, both those already established and others which came into large-scale use only during the present century. We left the story at the stage in which metals, or their alloys, were available in bulk. We have now to consider the ways in which they were worked into the multiplicity of parts – from armour-plating for battleships to hair-thin filaments for electric lamps – required by modern industry. Again, we shall see a mixture of old and new. Even at mid-century many metal-forming works would have contained nothing to surprise a nineteenth-century engineer; others would be using materials and techniques that would have been quite unfamiliar. Perhaps the most obvious general changes would have been a steady increase in mechanical handling to reduce heavy physical toil; the replacement of overhead drives and belts by individual electric motors; and greater attention to the health and comfort of the workers.

Metals can be shaped in a great variety of ways but most come under three main headings – those in which the molten metal is shaped in a mould; those in which it is mechanically deformed, as in forging and rolling; and those in which it is cut to shape by hard-edged tools. Among methods which do not conform to this classification we may mention the sintering of refractory metals in powder form; the deposition of thin films by hot-dipping in molten metal or by electrochemical methods; and the cutting of metals with an oxy-acetylene flame.

I. CASTING

As we have seen, the final stage of many extraction processes ends up with molten metal, which is cast into ingots. Much of this metal is remelted and recast in carefully prepared and shaped moulds made from master patterns. Carefully compacted green sand remained the traditional moulding material throughout our period, but during the Second World War shell moulding was introduced in Germany. In this process, the sand was mixed with a heat-setting resin. Where it came into contact with a pattern heated to around 250 °C the resin set and permanently fused the sand particles together; excess sand was removed and a thin shell was left around the pattern. This was further cured, split open to remove the pattern, and was

Traditional methods of casting iron in sand moulds. This photograph was taken in England about 1950, showing how long it took for improved methods to be adopted.

Madison Kipp pressure die-casting machine in operation in the 1930s: the casting here is of Mazak.

then ready to receive the casting. This method gives particularly close tolerance and good surface finish.

An alternative method, known as die-casting, came to be widely used for the greater variety of small metal parts increasingly required for motor-cars, domestic appliances, and so on. In this, a permanent steel or cast-iron mould – coated with soot or some refractory material to prevent surface attachment – was used to receive the molten metal. Under favourable conditions, such moulds might be used hundreds of thousands of times, and they were particularly successful with the new zinc alloys of the Mazak type (p. 124). In pressure die-casting the liquid metal is forced into the mould at pressures ranging from 1000 to 100 000 pounds per square inch. Another important innovation was centrifugal casting, in which the mould is rotated at high speed to spin the molten metal out to its edge, where surface speeds of up to 5000 ft per minute are attained. This proved particularly useful for cylindrical castings such as pipes and kitchen utensils. It is suitable for small experimental production runs.

Casting technique was greatly improved by growing understanding of the properties of metals and ways of modifying them appropriately. For example, careful control of the cooling of the mould ensured more uniform shrinkage of the casting and reduced the risk of formation of gas bubbles and cracks. It also made possible the development of a micro-crystalline structure appropriate for the ultimate use of the casting.

II. MECHANICAL SHAPING

The forging of metals, as practised by the blacksmith, is one of the oldest metal-shaping techniques and is based on the fact that, when strongly heated, many metals become malleable and can then be shaped by hammering, by passing them through rollers, or subjecting them to other forms of mechanical stress. Some metals can be forged cold. By the end of the nineteenth century the steam-hammer, hydraulic forging presses, rolling-mills, and other heavy equipment made possible the preparation of very large items. Krupps, for example, could roll a 130-ton steel ingot into a massive sheet a foot thick, 11 ft wide, and 43 ft long, destined to serve as naval armour-plating. Steel rail was being produced by the mile for the railways, acres of tinned steel sheet fed the canning factories, and vast quantities of rolled copper sheet were made into locomotive fire-boxes. Where a fine surface finish was required, the last stages of copper rolling were carried out with cold metal.

Up to this point, reduction in size was effected by repeatedly passing the metal plate to and fro between banks of rollers, just as clothes are wrung out in a mangle. With growing demand and rising costs, however, attention was increasingly directed to the possibility of continuous rolling of strip. This was first practised at Toeplitz, in Austria, in 1902. After preliminary rolling by conventional methods the sheet was presented to a series of five rolling-mills, 9 ft apart, rotating at different speeds to accommodate the diminishing thickness of the sheet as it progressed. It finally emerged at a speed of 6 ft per second with its thickness diminished to one-tenth of an inch. Unfortunately, this mill was not a success, partly because it was under-powered. The strip was not sufficiently uniform and operations stopped in 1907. A semi-continuous mill for steel strip was set up at Ashland, Kentucky, in 1922 but this, too, was unsuccessful. However, the growing demands of the motor-car and food-canning industries encouraged experiment, and in 1926 the Columbia Steel Company of Pennsylvania successfully introduced a continuous mill producing 48-inch strip. By the outbreak of the Second World War, twenty-eight continuous rolling-mills had been built in the USA, at an average cost of around $20 million, and such mills had also been established in Britain and elsewhere in Europe. An unusual development was the Sendzimir mill in which some twenty small working rolls surround a large back-up roll; in this, 2-inch steel stock could be reduced to one-tenth of an inch in a single pass.

The processes so far considered start from a metal ingot, which is gradually worked down to the desired size. As early as 1856, however, Bessemer had visualized the possibility of continuous casting of steel between rolls, thus eliminating the early stages; trials, however, were unsuccessful. No real progress was made until the 1930s when the machine invented by S. Junghans and I. Rossi for the continuous casting of copper and copper alloys between rolls was introduced. Here the problem was easier because of the lower melting-point and higher thermal conductivity of the metal. Working in Germany, Junghans then turned his attention to steel casting in 1939, and eventually achieved success in 1943. In Russia, there had been interest in the process since 1931, using the principles proposed by Bessemer; a continuous casting plant was commissioned in 1951 in the Krasny Octyabr works. It was followed four years later by a unit at Krasnoe Sormovo, then the largest in the world, capable of producing 50 tons of cast steel slabs an hour. In the USA and Britain small Junghans-Rossi plants were established around 1950, but it was ten years before the process was firmly established in the West.

Lead tubing had been made since 1820 by subjecting the cold metal to high pressure and forcing it out through a circular orifice with a central mandrel. Given sufficient pressure, such as was available in the twentieth century, most other metals can be shaped by extrusion – much as toothpaste is expelled from a tube – though the process may not be viable, or viable only for relatively small and special items, in the case of the harder metals. Extrusion may induce changes in the metal. Scme, like magnesium and aluminium, must be extruded slowly, but others – such as steel and copper alloys – can be extruded at speeds up to 1000 ft per minute. The most widely used method is direct hot extrusion, using hydraulic pressure to force the metal through the die. In cold extrusion – particularly useful for making small hollow cylindrical items having thick bases – the blank is punched, often by successive blows, into the space between a plunger and a die. For relatively soft non-ferrous metals – such as are used, for example, in making cartridge cases – this method was in use at the beginning of the century, but about 1930 it began to be used in Germany to manufacture small steel articles. The process is simple and quick, and lends itself to mass production of small articles.

Extrusion is, of course, the reverse of the drawing processes long used to manufacture metal wire and rod. In the drawing process, however, the force that can be applied is limited by the breaking-strength of the product. In 1900 wire was, of course, demanded in enormous quantities for established uses such as netting, rope, barbed and chain-link fencing, and nail manufacture. To this was added, in the twentieth century, an enormously increased demand for electrically conducting wire – largely copper-based – for the expanding electrical industry. As we have noted, however, reinforced aluminium wire came into use for long-distance transmission cables and at the other end of the scale very small quantities of special fine wires were required for the electric lamp industry to replace the original carbon filaments. The metals involved in this last use were chosen primarily because of their very high melting-points and this inevitably presented difficulties in fabrication. Osmium filaments were introduced in 1898 by Auer von Welsbach, well known for his invention of the incandescent gas mantle, and tantalum by Siemens and Halske in Berlin in 1905. The major development in this field, however, was William Coolidge's use of tungsten in the USA in 1908. This melts at 3400 °C and is thus very intractable. Coolidge succeeded in making it ductile by first compressing the powder under exceedingly high pressure in a hydraulic press. The resulting rod was suspended in an inert atmosphere and the particles were fused together by passing a heavy electric current through it. This metal rod was then made ductile by repeated heating and hammering, after which it could be drawn into very fine wire through diamond dies. Molybdenum metal (melting-point 2600 °C) was made similarly and used to make the filament supports. This technique of powder metallurgy – which William Wollaston had used more than a century earlier to make malleable platinum – did not make a major over-all contribution to the metal industries, but was widely used for specialist purposes as, for example, in fabricating beryllium (p. 128). It was also used, as we shall see shortly, for fabricating some of the cutting edges required for the new generation of machine tools.

III. MACHINE TOOLS

In 1776, early in the Industrial Revolution, Matthew Boulton was delighted when John Wilkinson bored him a cylinder for a large steam-engine 'almost without Error . . . [it] doth not err the thickness of an old shilling in no part'. By the middle of the nineteenth century Joseph Whitworth was deriding engineers who worked to such sizes as 'a full thirty-second' and, to standardize his own workshop equipment, had built a machine capable of measuring to one-millionth of an inch. Such precise control of size made possible the production of parts for elaborate mechanisms on a completely interchangeable basis, as exemplified by Samuel Colt's manufacture of his famous revolver in 1849–54, and by the adoption of his methods in Britain for the manufacture of rifles at Enfield, where 2000 rifles a week were made with all parts completely interchangeable. At the very end of the century the Swedish technologist C. E. Johansson introduced an ingenious collection of gauge blocks for use in precision engineering. This pointed the way to mass production of precisely machined engineering parts for many purposes, most especially for the motor-car and, later, aircraft industries.

By 1900 a formidable array of machine tools had developed from their

Johansson gauge blocks; this set was made in 1919.

nineteenth-century predecessors. They were capable of dealing not only with the relatively small items encountered in normal workshop practice but also with the very much larger ones required for heavy machinery such as ship's propulsion units and turbo-alternators. They covered such basic metal-shaping operations as turning, milling, grinding, planing, and gear-cutting. By the turn of the century there was, predictably, a clear tendency towards higher production rates and automatic control. Individual electric motors, often an integral part of the machine, were becoming the usual source of power. Improved casting methods reduced the amount of machining necessary to produce the finished article.

During the first half of the twentieth century the machines themselves were improved in many ways; by far the most important advance was in the materials used for the cutting tools. Up to the middle of the nineteenth century the best material available was carbon steel, and if this was to have a reasonable working life cutting speeds had to be limited to 40 ft per minute. This was increased by half with the advent of Mushet's tungsten/vanadium/manganese steels in 1868, which were almost universally used up to 1900. In that year F. W. Taylor and M. White of the USA demonstrated at the Paris Exhibition a tungsten/chromium/vanadium steel that would retain its cutting edge even at red heat. Such steels increased cutting speeds to 120 ft per minute over the next decade, obliging designers of machine tools to make them more rigid in order to retain precision. These new steels contained a little (0.7 per cent) carbon, which resulted in the formation of small particles of tungsten carbide, an exceedingly hard material. This led, in the 1920s, to tungsten carbide cutting tools. In making these, granular tungsten carbide was mixed with roughly three times its weight of powdered cobalt and the whole was compacted and fused by the technique of powder metallurgy. The new material still further improved cutting speeds but it was expensive. For this reason, it was customary for carbide cutting edges to be brazed to steel bases. Finally, we must note the arrival of the so-called cermets (ceramic/metal), which are mixtures of metals with various oxides and carbides. These have high strengths at high temperature, and are also produced by a sintering technique.

The development of new cutting tools was matched by advances in abrasives. In Ancient Egypt craftsmen ground stone surfaces with sand, and the grindstone was a traditional device for sharpening weapons. Nevertheless, up to 1900 grinding was not an economic way of shaping materials except in special circumstances, such as the making of optical lenses. At the turn of the century, however, the situation changed. E. G. Acheson, in 1898, introduced silicon carbide as an abrasive under the name carborundum, and almost at the same time C. B. Jacobs introduced an artificial form of aluminium oxide – second in hardness only to diamond – under the trade name alundum. The availability of these new products led Charles Norton in the USA to promote grinding as a basic engineering process, and in 1900 he completed two successful prototype grinding machines. Ten years later he had devised grinding machines capable of producing motor-car crankshafts in a small fraction of the time required for manual work. By the 1920s grinding was a widely used process in precision engineering.

During this period the mass production of accurately cut gearwheels expanded, first for bicycles and then for motor cars and aircraft, and it was an important field of application of the grinding machine. In such applications the wear was concentrated on the intermeshing surface, rather than on the gears as a whole; it was, therefore, uneconomic to make all the components of a gearbox from high alloy steels. A compromise was found in the technique of case-hardening in which carbon was diffused into the surface layers by heating the parts in charcoal in a closed box for two days at around 900°C. Later, with the use of steels containing a little aluminium, case-hardening with molten cyanide salts was introduced.

IV. ASSEMBLY

By the end of the nineteenth century a wide range of constructional materials was available in the shape of girders (in a variety of sections), sheet, rod, tube, and so on. From these basic units, available from stockholders, a great variety of metal structures could be made, from railway trucks to steel-frame buildings, bridges, and ships.

In 1900 much construction work of this kind was done with rivets. The rivet is passed through holes punched or drilled in the parts to be joined and the protruding end is then mushroomed out with a hammer to prevent it slipping out. For a more elegant finish, the end of the rivet may be given a domed shape with a special punch, or the hole may be countersunk so that the end of the rivet can be struck off flush. Small rivets can be worked cold, but larger sizes are inserted red-hot so that the metal is more malleable. An additional advantage of hot working is that the rivet contracts on cooling, giving an extremely tight joint. Most rivets were of mild steel but other metals – brass, copper, and later aluminium – were also used.

Riveting was carried out on a tremendous scale. For example, the Britannia Bridge over the Menai Straits – carrying the railway from London to Holyhead – contained more than two million rivets. Hundreds of bridges constructed by riveting plate girders together are still in use all over the world. In the shipyards, the riveter – and his mate carrying a portable forge – was a key operator. Work on this scale was arduous, and pneumatic and hydraulic riveting hammers were introduced in 1865 and 1871 respectively, but they were not widely used until the twentieth century.

By the 1920s, however, riveting was rivalled by welding, another ancient process, which was soon to eclipse it. In this, the two parts to be joined are fused together by intense heat, additional metal usually being supplied, as in soldering, in the form of a consumable welding rod. The source of heat may be an intensely hot flame, such as an oxy-acetylene flame, or an electric arc. The weld may be continuous or at a succession of points (spot welding). Although the material most commonly welded is steel, other metals can also be used, though the technique may then have to be modified. In the case of aluminium, which is easily oxidized, the joint must be formed in an inert atmosphere such as helium or argon. It is estimated that by 1950 world consumption of welding rods of all kinds was around 250 000 tons.

In theory a welded joint is stronger than a riveted one because it is continuous metal, whereas rivet holes are potentially a source of weakness. In reality, however, the local heating and cooling inevitable in welding can cause local stresses and changes in the metal which weaken it. Experience showed that the first sign of trouble in riveted structures was local and obvious, and total failure was rare. In welded structures, however, failure was liable to be rapid and extensive. This left a balance of practical advantage with welding but a major factor in its widespread adoption was economic. No rivet holes had to be made and aligned, and no stocks of variously sized rivets had to be held. Less labour was required. More important, perhaps, was that design was more flexible and new scientific principles of construction could be employed. A very important development was the introduction of the welded box girder which has great resistance to torsional forces.

The intense heat of the oxy-acetylene welding flame was turned to a different use, the cutting of metals. Heavy steel sheet, girders, and so on could by this means be easily cut to shape *in situ*.

V. PROTECTION OF METALS

Metal structures may fail from mechanical causes, of which sheer overloading is the most obvious. In 1854 J. Braithwaite used the word fatigue for the failure of metal subjected to repeated stress. In the twentieth century other causes of failure were identified. Just before the First World War E. N. da C. Andrade discovered the phenomenon of creep, a deformation that occurs as a consequence of continuous, rather than repeated, stress. In 1917, B. P. Haigh identified corrosion fatigue – particularly common in non-ferrous metals such as brass and copper – as a cause of failure associated with environmental conditions. Understanding of the causes of these sources of

The motor-car industry provided an important new outlet for paints. This picture shows a refinish training centre for spraying and painting; ICI, Chiswick, 1926.

failure led to improvements in design to reduce their incidence and to inspection and maintenance procedures designed to identify them before they became dangerous.

Overall, however, the biggest enemy of metal is probably corrosion. The traditional method of preventing corrosion, and often of simultaneously improving appearance, was to use paint, which provides a thin barrier between the metal and the atmosphere. Paints based on drying oils, such as linseed oil and, increasingly, tung oil, continued to be extensively used, but new pigments came into use. Among these was yellow zinc chromate which has important rust-inhibiting properties; this was an important addition to the traditional red lead. Another newcomer was titanium dioxide, which almost completely displaced white lead between 1920 and 1950. Although more expensive than white lead, it is non-toxic and does not blacken in sulphurous atmospheres. Natural resins, used to give a harder finish, were increasingly replaced by synthetic products of the chemical industry, especially the alkyd resins introduced in the USA about 1928. Synthetic resins were also incorporated in stoving finishes which give an exceptionally hard, heat-resisting and durable enamelled surface to a wide range of products such as baths, cookers, and kitchen-ware.

The motor-car industry was an important new customer for paints. With the introduction of mass production, cars could be manufactured more quickly than they could be painted, and this produced a serious bottleneck: at the beginning of the century painting the bodywork could take as long as ten days, allowing for complete drying between successive coats. Stoving enamels reduced this to 2-3 days, but the biggest change came in the 1920s with the advent of lacquers based on nitrocellulose, for the manufacture of which there was surplus capacity with the closing of wartime explosives factories. So important was this market that Dupont in the USA and Nobel Industries (later ICI) in Britain acquired a financial interest in General Motors. Nitrocellulose lacquers enormously accelerated the finishing process, for they could dry completely in half an hour. Emulsion paints, introduced in Germany during the Second World War to reduce dependence on drying oils, were widely used afterwards, though more for household decoration than as a protection for metals.

Up to the First World War paints were almost universally applied with a brush, though in 1907 the first spray-gun appeared. After the war, when the scale of operations and the need for speed in dealing with the new paints both increased, spraying was widely used in industrial applications. An important development, introduced in the USA in the 1940s, was electrostatic spraying. In this, the spray droplets are given an electric charge, and are attracted electrostatically to the object to be painted. Another method of speeding up the painting process was to use a dip-tank. In flow-coating, used extensively during the Second World War, paint was hosed on to the article to be coated.

An alternative method of protecting a vulnerable metal surface is to coat it with a thin layer of a more resistant one. Iron, or steel, for example, may be coated with tin (tin plate) or zinc (galvanized iron). The role of the zinc in the latter case is not merely to provide a mechanical barrier but to take the brunt, as it were, of corrosion: as a result of electrochemical processes occurring in the presence of corrosive solutions the zinc is preferentially dissolved. Plating was not always solely protective: in the motor-car industry, for example, plating with chromium, nickel, or cadmium gave a bright and attractive finish. Silver plating gave domestic ware the appearance of silver without the cost.

Up to 1941, tin plate was made by the traditional hot-dipping process, but thereafter continuous electroplating processes became increasingly important; by this means exceedingly thin films, down to 0.000015 in, could be rapidly produced. By 1957, little more than 10 per cent of American tin plate was produced by hot-dipping. By then the USA was by far the largest producer – 5 per cent of all her steel went into tin plate, total production of which was 5 million tons in 1957, requiring 33 000 tons of tin. The biggest customer was the tin-can industry, which produced 46 000 million tinned-steel containers: of these, 60 per cent went to the food-canning industry (p. 206).

Galvanizing, too, increased greatly in scale: by mid-century the USA alone was producing $2\frac{1}{2}$ million tons of galvanized iron sheet, requiring about 200 000 tons of zinc. Roughly the same amount of zinc was used for galvanizing other products. Sheet was still made by the hot-dip process, but small items – such as nuts, bolts, nails, and washers—were made by a process known as sherardizing. In this, the parts to be treated are tumbled in a drum for several hours with hot zinc dust (350 °C). This gives a dull but otherwise satisfactory finish, with a final zinc coat about 0.003 in thick.

Hot-dip or electrolytic processes give relatively thin metal coatings, measured in hundredths or thousandths of an inch or less. In the process known as cladding, which appeared about 1930, a much thicker coating is applied; this is usually specified in terms of the total thickness and commonly amounts to about 10 per cent. The cladding layer is welded or otherwise bonded to the base metal and the sheet is then rolled to the desired thickness; both sides may be so treated if desired. One of the earliest applications was the nickel cladding of steel plate for the construction of railway tank cars to carry caustic soda. Duralumin was clad with aluminium and, later, steel was clad with titanium (p. 128) to make parts for jet engines.

Another important method of preventing corrosion of metals is by alloying with relatively small proportions of other metals. In this respect, the twentieth-century metallurgist had something in common with the plant breeder. Growing understanding of the nature of metals, and of the relationship between composition and properties, enabled him increasingly to tackle his problems on a logical basis but, even so, it was not easy to combine a number of desired properties. Thus an alloy resistant to corrosion might be too soft to take a cutting edge, be unduly prone to fatigue, or have an excessive electrical resistance. Like the plant breeder, who must be ready to take advantage of spontaneous mutants, the metallurgist must be quick to notice and exploit useful qualities discovered by chance, as in the age-hardening of aluminium alloys (p. 127).

As ferrous metals are by far the most widely used, we may usefully include in this section a brief account of the development of corrosion-resistant steels. The key discovery here may, with some qualification, be said to be that of H. Brearley in Britain. In 1912 he was investigating the suitability of high-chromium steels (9–16 per cent) for making rifles, and observed their high resistance to corrosion; in the event, such steels found their main application in the table cutlery trade and later, to some extent, in the exhaust valves of aeroplane engines. It must be noted, however, that similar alloys had been investigated at least twenty years earlier, but their rust-resisting properties were obscured by the fact that their resistance to corrosion was measured in terms of their reaction with dilute sulphuric acid. Early in the century the addition of nickel to steel was regarded as essential to the prevention of corrosion; in Germany B. Strauss and E. Maurer in

1912–15 – in association with Krupp – developed rustless steels containing as much as 20 per cent chromium but also 6 per cent nickel.

Finally, as an example of the development of corrosion resistance in non-ferrous metals, we may consider the history of the development of the metal used for making condensers for steam-engines. For some applications fairly pure water might be available for cooling, but ships were dependent on sea water, and power-stations and many other installations often had to use river, sometimes estuarine, water. The result was severe and rapid corrosion, and repair and ultimate renewal of the condensers – long lengths of tubing contained within a shell – was an inescapable charge. In 1900 brass containing 70 per cent copper and 20 per cent zinc was, on balance of effectiveness and cost, the best alloy available. After the First World War it was found that resistance was much increased by addition of a very little arsenic (0.05 per cent), and still further if aluminium (2 per cent) was present too. Later, with nickel more readily available, a 70:30 copper:nickel alloy was used. Between 1914 and 1940 the average life of condensers in British warships had increased from barely six months to seven years. Later, costs were cut by reducing the nickel content to 10 per cent and including 1 per cent of iron.

BIBLIOGRAPHY

Althin, T. K. W. *C. E. Johannson 1864–1943: the master of measurement.* Stockholm (1948).
Beck, A. *The technology of magnesium and its alloys.* F. G. Hughes, London (1941).
Carpenter, H. and Robertson, D. *Metals.* Oxford University Press, London (1939).
Davies, R. C. The history of electrodeposition. *Sheet Metal Industry,* **28,** 477 (1951).
Dennis, W. H. *A hundred years of metallurgy.* Duckworth, London (1963).
Edwards, J. D., Frary, F. C., and Jeffries, Z. *The aluminium industry.* McGraw-Hill, New York (1930).
Evans, U. R. *Metallic corrosion, passivity and protection.* Edward Arnold, London (1937).
Field, Foster P. *The mechanical testing of metals and alloys.* Pitman, London (1948).
The making, shaping and treating of steel. Carnegie Steel Company (1925).
Morley, A. Strength of materials. (9th edn.) Longmans Green, London (1940).
Morral, F. R. A chronology of wire and wire products. *Wire and Wire Products,* **20,** 862–85 (1945).
Newman, R. P. and Houldcroft, P. T. Welding-engineering and metallurgical aspects. *Chartered Mechanical Engineer,* **8,** 214 (1961).
Pearson, C. E. *Extrusion of metals.* Chapman and Hall, London (1953).
Roe, J. W. *English and American tool builders.* Yale University Press, New Haven (1916).
Rolt, L. T. C. *A short history of machine tools.* MIT Press, Cambridge, Mass. (1965).

Seferian, D. *Metallurgy of welding*. Chapman and Hall, London (1962).

Steeds, W. *A history of machine tools 1700–1910*. Clarendon Press, Oxford (1969).

Street, A. C. and Alexander, W. O. *Metals in the service of man*. Penguin Books, Harmondsworth (1972).

Timoshenko, S. P. *History of strength of materials*. McGraw-Hill, New York (1953).

Warren, K. *The British iron and steel sheet industry since 1840* G. Bell, London (1970).

Woodbury, R. S. *Studies in the history of machine tools*. MIT Press, Cambridge, Mass. (1972).

16

POTTERY AND GLASS

In American usage pottery and glass are both regarded as ceramics, whereas in Europe the term is usually restricted to products made from clay. However, with the advent of electroceramics such as steatite – an important electrical insulator which contains no clay – the distinction has become more apparent than real. For present purposes we can regard them both as products made by subjecting relatively simple inorganic materials to high temperatures. In the case of pottery, the shaping of the material is done before firing, whereas glass is fired before it is shaped. Both pottery and glass are, of course, among the most ancient of man-made materials and they, especially pottery, figure prominently among the surviving artefacts of the ancient civilizations. By the beginning of the twentieth century the manufacture of both was being carried out on a very large scale, with mechanization of most of the old hand processes. In the present work we can give only a very general account of recent developments, as the variety of products is so great. Pottery, for example, ranges from building bricks and drain-pipes to delicate ornaments and fine tableware; glass from large plate-glass windows to the very precisely worked lens components of optical instruments.

I. POTTERY

The move from the traditional local potteries into large factory units concentrated in relatively small areas was as characteristic of the Industrial Revolution as the similar reorganization and growth of the textile industry. By 1900 the great majority of pottery products were made in factories by mechanized labour-saving processes though there was, and still is, a substantial demand for high-quality products, especially tableware, involving a great deal of individual craftsmanship and artistry, especially in their decoration.

Like other industries, the manufacture of pottery influenced, and was influenced by, scientific and technological developments in other fields. For example, the important new industries of the twentieth century – the electrical and the motor-car (and later aircraft) industries – created new demands. The electrical industry found in ceramic products a source of useful and versatile insulators, suitable equally for the massive insulators of high-tension overhead cables and the inner parts of electric-lamp fittings.

No less important, their high heat resistance made them ideal insulators for the spark-plugs which by mid-century the motor-car industry was requiring by the hundreds of millions. Their resistance was not limitless, however, and the growing use of strong detergents and washing-up machines demanded improvements in the glazes used. The increasingly exacting requirements of the steel industry – and particularly the introduction of oxygen for smelting – led to a demand for improved refractories. Underlying and facilitating all these new developments was a growing understanding of the scientific principles involved, especially the physico-chemical changes that occur during the mixing and firing processes.

These new developments were, however, superimposed on a huge traditional output. The requirements of the building and allied industries for roofing tiles, bricks, and piping still represented a major part of the industry's sales, though not a wholly predictable one. The demand for bricks rose rapidly in the years immediately after the First World War, as arrears of building were made good, only to fall again in the subsequent years of depression. The Second World War, too, was followed by a boom, which was accentuated by the massive destruction of property in the aerial bombardments which had been a negligible factor in 1914–18 when artillery fire along the lines of battle had been the main destructive agency. There was growing competition from rival products, such as reinforced concrete for building construction, asbestos tiles for roofs, and plastic piping for water supply and drainage.

Important developments took place in the first stage of manufacture, the mixing of the clay body from which the finished articles are made. With increasing use of mechanical moulding machines, capable of producing up to a thousand small articles – such as cups and plates – an hour, it became more and more important that the rheological (flow) properties of the plastic material were absolutely uniform and reproducible. It was particularly necessary to develop pug-mills that would produce a product free of entrapped air; these were introduced in the USA in the 1920s. Traditionally, moulded clay was allowed to dry naturally before firing, but it was impossible to do this and keep up with the higher production rates made feasible by mechanization at other stages. Artificial drying methods were therefore introduced, employing hot air or radiant heat. With the advent of more powerful machinery it was possible to mould semi-dry mixtures, with correspondingly less water to be expelled and with reduced porosity.

Some changes in practice reflected growing realization of health hazards. The British practice of wet mixing diminished the dust hazard, but in the USA air-flotation methods were widely used, and in these dust control was

much improved. In the firing process itself the ware was traditionally placed
on a layer of ground flint but this created the risk of silicosis; from about 1930
alumina was substituted increasingly for flint, and in Britain it was made
compulsory in 1947. At about the same time (1949) Britain forbade the use
of lead in glazes, on the grounds of its toxicity; some other countries were
content to strengthen legislation safeguarding the workers concerned. Up to
the end of the nineteenth century pottery was fired in coal-burning bottle-
kilns which were inefficient and created much smoke: it was not without
reason that in Britain the pottery area of Staffordshire was known as the
Black Country. In the twentieth century increasing use was made of electric
or gas-fired furnaces, partly for environmental reasons and partly because
they were cleaner and more convenient. The potters could dispense with the
fire-clay saggars necessary to protect ware from smoke in coal-fired kilns

Contrasting views of Longton, in the Potteries, in 1910 and 1970.

and, by using improved tunnel kilns, first introduced in Denmark in the
1870s, the firing process could be made continuous, in contrast to the old
bottle-kilns which had to be cleared and refuelled between each batch.

Refractories were required for a variety of purposes, ranging from the
firebacks of domestic fires (a diminishing market as the century progressed)
to the highly resistant bricks used to line steel-making furnaces of ever
increasing size. The biggest market was for traditional fire-clay products
based on aluminosilicate minerals; in these, the higher the alumina content
the more refractory is the product. In the first decade of the century,
however, refractories based on almost pure silica appeared. These were
made from crushed and ground ganister (quartzite) to which about 2 per
cent of lime was added to promote binding. In these bricks, in direct contrast
to those made from fire-clay, the presence of alumina, in even quite small
amounts, reduced the refractoriness considerably. Normally, less than 1 per
cent of alumina would be tolerable. The great advantage of these bricks is
that, unlike fire-clay, which softens at a comparatively low temperature, they retain their strength to within 50 °C of the melting-point of silica
(1725 °C). This allows them to be used, for example, in furnace roof spans of
up to 10 metres.

Developments in metallurgical processes created new demands for refractories. Here the
bottom of a 360-ton tilting furnace is being rebricked at the Lackenby Works of Dorman
Long: United Kingdom, 1930.

II. GLASS

As with pottery, glass manufacture was greatly influenced by developments in other fields of industry, especially, again, the electrical and motor-car industries. The manufacture of electric lamps, and later of thermionic valves, demanded glasses possessing special characteristics, as did the making of the optical components of cameras, cine cameras, and projectors. The motor-car industry was a rapidly growing market for high-quality flat glass for windscreens and side and rear windows.

By the beginning of the twentieth century a variety of special glasses had been developed, notably by Ernst Abbe and Otto Schott in Germany, in order to provide the wider range of refractive indices necessary to improve the optical systems of microscopes, telescopes, binoculars, and, later, cameras. In the main, however, only two kinds of glass were manufactured: glass made from soda, lime, and silica for common objects like window panes and bottles, and glass made from potash, lead, and silica for higher-class products, especially tableware.

The first important advance for the mass market was the development of borosilicate glass by the Corning Glass Company in the USA in 1915. As its name implies, this consists largely of boron oxide (12 per cent) and silica (80 per cent). Its important characteristic is that its coefficient of thermal expansion is only one-third of that of soda-lime glass and it is, therefore, much more resistant to sudden changes of temperature. It was this characteristic that led to its widespread use in domestic ovenware, best known under the brand name Pyrex.

We have already noted some of the problems of the electric-lamp industry, particularly the manufacture of very refractory metals such as tungsten in the form of fine wires. Mounting these wires in a glass bulb presented other manufacturing problems, because the seal had to be absolutely gas-tight. Initially, the incandescent filament was surrounded by a high vacuum, but later inert gases were used (p. 76): at all costs, air had to be excluded or the delicate filament would instantly burn away. To meet these requirements the glass base of the lamp was fused round the electrical inlet leads, but two problems were encountered. At first, the only satisfactory metal was platinum, because its change of dimension with temperature was roughly equal to that of glass; its disadvantage was its cost. Later, an iron-nickel alloy was developed that had the appropriate characteristics. Secondly, at the high working temperatures that were quickly reached, some short-circuiting occurred between the two leads, resulting in slow decomposition of the glass. Potash-lead glass was found to be more resistant but also more expensive. A compromise was reached by making the base of the lamp from potash-lead glass and fusing this on to a soda-lime bulb. For powerful lamp bulbs and for large thermionic valves, it was necessary to use borosilicate glass. With the advent of gas discharge lamps in the 1930s, new problems arose owing to corrosion by the vaporized sodium or mercury; this was overcome by developing new glasses containing much less silica and much more alumina and lime.

Once prepared, molten glass is fabricated into two main forms: sheet glass and containers such as bottles and jars. Save that they both involve manipulating molten glass, the techniques involved are quite different and must be considered separately.

III. SHEET GLASS

In the second half of the nineteenth century most sheet glass was made by a process developed by Chance Brothers of Birmingham. In this a large glass cylinder – 6 foot long and more than a foot in diameter – was blown and then split with a diamond when cold. It was then heated in a kiln until the glass softened and the cylinder gradually unrolled and flattened itself. In 1903 John Lubbers in the USA introduced a modified version of this in which a much larger cylinder – up to 40 foot in height and 3 foot in diameter – was drawn upward from a large pot of molten glass.

Long before this, the idea had been conceived of making sheet glass by continuous drawing from a reservoir of molten material, and a patent for such a process was taken out by W. Clark of St. Helens as early as 1857. Bessemer, too, conceived the idea of making sheet glass by tapping molten glass from a slit at the bottom of a tank and passing it between rollers. In 1884 Chance Brothers developed a process for making sheet glass by rolling it. Nevertheless, such processes presented great technological difficulties, and only in the twentieth century were these overcome. In 1901 Fourcault patented a process for dipping a 'bait' made of sheet metal into molten glass and drawing off a ribbon of glass which was then cooled to stabilize it. At much the same time I. W. Colburn was working on a similar process in the USA. However, all such processes suffer from the difficulty, among many others, that molten glass tends to form a 'neck', like treacle drawn from a tin with a spoon. Not until 1913 was Fourcault's process working satisfactorily. In the early years of this century, therefore, virtually all plate glass was still being made by the traditional method of casting glass on a flat iron table and subsequently grinding and polishing it.

After the First World War, however, the growing demand of the motor-car industry made it imperative that good quality plate glass be produced more rapidly and cheaply. The Ford Motor Company, as an interested party, developed a process in which molten glass was poured on to a moving table to form a continuous strip which was subsequently rolled, ground, and polished. This process was developed in collaboration with Pilkington Bros. in England. It was successfully operated from about 1925, and was the standard method of making plate glass for about twenty-five years, until Pilkingtons invented the float glass process in 1952. During the next twenty years this was almost universally adopted. In this process, molten glass is floated continuously along the surface of molten tin, which imparts a fine finish to the lower surface. To prevent oxidation, the process, in which the upper surface of the glass is simultaneously fire-polished, is carried out in an inert atmosphere. The glass strip – the thickness of which can be controlled in the 2–7 mm range – is then passed through an annealing furnace and cut to size.

While very satisfactory under normal conditions, glass can be exceedingly dangerous in the event of an accident. As early as 1905 safety glass for windscreens was made by gluing two sheets of glass to a middle sheet of celluloid, but by 1950 this process had been replaced by polyvinyl acetate, which forms a very tough translucent film. For special purposes – such as bullet-proof glass or the windows of some pressurized aircraft – multilaminated glass was developed. Glass incorporating wire mesh was made to prevent theft and the possible shattering of glass in the event of a fire. Safety glass is now also made by forms of heat treatment that create stresses which cause the glass to crumble rather than splinter in the event of accident.

IV. CONTAINERS

The traditional method of making a bottle was to form a thick-walled bulb of glass at the end of a blowing-iron and to give this its final shape by further blowing within a hinged iron mould. Four men, apart from a man tending the furnace, were required for the operation: the gatherer, whose work has just been described; the wetter-off, who took the blown bottle from the mould, and separated it from the iron; the finisher, who applied a rim of glass to form the neck; and a taker-in who took the hot bottle to an annealing oven where it could cool slowly.

The Ashley machine of 1887 eliminated the wetter-off and the finisher but was only semi-automatic. Its output was roughly the same as that of the manual method, around 150–200 bottles an hour, but with half the labour. In this machine, a plunger formed the neck in a blob of molten glass and an injection of compressed air then blew the bottle to the required shape in a mould. Thus the neck was formed before the bottle itself was shaped, in direct contrast to the old manual method, and this was a feature of all later machines.

The first completely automatic machine was that designed by M. J.

Process for making bottles by hand.

Owens in the USA, which went into production in 1904. In this, fixed amounts of molten glass were sucked up by blow-guns mounted on a rotating head. A 10-head machine could make about 3600 bottles an hour. The machine was demonstrated in 1907 to the leading European glass manufacturers, who were offered the European patent rights with the clear implication that if these were not acquired the Owens Company would itself develop the machine. In the event, a consortium paid £600 000 for the European rights and by 1913 164 machines were in operation.

The Owens machine was a great step forward, but its supremacy was soon challenged by feeder-fed machines – of which the IS (individual section) was the most successful – in which the standard molten glass gobs were formed not by suction but by gravity feed assisted by a plunger. This eliminated the heavy rotating turntable, an inherent feature of the Owens machine, and was faster to operate. This type of machine, too, was of American origin and was introduced by Hartford Fairmount. By 1920 two billion bottles a year were being produced by feeder-fed machines, compared with three billion on Owens machines. By 1927 the two types of machine were making roughly equal contributions but thereafter the

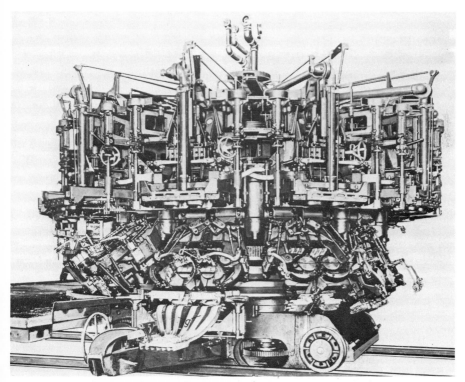

Modern Owens bottle-making machine.

Owens machine declined rapidly. By mid-century output was enormous: about 15 billion glass containers were made annually in the USA alone. Changes in design resulted in substantial economies in glass. In 1932 a US quart bottle required 700 g of glass; but by 1940 only 500 g. Additional strength was conferred by very thin surface coatings, often of tin oxide.

Up to the First World War electric lamp bulbs were blown by hand, but subsequently automatic machines, notably the Westlake, were introduced. By 1950 a single machine could produce more than 100 000 automatically blown bulbs an hour with the aid of a single operator.

The development of cheap mass-production methods for a wide range of glass containers had important social consequences, for they provided a convenient means of packing a great variety of products, thus displacing established materials. Food and drink were the commodities most affected. Until after the First World War, for example, milk was still delivered to the door in churns and there dispensed into the housewife's own container. By 1939 this practice – deplorably unhygienic by modern standards but taken as a matter of course at the time – had virtually disappeared, and milk was delivered in sterilized bottles. Bottled beer, too, increasingly displaced that drawn from the wood.

BIBLIOGRAPHY

Angus-Butterworth, L. M. *The manufacture of glass*. Pitman, London (1948).

Chandler, M. *Ceramics in the modern world*. Aldus Books, London (1967).

Dickson, J. H. *Glass*. Hutchinson, London (1951).

Douglas, R. W. and Frank, S. *A History of glass making*. Foulis, Henley-on-Thames (1972).

Dralle, R. *Die Glasfabrikation*. Oldenbourg, Munich (1911).

Hodkin, F. W. and Cousen, A. *A textbook of glass technology*. Constable, London (1925).

McGrath, R. *Glass in architecture*. Architectural Press, London (1937).

Maloney, F. J. T. *Glass in the modern world*. Aldus Books, London (1967).

Meigh, E. The automatic glass bottle machine. *Glass Technology*, **1**, 25 (1960).

Scoville, W. C. *Revolution in glass making*. Harvard University Press, Cambridge, Mass. (1948).

Taylor, W. C. The effect on glass of half-a-century of technical development. *Bulletin of the American Ceramic Society*, **34**, 328 (1951).

17

FOOD TECHNOLOGY

The development of the food industry followed a trend that we have already noted in other contexts – a trend towards centralized production. In 1900 most food was still processed locally, either in the home or in small establishments, like bakeries, supplying a limited market. Food-processing factories were already well established for certain products, but their contribution was, in relative terms, much smaller than it was to be by mid-century. To some extent this can be associated with social factors: the middle-class housewife, at least in Europe, prided herself on her household management and disdained ready-prepared food. With domestic help still easily available she could afford this prejudice; moreover, while the prejudice existed, the food manufacturers had no great incentive to improve quality. With the poorer working-class housewife the situation was different: quantity was at least as important as quality, and she was quite happy to see some of her domestic chores translated to the factory.

In the USA, with its predilection for labour-saving techniques of all kinds, there was a different attitude, at least in the rapidly growing urban communities. Processed food was welcome also in the areas of pioneering activity where ordinary sources of supply did not exist and the acquisition and storage of fresh food presented considerable problems. We must recall, too, that new food-producing areas were developing far from the main areas of consumption and – so far as perishables like meat, fish, and fruit were concerned – their products could not be marketed at all unless processed in some way on the spot. Even in the case of non-perishable materials, such as wheat, the products of the new areas often could not be satisfactorily handled without the development of new technologies. The hard wheats of North America, for example, did not lend themselves to the traditional stone-milling methods of Europe.

I. FOOD TECHNOLOGY IN 1900

At the beginning of this century certain aspects of modern food technology were already well developed but others, which had become commonplace by 1950, had scarcely developed at all. We may, therefore, conveniently begin our study by outlining the situation in 1900.

By that time the canning industry had been firmly established for meat, fish, fruit, and vegetables. Although great improvements were to be made

At the beginning of this century refrigerated food was becoming commonplace. This picture shows the compressor and ancillary equipment for a cold-storage room, *c.*1901.

both in the technique itself and in the quality of the product, canned food was widely used for both its convenience and its cheapness. The fact that it would keep almost indefinitely was a most important point in its favour – as much for the housewife anxious to hold a reserve against unexpected demand as for the military victualler meeting the requirements of an army in the field.

Refrigeration, too, was firmly established as an important means of preservation, and large cargoes of frozen meat were coming into Europe from the new sources of supply in Australia, New Zealand, and North and South America. The quality left something to be desired, however, partly because of imperfections in the freezing process, which affected the taste, and partly because of the inferior quality of much of the product, reared under conditions very different from those on European farms. Nevertheless it was acceptable and, because of its cheapness, it found a ready market. As opposed to canned products, however, which would keep more or less indefinitely, refrigerated products had a relatively short life because the cost of fuel made it uneconomic to freeze for long periods. The magnitude of the trade is indicated by the fact that in 1900 New Zealand exported four million carcasses of mutton and lamb and the Argentine more than two million. There was also a rapidly growing trade in frozen beef: 20 000 tons from the Argentine in 1900. The freezing of fish was less successful, however, and in the fish trade refrigeration was generally on a short-term basis sufficient to cover transit from the increasingly distant fishing grounds to the ports. Fruit, vegetables, and eggs did not at that time lend themselves to preservation in this way.

The processing of dairy products, too, was being gradually moved from the farm to the factory. By 1850 cheese factories were established in both the USA and Australia, but in Britain and elsewhere in Europe they met fierce opposition from farmers and dairymaids alike: the one feared for his position as the traditional cheese producer and the latter for her employment. Nevertheless, the first English cheese factory – at Longford, Derby – was in operation in 1870, under the guidance of American advisers, and within five years there were ten such factories, processing the milk of 8000 cows. Production of factory-made cheese increased steadily, but the larger farmers, able to buy modern equipment and assimilate new techniques, contrived to hold their own.

In butter-making the move to the factory was fostered by the invention of the centrifugal cream-separator by the Swedish engineer de Laval in 1877, which made possible great savings in labour and space. While the traditional dairy-farming countries continued to be major suppliers, the new producers, notably Australia and New Zealand, made rapidly growing contributions. All producers found a rival, however, in margarine. Invented in the 1860s by the French chemist H. Mège-Mouriès, its large-scale manufacture had been firmly established, especially in Holland, by 1900. An important discovery here, made in 1899, was that certain oils, too soft for margarine manufacture, could be hardened by hydrogenation in the Sabatier-Senderens reaction. This process began to be worked on an industrial scale about 1910.

Milk is another important perishable food whose handling changed rapidly. With the extension of the railways and the growth of the urban areas, the towns increasingly drew their supplies from the countryside in bulk. The milk travelled in tinned-steel churns and was delivered from smaller churns at the doorstep. Some surplus milk was condensed or dried. Pasteurization was introduced about the turn of the century, originally with the commercial objective of prolonging the life of the milk and later it was used more widely for the sake of health in general and for the prevention of tuberculosis in particular.

In the first half of this century the trends indicated above continued, with emphasis on preservation and factory processing. Development was facilitated by a better understanding of the scientific principles involved. In addition, some quite new processes came into general use, especially after the Second World War. The most important of these were freeze-drying, which allowed easy reconstitution of many products, and deep-freezing, which went hand in hand with the availability of domestic cold-storage lockers. Before studying these innovations, however, we may conveniently consider progress in established food technology.

II. MILLING AND BAKING

We have already mentioned the new pattern of wheat production and the problems arising from the introduction of new types of wheat having physical properties different from those to which millers were accustomed. A compensation, however, was that increasing use of the combine harvester – by 1938 50 per cent of American wheat was harvested in this way – gave cleaner grain. At the same time, the increasing mechanization of dough-making for bread – and of the making of cakes and biscuits too – demanded production of a uniform product. New milling processes had to take account, in particular, of two characteristics of the grain: its hardness and its shape, especially that of its longitudinal crease.

The steel rollers that replaced the traditional mill stones served a double purpose. First, a set of fluted rollers crushed the grain and effected some separation of bran from flour. Then the grain was put through a series of

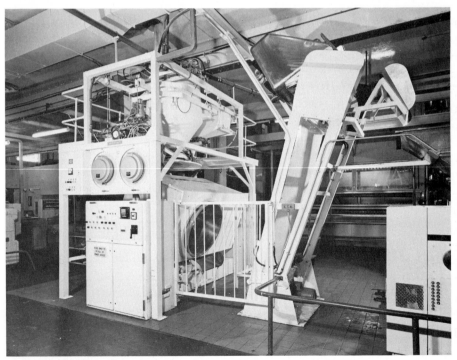

'Tweedy' high-speed dough mixer.

pairs of reduction rolls, one rotating at a slightly different rate from its companion in order to introduce a shearing movement, before being sieved. The most important changes in this process as the century advanced were a great reduction in the size of the rollers, relative to throughput, and the introduction of air-flotation techniques, rather than sieving, for separating flour grains of different degrees of fineness.

The proportion of the wheat that could be converted into flour varied, and so, other things being equal, there was a tendency to favour new varieties offering a high extraction rate. Other things were not always equal, however; thus many of the more productive strains proved to contain excessive amounts of amylase, an enzyme that breaks starch down into sugar.

Much bread is still made by the traditional method of setting the dough aside for two or three hours to let it ferment or rise. Shortly after the Second World War, however, American bakeries adopted a dough development process based on mechanical manipulation, which cut this preliminary stage to a few minutes, and saved much space and labour. This development made it possible to bake continuously, instead of in batches, in tunnel ovens.

Production was also speeded by the use of improvers. Flour, like wine, improves with keeping, as a result of spontaneous chemical changes. In the nineteenth century, when flour was imported from abroad, the maturing process took place during the voyage but when Britain and other European importers bought grain from North America and did their own milling this no longer happened. To overcome this difficulty, certain chemical improvers were developed which accelerated the ageing process; they also had the advantage of producing a more uniform type of flour and one that led to a more easily workable dough.

III. DAIRY PRODUCTS

Throughout the first half of this century milk and products derived from it remained of major importance. At mid-century world milk production was around 180 million tons annually, and from this were made some three million tons of butter and two million tons of cheese. Most countries remained more or less self-supporting in respect of these commodities: few imported more than 10 per cent of their over-all needs. The position of the United Kingdom was quite different, however: in 1939 she imported 80 per cent of her butter and 50 per cent of her cheese. The main exporters were New Zealand, by far the largest, followed by Australia, Denmark, and Holland.

While technological developments, notably mechanization of existing methods and centralization of manufacture, ensured that quantities were sufficient, this was not always the case so far as quality was concerned. Factory-made butter was the least affected, in that the process of its manufacture is relatively simple. With cheese, however, the decline was very evident, and mainly the result of commercial factors. The full flavour of a cheese is realized only as the result of a long ripening process governed by a complex system of enzymes. If this process is curtailed, the product , though unimpaired from the nutritional point of view, may be relatively tasteless and of poor texture; such curtailment did occur in the cheese factories and stores, though it would be idle to suggest that all farmhouse cheese was perfect.

Much of the decline was promoted by changing demand. Bread and cheese was the traditional meal of the industrial worker in the factory and the demand was for quantity rather than quality. To satisfy this, quick-ripening moist cheeses were produced on the basis of rapid turnover and larger profits. Less cheese was stored in lofts, to ripen, but more in cold-stores where microbiological action was suspended. Worse, means were discovered of making cheese from pasteurized milk. At first the process denatured milk proteins and prevented coagulation with rennet, but this was overcome by using a somewhat lower pasteurization temperature. The result was 'processed' cheese, first manufactured in Switzerland at the beginning of the century and later made elsewhere in Europe and in North America. From the purely practical point of view it has much to commend it: although its moisture content is higher than that of real cheese, thus making it poorer value for money, it keeps indefinitely and there is no wasteful rind. Its disadvantage is its taste and texture.

We have already noted the advent of margarine as a rival to butter. The original product was in effect little more than an emulsion of purified beef tallow with milk, from which the fatty fraction was then precipitated by addition of ice and worked to a buttery consistency. It was not very palatable but it was cheap and nutritious. In the twentieth century vegetable fats and oils were introduced and the skim milk used was subjected to lactic fermentation. Great improvements were made in the emulsification process, and incorporation of vitamins A and D (compulsory in Britain during the Second World War) was common. By 1950 the palatability of margarine and its nutritional value had been greatly enhanced. By careful control of the manufacturing process, products could even be made suitable for different climatic conditions. World production in 1938 amounted to more than one million tons, roughly one-third that of butter.

Milk contains approximately 87 per cent water; it is not surprising therefore that there has long been an interest in drying (or condensing) it, not only as a means of indefinite preservation but also to reduce its bulk.

Most early processes depended on direct heating: the milk was allowed to run over a hot roller and the dry residue was then scraped off. Unfortunately this causes considerable change in flavour and the lost water is not easily reabsorbed when desired. Nevertheless, it provides a product useful for various purposes in the food industry. During the Second World War spray-drying processes were developed in which a spray of skimmed milk was allowed to fall through a rising current of hot air; by the time the droplets reached the bottom of the plant they have turned to powder. Again, however, there was difficulty in reconstituting the product. In 1946 this was overcome by an agglomeration process developed in the USA by the Instant Milk Company. Spray-drying was also used to dry eggs.

There were some important changes in the process of pasteurization. From the 1920s this was done in bulk rather than in bottles and consisted in holding the milk for half an hour at about 65° C; it was then rapidly cooled. During the Second World War, however, this method began to be replaced by the so-called HTST (High Temperature Short Time) process in which the working temperature was about 72° C but the operating time only a quarter of a minute. It was followed by the UHT (Ultra High Temperature) process in which the milk is superheated to about 150° C for only a few seconds. The short heating periods of these last two processes demanded the development of elaborate heat exchangers.

IV. MEAT AND FISH

For meat and fish, and for fruit, vegetables, and condensed milk, canning continued to be the most important means of preservation. The basic principle remained unchanged but there were significant developments in practice, organization, and, of course, in scale; in 1950 the USA, which had become the world's greatest producer of canned foods, used some 10 000 million cans. At the same time developments in bacteriology – a science which had not even been born when Nicolas Appert introduced his preservation process in 1810 on a purely empirical basis – made it possible to conduct the process more efficiently and, above all, more safely. An essential requirement was proper understanding of the heat-transfer process: it was essential that the whole of the food mass was held for long enough at the required temperature.

The essence of canning is to destroy by heat all the micro-organisms that cause food to go bad and then to seal the container completely so that there can be no subsequent contamination until it is opened. Contamination, should it occur, is usually harmless but certain organisms can produce severe, and sometimes fatal, food poisoning. This was not uncommon in the early days of canning but is now very rare.

In Appert's original process the food to be preserved was put into loosely stoppered glass bottles which were then immersed in boiling water for what experience indicated was 'long enough'. The stoppers were then driven home and luted round the edge. This procedure presented two hazards: some organisms can survive the temperature of boiling water and there might be leakage round the stopper. The first hazard was overcome by replacing the water by a solution of calcium chloride, which boils at a much higher temperature ($240°$ F), and later, in the 1870s, by using autoclaves in which the boiling-point of water is raised by application of pressure; this was a technique widely adopted for sterilizing surgical instruments. By the end of the century large reliable autoclaves were in general use. They remained in normal practice until the 1960s, when the process of flame sterilization began to be used. In this, the cans are spun rapidly (two revolutions per minute) while a gas flame is applied to their surface. Under these conditions heat transfer is so rapid that there is no local overheating.

The risk of contamination after sterilization was largely overcome by using cans made of tinplate instead of glass containers. The end discs were soldered on and a small hole was left in the upper disc to be closed by soldering when heat treatment was complete. It was a slow and laborious process: even a good artisan could make no more than about 600 cans a day. The development of can-making machinery was, therefore, essential to the expansion of the industry: by mid-century such machines were capable of making 600 cans in two minutes. The major development here was the advent of the sanitary or open-top can about 1905. This was made from seamed tubing double-seamed at top and bottom; the need for soldering was eliminated.

For most of the period in question, tinplate (p. 185) was the standard material for making cans, the thickness of the tin, which is expensive, being progressively reduced: in 1950 a thickness of 0.0001 inch was normal. Such thin films were easily damaged and any imperfection could be a site of corrosion; to avoid this it became customary to lacquer the tinplate. After the Second World War the cheapness of aluminium led to its use in making cans, especially for vegetables and drinks; for the latter, the pull-ring opener was introduced. At the same time, steel coated with an exceedingly thin film of chromium/chromium oxide began to replace tinplate; this, too, was lacquered.

The storage and transport of refrigerated food was well established by the end of the nineteenth century, but it was generally believed that if food was cooled below $-2\,°$ C it would suffer irreversible and deleterious changes. In 1929, however, Clarence Birdseye showed that quick freezing to lower temperatures is satisfactory for many products, mainly because the ice

(*Top*) Early vegetable cannery, with food handled under primitive conditions. (*Bottom*) Dole aseptic canning system, developed in the USA after the Second World War; sterilizing unit on left, filling and sealing unit on right.

crystals formed are much smaller. This opened the way to the enormous frozen-food· industry which, as a complement, demanded the general adoption of domestic deep-freeze lockers in addition to refrigerators.

A very important innovation in this area was accelerated freeze-drying, originally developed in Sweden in the 1930s for drying biological materials of pharmaceutical importance. It is based on the fact that at low pressures ice will sublime – that is to say, it passes directly into the vapour phase without melting. The dehydrated residue is left in a powder form which readily reabsorbs water to reconstitute the material in something very like its natural form. Freeze-drying has been used for many vegetables; for meat and fish; and for coffee. During the Second World War it was used for the purification of penicillin, but the first commercial plant for food preservation seems to have been opened in Russia in 1954. It is an expensive process in terms of energy consumption but against this it can be argued that the product can be stored indefinitely without further demand for energy, such as is necessary for refrigerated goods.

V. FOOD ADDITIVES

In the nineteenth century food adulteration was widely practised, for a variety of reasons. Sometimes the motive was simply greed, as when an expensive commodity was 'extended' by mixing in a cheap one. Sometimes the aim was to improve the appearance or keeping qualities of the food, as when vegetables were kept green by adding copper salts to the water in which they were boiled. In some cases the adulterants were harmless and no more than a cheat, but some were dangerous and might even be fatal. By 1900 most western countries had introduced restrictive legislation to deal with the problem. In Britain legislation dates effectively from the amended Adulteration of Food and Drugs Act of 1875; in the USA the Food and Drugs Act was introduced in 1906.

The legislation was designed not to forbid food additives altogether but to restrict their number to those regarded as safe and to lay down the concentrations at which they might be used. In the first half of the twentieth century some additives were forbidden, on the ground that they had shown previously unsuspected toxic properties, and others were added as research and experience showed them to be both safe and useful. After the Second World War there was particular emphasis on the elimination of potentially carcinogenic substances. There were, however, few internationally agreed standards, and this inevitably had some effect on the growing international trade in processed foods: a product quite acceptable in one country might be forbidden in another. The USA tended to be more nervous than Europe, sometimes justifiably.

Among widely used additives were sulphur dioxide (and other closely related chemicals), benzoic and propionic acids, certain antibiotics, and esters of gallic acid. One of their most important roles was as an anti-oxidant, preventing fats and oils from becoming rancid. Such rancid materials are not only unpalatable but many also cause more or less serious digestive disorders. The ageing of flour by means of additives has already been mentioned.

In the early years of this century, but more particularly in the 1920s, it was realized that the nutritional value of foods could not be expressed simply in terms of their content of protein, fat, and carbohydrate. They also contained accessory food factors, known as vitamins, which – even though present only in minute quantities – were essential for health. In a good mixed diet containing fresh food the body normally obtains sufficient vitamins, but heavy reliance on processed foods can lead to deficiency and ill health. A classic example was Christiaan Eijkman's discovery in 1890 that beriberi, a serious disease widely prevalent in the East, was due to lack of vitamin B; in rice-eating countries intake of this is normally quite adequate because it is contained in the husk of the rice grain. When polished rice is eaten, however, this source of the vitamin is lost and more or less serious deficiency symptoms will appear.

Many vitamins are heat-sensitive substances and may be destroyed in cooking or in preservation processes such as canning. As this hazard became generally recognized, many manufacturers began to add vitamins, especially vitamins B and C (the anti-scurvy factor), to make good the loss. We have already noted the addition of vitamins A and D to margarine to make it comparable in this respect with butter.

VI. ORGANIZATION

In conclusion, we must mention two major organizational changes in the food industry. Apart from the steady trend towards centralized production, there developed an increasingly close association between the food processor and the producer. Many farmers, for example, would grow crops such as peas or rhubarb under contract to a particular processor. Similarly meat and fish factors would have long-term supply contracts. Such arrangements were beneficial to all concerned, including the customer, for the contracts normally embodied strict clauses about the quality of the product. Very often the precise variety of crop to be grown, and the stage at which it was to be harvested, was laid down.

The second organizational change also directly concerned the customer. In the 1930s European food shops still followed traditional lines; such commodities as butter, tea, bacon, meat, and fish were weighed out from

New methods of marketing radically altered the design of shops. (*Top*) Typical large food shop in the 1920s. (*Bottom*) Post-war supermarket largely stocked with pre-wrapped food for self-service.

bulk to the customer's requirements by assistants in the open shop. But in the USA signs of change had appeared as early as 1912, when the first self-service store opened in Memphis; by 1929 there were at least 3000 such shops. In Europe, the self-service shop made rapid headway after the Second World War. In these shops food offered for sale is exposed to constant handling by customers, and for reasons of hygiene virtually all items must, therefore, be wrapped. This development had profound consequences for the packaging industry.

BIBLIOGRAPHY

Bitting, A. W. *Appertizing or the art of canning: its history and development.* Trade Pressroom, San Francisco (1937).

Coppock, J. B. M. The evolution of food science and technology in the UK. *Chemistry and Industry*, 455 (1973).

Fannema, O. R., Powrie, W. D. and Marth, E. H. *Low temperature preservation of foods and living matter.* Marcel Dekker, New York (1973).

Filby, F. A. *History of food adulteration and analysis.* Allen and Unwin, London (1934).

Forbes, H. Rise of food technology 1500–1900. *Janus*, **47**, 101, 139 (1958).

Francis, C. A. *A history of food and its preservation.* Princeton University Press (1937).

Holdsworth, S. D. Dehydration of food products. *Journal of Food Technology*, **6**, 371 (1971).

Kaydlereas, S. A. On the history of food preservation. *Scientific Monthly*, **71**, 422 (1950).

McLachlan, T. *History of food processing.* Pergamon Press, Oxford (1975).

Sturck, J. and Teague, W. D. *Flour for man's bread: a history of milling.* University of Minnesota Press, Minneapolis (1952).

Tannahill, Reay. *Food in history.* Eyre Methuen, London (1973).

Van Stuyvenberg, J. H. (ed.), *Margarine: an economic, social and scientific history*, Liverpool University Press (1969).

18

TOWN PLANNING

It is generally acknowledged that almost the first manifestation of civilization is transition from a pastoral, sometimes nomadic, way of life, to life based on settled urban communities. The degree of transition may, of course, be very variable, ranging from simple townships serving mainly as a market for a scattered rural population to the sophisticated city states of the ancient world. Many of the great cities of that world were carefully planned in respect of both the siting of buildings – as in the rectilinear plans of Roman towns – and the provision of basic services such as water and drainage. It must always be remembered, however, that what we see now are mostly the remains of public edifices built of durable materials, and impressively sited to satisfy the vanity of rulers and impress outsiders. Humbler citizens had to be content with far simpler homes, built of wood, wattle, mud bricks, and similarly perishable materials. Generations of these, collapsed layer upon layer, form the foundations of many modern cities. For them, town planning doubtless had quite a different meaning. Of the medieval towns of Europe much the same may be said. Their cathedrals and churches, palaces and mansions, barracks and castles, and other public buildings were well sited and amply provided with amenities but the houses which crowded round them were, as so many survivals attest, closely crowded and ill-supplied. Even in great capital cities like London, sewage was, almost within living memory, discharged into open drains and water was drawn from polluted wells and rivers.

The Industrial Revolution, in both its first manifestation in Britain and its subsequent extension to Europe and North America, accentuated the move from the country to the towns, where workers could live close to the factories in which they worked. In many cases they lived in uncompromising rows of back-to-back houses laid out on strictly utilitarian lines with a minimum of amenity in the way of gardens or parks. It is interesting to note that, while socialists have regarded these as almost synonymous with capitalist oppression, some such properties today command high prices where the fashionable residential tide has flowed round them. While many of the new industrial towns had long histories, almost all were quite small and geared only to the needs of a modest population, when they were engulfed by the population wave. Birmingham is a fairly typical example. Although mentioned in the Domesday Book, and well known for its arms

Preston, England, a typical industrial town dating largely from the nineteenth century.

manufacture, its population in 1770 was only 30 000. It doubled during the next thirty years and almost exactly doubled again in each of the subsequent three half-centuries. At the beginning of the twentieth century its population was half a million; by 1950 it was well over a million. The growth of Glasgow closely paralleled this.

In the face of such population pressures the problems of town planning were enormous. Niceties tended, inevitably, to give way to the supplying of basic needs such as drainage, water supply, roads, and schools. By 1900, however, the pressure was slackening, and in Britain and in Europe generally, a new and relatively stable population pattern had emerged, though this was not so in North America, Australia, and other still expanding countries. The qualification 'relatively' is used advisedly, however, for, although planners could at least begin to discern what they had to plan, new factors were emerging and complicating the situation, and population shifts had by no means disappeared. Better understanding of the nature of disease and its transmission, and a steadily rising standard of life, threw an added burden even on the basic services of water and drainage, and to improve these in already densely populated areas was a major task in itself. The advent of the motor transport, which led – first in the United States but with Europe not far behind – to a situation in which private transport tended to be the rule rather than the exception, created great urban traffic problems, especially as public road vehicles also appeared in great numbers. The growth of cities pushed the population out from the centre, with corresponding new demands on public transport. Although the cities remained the magnet as a place of work, a drift back to the country as a place for living began.

The gas and water undertakings had rights to dig up the highway, and in this they were joined by the new electricity companies; every major urban development was complicated by the need not to disrupt a bewildering network of underground services, not all of which were very precisely charted. Towards the end of our period the demands of air travellers had seriously to be considered: space had to be found for airports as well as means of conveying passengers to them. The physical destruction of large urban areas – considerable in Europe in the First World War and enormous in the Second – posed major reconstruction problems, especially as it was coupled with a long pause in private building. It also provided opportunities, not always seized, to correct past planning mistakes. Increasingly, planning decisions began to be influenced by environmental considerations which, in the second half of this century, were to become so powerful an influence that many desirable planning developments were completely stultified.

Today, it is fashionable to decry the work of earlier town planners, but perhaps we should remark how much was achieved rather than how little. Those responsible for controlling urban development had little public support. In the United States, Australia, and other countries where the pioneering tradition was still strong, and in western Europe where the concept of every man's home being his castle remained deeply entrenched, planning in any positive sense was not popular. Its unpopularity was reflected in the lack of statutory powers enabling local authorities to plan in anything more than general terms; much legislation was persuasive rather than obligatory. Even when they had statutory powers, authorities often could not afford to make full use of them because of the need to pay substantial compensation to property owners affected. One of the first major pieces of legislation that confirmed effective town planning authority was Britain's Town and Country Planning Act enacted in 1947, almost at the end of our period.

Thus twentieth-century planners found themselves in something of a dilemma. On the one hand they had more facilities at their command to improve the amenities of the areas for which they were responsible but the very existence of these increased the complexity of the situation, and changes that seemed desirable when introduced had far-reaching consequences that were not foreseen, and indeed were probably not foreseeable. A classic example was the building of high-rise flats which seemed to combine the virtues of conserving valuable land – a point dear to the hearts of environmentalists – and avoiding dispersal of communities. In the event they created grave social problems. On the other hand, the launching of major planning projects demanded so much preliminary inquiry and legislation that situations might materially change before the necessary authority was obtained so that major modifications, or even a fresh start, had to be authorized. Over and above all this was the sheer magnitude of the problem. In Britain, for example, a few garden cities with a population of around 50 000 each could make little impact on urban sprawls housing tens of millions. Grandiose schemes could be formulated but, whatever their intrinsic merit, they could not be put into practice in the face of opposition by ratepayers and taxpayers. The ambition of many local authorities was often no more than to survive and remain solvent in the face of demands that stretched their resources to the limit: in the 1970s some great cities, such as New York, were not even to achieve this. At mid-century one of the acknowledged social problems of the western world was an ageing population. No less important was an inheritance of aged and decayed buildings.

New York, USA, in the mid-1960s, a classic example of over-building.

Canberra, Australia, as it appeared in 1970; it was built on a plan conceived shortly before the First World War.

Though any traveller on the rail and road networks of the western world could see abundant evidence of planning failure, this is not to say that there were no successes; merely that they were less conspicuous.

I. SOME PLANNING SUCCESSES

Acknowledgement of the extent of failure should not blind us to some projects which were undoubted successes in themselves and served as models for others. It is always a mistake to suppose that because everything is not right, all is wrong. Not surprisingly, many of the successes were achieved in green-field sites where planners were untrammelled by existing works. Some were on a grand scale, among them Canberra and Brasilia. Canberra, the Federal Capital of Australia, was founded by Act of Parliament in 1908, on the site of a small hamlet, and it followed the cobweb design of W. B. Griffin. Growth was slow, however, and Parliament House was not completed until 1927. In 1946 the Australian National University was established there, but even at mid-century it was a relatively small township with a population of about 25 000. Brasilia, founded in 1960, is different in many ways: in particular, it is located on a remote inland site and its functioning as a capital city is dependent on access by air, a possibility which could not have been considered in the case of Canberra, which was planned for traditional modes of transport. Brasilia grew far more rapidly than Canberra, attaining a population of more than half a million within ten years.

Most plans, however, were necessarily less ambitious. Important pioneer work was done in Britain by Ebenezer Howard, founder of the garden-city movement, exemplified by Letchworth (1903) and Welwyn Garden City (1919), which was designed to move population away from urban centres into new mixed townships distributed throughout the country. A typical community-owned garden city would have a population of 30 000 occupying some 6000 acres, of which 5000 would be agricultural land. After the Second World War the New Towns Act (1946) provided for the establishing of larger communities under government agencies known as Development Corporations; these were to include new industries. By 1950 ten had been established, with a target population of 700 000.

In the United States the garden-city movement was paralleled after the First World War by the Regional Planning Association (1923). This led in the inter-war years to such communities as Chatham Village (Pittsburgh) and Sunnyside Gardens (Long Island). In Europe a notable success was Vällingby, nine miles from central Stockholm, which forms the hub of four peripheral communities. The land was purchased by the City of Stockholm in 1930, and building was completed in the 1950s. In Stockholm itself the city centre was imaginatively rebuilt in accordance with a plan – now

City-centre development in Stockholm, Sweden, designed to separate traffic from pedestrian areas.

generally followed when feasible – of separating pedestrian and motor traffic.

Germany was a special case after the Second World War: destruction was so extensive that rebuilding largely solved the old problem of slum clearance. In Cologne, for example, pedestrian precincts were separated from the road system, as in Stockholm.

II. ACCESS

New technological developments posed many problems for planners, but perhaps the greatest was that arising from the dramatic growth of motor transport. In the United Kingdom, for example, there were 8465 registered motor vehicles in 1904, and 4.4 million in 1950; in the next decade the number was to double. In 1958 the world population of motor vehicles exceeded 100 million, 80 per cent of them being in the USA. To keep them moving in urban centres, especially during rush-hours, proved increasingly difficult and, ultimately, insoluble. Cross-country transport was relatively – but only relatively – more easily controlled: main roads could be widened and improved to take heavier traffic but innumerable bottle-necks developed where they passed through towns which, for reasons of finance or geography, could not readily be bypassed. Beginning in Germany with the Avus *Autobahn* of 1921 and the Dusseldorf–Bonn link of 1929, roads specially designed for heavy fast traffic were built *ab initio*. By 1939 over 4000 km of *autobahns* had been built in Germany, much of which was, of course, dictated by military considerations. After the war, plans were made for a European network ten times greater than this: work began on Italy's 725 km *autostrada* in 1954. In the USA, in the 1920s and 1930s, work began on the construction of city parkways – such as the Hutchinson River Parkway south from New York – on which commercial vehicles were banned and private cars could move swiftly. At the same time a system of turnpikes – motorways financed by tolls – was established, beginning with the Pennsylvania Turnpike (1940). By 1960 some 3000 miles of toll roads had been built.

These systems proved difficult to integrate, for it was not easy to absorb the heavy fast traffic off the new custom-built roads into the still largely congested urban networks. The only workable solution was the building of complex junctions involving long smooth curves at varying levels; typical examples are the access points to the Californian Freeway in Los Angeles and Spaghetti Junction near Birmingham in Britain. Such junctions are not only very costly but they also occupy large areas of land just where it is needed for existing urban developments. Moreover their construction is a slow process and while they are being completed existing traffic problems are aggravated rather than ameliorated.

BIBLIOGRAPHY

Abercrombie, Patrick. *Town and country planning*, Butterworth London (1933).

Adams, Thomas. *Outline of town and city planning*, Churchill, London (1935).

Buchanan, C. *Traffic in towns* (Ministry of Transport Report). HMSO, London (1963).

Committee for Economic Development, Research and Policy. *Guiding metropolitan growth*. New York (1960).

Jacobs, Jane. *The death and life of great American cities*. Random House, New York (1961).

McKelvey, Blake. *The urbanisation of America 1860–1951*. Rutgers University Press, New Brunswick (1963).

Merlin, Pierre. *Les villes nouvelles*. Presses Universitaires de France, Paris (1969).

Mumford, Lewis. *The culture of cities*. Harcourt, New York (1938).

— *The city in history: its origins, its transformation, and its prospects*. New York (1961).

Sharp, Thomas. *Town planning*. Penguin, Harmondsworth (1940).

Unwin, Raymond. *Town planning in practice*. Benn, London (1909).

19

THE NEEDS OF TRANSPORT

In the previous chapter we singled out the difficulty of keeping traffic moving in congested urban areas as a major problem of town planning. This was, however, only one aspect of a much bigger problem. The changing pattern of transport created new demands of many kinds. Some of these were merely extensions of existing ones, as in the building of new railways, canals, docks, tunnels, bridges, and harbours; here the main advances lay in the use of new techniques – especially machinery in place of manual labour – to solve old problems. Some demands were essentially new, however; thus, while new roads largely, but by no means entirely, followed the lines of old ones, surfaces that had sufficed for slow horse-drawn vehicles were inadequate for fast motor transport. Similarly, in the earliest days of flying, almost any large open field sufficed for take-off and landing, but by mid-century the world's principal airports required not only large areas of land – by then not easily acquired within convenient distance of urban centres – but also highly sophisticated facilities for the handling of passengers and freight. These facilities were, however, so specifically linked with the aircraft they served that they are most conveniently considered in that context (Chs. 22, 23).

I. ROADS

In the nineteenth century the main innovation in transport had been the railways, powerful rivals to the older canals; their development diminished the incentive to maintain and improve the main roads. It has been argued that the first moves towards road improvement were prompted by the advent of the bicycle, which provided a novel means of transport that became widely popular in the 1880s and 1890s. There is certainly some truth in this, but it was the advent of motor vehicles – first light-weight cars and later increasingly heavy commercial vehicles – which revealed the inadequacy of the existing systems. This inadequacy lay primarily in the mode of construction rather than in their ramifications; by 1900 the European pattern of population was fairly well established – though this was not true of North America – and roads of a sort existed between the main urban centres and incidentally served many country areas.

At the beginning of this century France had probably the best road system in the world, and it is significant that it was in France that motor cars first became popular. By contrast, the USA – which was soon to lead the world in motoring – had one of the worst road systems: if we ignore the stone-paved urban streets, there were probably no more than 200 miles of properly metalled roads in the country.

Basically, the new roads for motor traffic had to meet three requirements. They had to be strong enough to withstand the constant pounding of fast, heavy traffic: the limit of a well-compacted dirt road is at most 100 light vehicles a day. They had to be, and remain, smooth enough to give a comfortable ride, even allowing for the cushioning effect of pneumatic tyres. Finally, they had to be as dust-free as possible; the traditional dress of early motorists is evidence of the importance of this aspect.

Whatever system of construction is used, it is the subsoil that ultimately supports the traffic, and a first condition for success is the proper preparation and compaction of this and its subsequent protection from the erosive action of water. At the beginning of the century, much of the preparatory work was done by pick and shovel, as it still is in parts of the world where labour is abundant and cheap; increasingly, however, heavy earth-shifting machines were introduced, such as bull dozers and graders, many depending on caterpillar traction – later used for army tanks – to keep them mobile in severely muddy and rough conditions. The steam road-roller was already in use in the latter half of the nineteenth century and was followed by similar machines driven by diesel engines. Two types of surface finish came into general use. The cheapest, and most widely used, was asphalt or tar applied hot and finally given a rough – and supposedly skid-proof – surface by rolling-in stone chippings. This could be applied to existing roads whose foundations were sound and it served the dual purpose of laying dust and keeping water from the foundations. Additionally, it was slightly plastic so that it resisted cracking if there was minor subsidence. At first the tar or asphalt was applied by hand from tar-boilers (the health hazard had not been appreciated), but mechanical spreaders appeared in the 1930s.

Concrete roads were no novelty in 1900. Not only had they been used by the Romans, though this had long been forgotten, but short stretches of concrete road had been built in the latter half of the nineteenth century in Austria, Scotland, France, and Germany. The first concrete road in the USA was built at Bel-Fontaine, Ohio, in 1892. This method of construction made no significant contribution, however, until two major advances had been achieved. The first was the development of machinery to mix and apply concrete, which seems to have begun in Germany in the 1890s. The second was the improved formulation of concrete which accelerated its rate of setting: as late as the 1930s a concrete road-base, in contrast to one of tar or asphalt, could not be used until a month after it had been laid. Originally, concrete – commonly reinforced with wire mesh – was used as a base, being subsequently coated with tar or asphalt. Later, however, it came to be used also for the final surface, especially with the advent of major city bypasses and motorways. For very heavy traffic, concrete has the advantage of being rigid and thus the load is spread over a much bigger area.

II. BRIDGES

The pattern of communications and the location of towns have always been powerfully influenced by geographical features. Mountain ranges and major rivers could be crossed only where circumstances were favourable – a natural pass in the one case or a ford or the possibility of bridging in the other. An important feature of the development of communications in the nineteenth century, and even more so in the twentieth, was that technological progress made these barriers more easily surmountable. It was in bridging that the most striking advances were made. Tunnelling, like the driving of mine workings, depended in the main on deploying enough labour for a long enough period of time. Eupalinos' tunnel in Samos, built in the sixth century BC, was nearly 3 metres square and more than 1000 metres long; it was excavated from each end, and the precision of the junction at the centre bears witness to the accuracy of the ancient surveyors' work.

Bridging was more difficult, a limit being set by the maximum distance that could be covered in a single span. In favourable circumstances, of course, a series of closely spaced piers enabled long distances to be covered. The Lower Zambezi Bridge (1935) consisted of thirty-three identical 80 m spans. This design was not practicable where deep water or gorges had to be spanned; for these, suspension bridges were often necessary, and there were major developments in their construction. One of the most famous was the Golden Gate Bridge in San Francisco which, with a span of 1280 m, was the largest in the world in 1936.

In 1900 iron and steel bridges were widely used; they were constructed, like ships, by riveting the members together. Prefabrication was common, the members being cast and drilled before dispatch to the point of erection. Many such bridges were built in India between the wars from parts shipped from Europe. In their simplest form, such bridges consisted of single trusses resting on piers at each end. In 1917 the largest span of this kind was in the Sciotoville Bridge over the Ohio River (236 m), but this was getting near the attainable limit; a bridge of this design built in Japan in 1966 just exceeded 300 m. Cantilever bridges, built outward from each end to meet in the middle, made larger spans feasible. The most famous was the massive Sydney Harbour Bridge (504 m) opened in 1932; it was the heaviest riveted steel structure ever built.

Golden Gate Bridge, San Francisco. When opened in 1936 it had the longest span in the world (1280 m).

For many years riveted bridges prevailed. Their advantage was that they never failed catastrophically: signs of weakness were apparent long before serious trouble occurred, for the rivet holes themselves put a stop to the development of cracks resulting from brittle fracture. On the other hand, riveting is a labour-intensive operation and skilled riveters became increasingly difficult to find as the twentieth century advanced. Moreover, any given structure demanded the holding of large stocks of various sizes of rivets and girders and very precise fabrication of parts to ensure that on erection the rivet holes were exactly aligned.

In such circumstances welding provided an attractive alternative, but two obstacles had to be surmounted before it was successful. First and foremost, welding techniques had to be perfected. Until the metallurgical principles were properly understood, welded joints were liable – unlike riveted ones – to fail without warning; this was a serious hazard encountered with welded ships, including the Liberty ships of the Second World War. For full advantage to be taken of welding in terms of lightness and elegance, new concepts of design had to be accepted; the earliest welded bridges were virtually identical in appearance with their riveted counterparts. Welded construction made slow headway, and plenty of riveted bridges were still being built at mid-century and after.

The first welded steel bridge seems to have been a modest footbridge, less than 30 m long, in a gasworks in Melbourne, Australia. Nevertheless, it is historically comparable with the first iron bridge that spanned the Severn at Coalbrookdale in 1779; this was a mortised structure containing no bolts or rivets. Bigger bridges were built in the USA and Europe during the next decade, but the most important new development did not occur until 1962 when the spans of Brunel's suspension bridge at Chepstow (1852) were replaced by welded box girders. Thereafter, welded box girders became almost universally used in all kinds of new bridges. For a time they came under suspicion owing to a series of widely publicized disasters, but the trouble proved to lie in the design and execution and not the principle.

In a sense, for the final product in both cases is effectively a single entity rather than an assembly of units, the prestressed concrete bridge resembles the welded steel one. First introduced in France in 1907, it opened up a new era of elegance and economy in construction. Outstanding examples are E. Freyssinet's bridge over the Marne at Esbly (74 m, 1949) and the much bigger Gladesville Bridge (304 m, 1964) over the Parramatta River near Sydney.

Many large suspension bridges were built in the nineteenth century, but in the twentieth there were improvements in design and longer spans were achieved. A recognized defect was that such bridges were liable, under certain wind or traffic conditions, to forced oscillations which, eventually, could lead to their total destruction. The most famous disaster of this kind – partly because the event was filmed and thus widely seen throughout the world – was the destruction of the Tacoma Narrows Suspension Bridge in 1940. Two remedial measures were adopted. In the earliest, the roadway was not attached directly to the cables by suspenders but was supported by a truss which conferred rigidity and distributed the load over a much greater length of the cable. A later measure, stemming from the improved knowledge of aerodynamics occasioned by the growth of the aircraft industry, was to give the structure an aerofoil section to minimize the effect of wind on it.

Other classes of bridge, built to meet special situations, can be mentioned only briefly. The most important are movable bridges, in which the roadway is temporarily moved to allow the passage of ships. In some the road swings horizontally, in others vertically. Outstanding examples of swing bridges are those in the USA across the Willamette River (158 m, 1908) and across the Mississippi at Ford Madison (160 m, 1927). The biggest in Europe was the Kincardine Bridge across the Forth in Scotland (111 m, 1936). The most spectacular example of the lift-bridge, in which one section is hoisted vertically upwards when necessary, is the Arthur Kill Bridge joining New Jersey and Staten Island in the USA; it was built in 1959 and has a platform 179 m long.

Very few transporter bridges have been built but they are, nevertheless, technically interesting. In these, passengers and vehicles are moved from

Rebuilding of Brunel's bridge
across the Wye at Chepstow
(1962).

Gladesville Bridge, Sydney, Australia (1964); this embodies the world's longest concrete arch
span.

Transporter bridge across the Mersey and Manchester Ship Canal, opened 1905.

one side of a river to the other on a carriage suspended from a beam spanning two tall piers. Like ferries, they have the disadvantage that traffic is considerably delayed during loading and unloading and during the necessarily slow crossing. European examples include the transporter bridge over the Tees at Middlesbrough (305 m, 1911) and one over Le Vieux Port at Marseilles (166 m, 1902).

The security of every bridge depends not only on the span but on the piers and abutments. Generally speaking the piers have to withstand mainly a vertical load and the abutments mainly a lateral thrust. Both may have to be sunk very deep to reach firm foundations. The foundations of the Oakland Bay Bridge in San Francisco (1936) and the Hawksbury River Bridge in Australia (1942) both go down more than 70 m; in the case of the latter, great difficulties were encountered in reaching bedrock. Depending on their situation, piers and abutments have to withstand not only the weight of the bridge itself but additional stresses arising from such causes as wave action, ice, log-jams and even, in some cases, earthquakes.

Foundations may be made by traditional methods of excavation, if necessary using caissons supplied with compressed air. The latter demand the use of decompression chambers to safeguard the men working in them. Alternatively, pile-driving is resorted to. An important development, made in Italy in 1938, was the ICOS system (*Impresa Construzioni Opere*

Spedializzate). In this, the excavation that is to receive the foundation –
usually steel-reinforced concrete – is filled with a slurry of bentonite clay
which prevents collapse of the walls and the intrusion of surface water. Into
this the foundation material is lowered, and the displaced clay is recovered
for further use.

III. TUNNELS

All civil engineering projects are fraught with the risk of encountering
unforeseen soil conditions that may not only pose severe, and occasionally
overwhelming, technical problems but also add disastrously to the cost. This
risk is greatest in tunnelling because, from the very nature of the operation,
it is not usually feasible to investigate with trial boreholes the whole of the
terrain to be encountered. Where there are serious doubts, a small pilot
tunnel may first be excavated, as with the Mersey Tunnel in England
(1935). The risk is particularly great in tunnels driven under water. Well-
known disasters of the nineteenth century were the flooding of the Severn
tunnel in Britain during its construction in 1879 and the story of the Hudson
River tunnel in the USA: begun in 1879 it was not finally completed until
1904.

Basically a tunnel is made by steadily attacking the face, removing the
spoil, and shoring up the bore thus created. Usually excavation is done from
both ends, but, where feasible, work may also proceed from shafts sunk
along the line of the tunnel. Traditionally this work was done by pick and
shovel, later reinforced by the use of high explosives, with heavy timber
shoring. Well before 1900, however, major innovations had been made. The
most important was the use of a shield – devised by J. H. Greathead in the
1870s – at the work-face, enabling several men to work simultaneously; as
the tunnel advanced the shield was periodically jacked forward and the new
section of tunnel was lined with iron or concrete. A second important
development, used when there was a threat of water bursting in, was to work
in compressed air, which would keep it out. Towards the end of our period
some use was made of liquid air or solid carbon dioxide to freeze watery soil
solid while the tunnel was extended. This freezing technique had been tried
in Europe as early as the 1880s but with limited success because of the
inadequate methods of refrigeration then available. In the course of the
Hudson River project it was found that in wet silt and similar very soft
material the shield could be forced forward by the hydraulic jacks – which
collectively applied a force of 2500 tons – without excavation. This 'blind
driving' became a recognized technique in such circumstances. In the
twentieth century it was developed by Eric Bridge into the rotating shield
system, in which the whole shield served in effect as a giant auger.

In all tunnelling operations, the excavation of rock and earth and its removal from the workings is one of the biggest of the engineers' problems. As in mining (Ch. 9), increasing use was made of mechanical devices for this purpose—pneumatic drills, mechanical haulage on tramways, bulldozers, dump-trucks, and the like.

Many shallow tunnels – around 25–30 ft below the surface – have long been made by a cut-and-cover technique, in which a deep trench is first cut and the upper part is subsequently filled in. In the twentieth century a somewhat similar technique was developed for underwater tunnels; in this, prefabricated concrete or steel tubes were laid in a dredged trench. The first such tunnel was laid under the Detroit River in 1910 – though the technique had been used in 1893–4 for the construction of a large sewage outlet in Boston Harbour – and twenty years later Detroit and Windsor were similarly linked. Subsequently, tunnels of this kind were built in Holland and Belgium. Naturally, the technique presupposes a reasonably smooth river bed capable of supporting the tunnel uniformly throughout its length.

Tunnels serve a great variety of purposes – carrying roads, railways, canals, water supply, sewage, etc. – and the twentieth century saw the completion of some very large projects, of which only a few can be listed here. As this chapter follows that on town planning, we may perhaps appropriately mention first the underground railway systems of London and Paris, which were designed primarily to relieve traffic congestion on the surface. The City and South London was completed in 1900, and by mid-century London had about 90 miles of underground railways. The first section of the Paris Metro (Porte Maillot to Porte de Vincennes) was completed in 1898, and by 1950 the system totalled about 60 miles. Roughly contemporaneous was the 12-mile Simplon Tunnel, joining Switzerland and Italy (1895–1906); the 9-mile St. Gotthard Tunnel and the 8-mile Mont Cenis had both been completed in 1881. Much longer tunnels were built in the USA: for example, the Catskill water supply system for New York, constructed in 1905–26, includes an 18-mile tunnel beneath the Shandaken Mountains. One of the most spectacular tunnels is the Rove Tunnel (1927) connecting the Port of Marseilles with the Rhône; it is 72 ft wide, 50 ft high, and 4½ miles long, and took fifteen years to construct.

The earliest major tunnels for transport were largely for canal boats and railways but, with the arrival of motor transport, the density of traffic soon made the construction of major relief tunnels economically feasible. Among these may be mentioned the Holland Tunnel (1927) in New York and the Mersey Tunnel linking Liverpool and Birkenhead (1934). All tunnels present ventilation problems, but these were accentuated by the need to disperse large volumes of poisonous exhaust fumes from motor vehicles. The

Holland Tunnel, for example, was designed to carry 1900 vehicles per hour in each tube, and the Mersey Tunnel carries an average of 5000 vehicles a day.

IV. CANALS

Canal building, like tunnelling, dates from antiquity, but the Industrial Revolution, which engendered the need to move large heavy loads smoothly and cheaply over considerable distances, provided an incentive for the expansion of the canal system. In the nineteenth century the railway became a rival to the canal, and, in the twentieth, motor transport rivalled both. Nevertheless, not only was much of the existing canal system maintained and improved but some major new projects were also embarked upon.

Generally speaking, we are concerned with two types of canal: barge canals carrying relatively small vessels along inland waterways and ship canals directly accessible to ocean-going vessels of considerable size. Ship canals are usually constructed either in order substantially to shorten ocean voyages – as with the Suez, Kiel, and Panama canals – or to convert important inland towns into ports – as with the Manchester Ship Canal and the St. Lawrence Seaway. Access to ship canals is usually by way of normal port facilities, which we will consider later.

As with roads, tunnels, and other major civil engineering projects, canal construction was greatly influenced by the increasing availability of mechanical means of excavation and of removing spoil. The quantities involved could be enormous: in the case of Panama, the spoil removed amounted to 305 million tons; at the time (1915) this probably represented the biggest civil engineering enterprise ever undertaken.

The trade effects of the Panama Canal were profound. It shortened the sea route from New York to San Francisco by 5000 miles and that from western European ports to the west coast of the USA by nearly 3000 miles. At mid-century it was carrying more than 5000 ships annually (a figure that was more than doubled by 1965), but it was already outdated and the US Congress was debating the possibility of constructing a canal capable of taking larger vessels. The main problem was that, although 305 m long, the locks could not accommodate really large ships, notably the largest US Navy vessels. The Suez Canal was encountering the same difficulties: by 1960 oil-tankers already exceeded 100 000 tons and far larger ones were in sight. For them, the only route was round the Cape.

The construction of the Panama Canal was closely identified with the use of railways and steam-power. Steam-shovels weighing nearly 100 tons – enormous by the standards of the day – loaded huge bites of earth and rock on to wagons running on a complex railway system built for the purpose. The wagons were subsequently unloaded by steam ploughs capable of discharging more than 500 cubic metres of spoil in five minutes; the discharged spoil was then spread mechanically.

Panama was an outstanding example of a canal designed to link two oceans. The St. Lawrence Seaway, built between 1954 and 1958, was an example of a project designed to enable large vessels to proceed inland. Linking Montreal with Lake Ontario, it brought sea-borne traffic into the heart of North America. In addition it incorporated a major hydroelectric scheme serving both Canada and the USA. Its construction was a highly complex operation, made so by the need to avoid any disturbance of existing canal facilities; any diminution of water flow to existing establishments; and any lowering of the level of Lake Ontario. Some 1500 hectares of land were permanently flooded; 35 miles of highway and 40 miles of railway had to be re-routed; and the rehousing of displaced population involved bodily moving more than 500 houses on specially constructed transporters.

The huge tunnelled canal joining Marseilles and the Rhône has already been mentioned. Another major development in western Europe was the 20-mile Princess Juliana Canal in Holland (1935). In the inter-war years the Soviet Union embarked on an ambitious Four Seas canal project, designed to link the White, Baltic, Black, and Caspian Seas. The White Sea Canal

Construction work on lock-walls for the St. Lawrence Seaway (1958).

was completed in 1933 and the Volga–Moscow Canal four years later, making Moscow accessible from the Baltic and the White Sea. The Volga–Don Canal was started in 1938 but not completed until 1952. In Germany, construction of the Mittelland Canal was completed in 1938 after more than thirty years' work. It was the last link in a great series of waterways connecting the North Sea, the Ems, and the Rhine with the Elbe, Oder, Vistula, and the Baltic.

V. DOCKS AND HARBOURS

The basic principles of the construction of docks and harbours changed little in the first half of this century, but, as in all civil engineering projects, there was a steadily increasing reliance on the use of various kinds of heavy machinery in place of manual labour. Dock installations, however, changed rapidly to meet new demands. Thus the need for coaling facilities declined as oil became the ascendant marine fuel; tankers of ever-growing size required to discharge liquid rather than solid cargo; commercial vehicles and private motor cars required roll-on facilities, already accorded in some ports to railway trains; and new and improved navigational aids had to be installed. The pattern of passenger travel changed rapidly as civil aviation developed, and the ocean terminals built to receive giant liners declined; in 1957, for the first time, more passengers crossed the Atlantic by air than by sea, and by the end of the 1960s more than 300 million passengers a year were travelling on scheduled airline flights.

Broadly speaking, docks are of three kinds: tidal docks, communicating directly with the open sea and thus subject to the full rise and fall of the tides; wet docks in which water can be enclosed at constant depth by the use of lock gates; and dry docks which can be pumped free of water for the inspection or repair of vessels. Where the tidal range is small, as on much of the east coast of the USA and in the Mediterranean, tidal docks are the rule, but where the range is large – as with the 30-ft tides at Liverpool or Bristol – wet docks become essential for two reasons. Apart from the inconvenience of working a ship that is rising and falling through a considerable distance twice a day, the cost of building a quay of the necessary height becomes uneconomic in comparison with the longer, but much lower, walls required for a wet dock. A disadvantage is the time required to work vessels in and out through locks.

Virtually all ports require dry docks, of which there are two main kinds. The first is built as a fixed and permanent part of the port facilities. The second is a floating dock which can be towed to any desired position. As it carries extensive workshop facilities, it is theoretically possible to work such a dock wherever suitable moorings can be found, but in practice it is difficult to operate in isolation from shore facilities. A fixed dry dock at Le Havre (1927) was something of a hybrid: it was floated into position and then sunk in a dredged basin.

Apart from increases in size, dictated by the growing length and draught of the ships they served, perhaps the most striking change in the construction of docks was in the material used. Almost up to the end of the nineteenth century masonry construction was virtually universal, but after 1900 rapidly increasing use was made of concrete, reinforced with masonry only where exposed to exceptional stress. As has been noted in the context of roads, the time necessary for the concrete to set was greatly reduced. A typical example is Rosyth Dockyard, built in 1909–18. Hitherto all British naval dockyards had been built of dressed stone, but at Rosyth concrete was used throughout. Dover Harbour (1909) was another early British project making extensive use of concrete, largely in 42-ton blocks; in the course of a year no less than 575 linear metres of breakwater were laid. By mid-century French and Italian engineers were using blocks nearly ten times as heavy.

The first entrance locks to attain a length of 1000 ft were those for the Alexandra South dock at Newport, Wales (1914); the biggest built up to 1939 was the 1312-ft lock at Ymuiden on the North Sea–Amsterdam Canal (1930). Increasingly, lock entrances were made square in section, rather

Dover Harbour under construction 1907. The basic units were 42-ton concrete blocks.

than being deeper at the middle, to accord with the midship cross-section of modern vessels. To facilitate the entrance of large ships, locks were inclined to the channel instead of being at right angles to it. Until the Second World War lock gates were still commonly made of timber, though some use had been made of wrought iron even in the nineteenth century. As it became increasingly difficult to obtain timber (usually greenheart) of the necessary dimensions, the use of mild steel became general. One advantage of this was that the gates could be built with buoyancy chambers to reduce their effective weight and make their opening and closing easier. This was an important consideration, as indicated by the fact that one of the lock gates for the Gladstone Dock (1927) at Liverpool had a gross weight of 496 tons.

By the start of the Second World War dry docks exceeding 1000 ft in length had been built at various ports around the world – including Liverpool (1913), Bombay (1914), Boston (1920), Durban (1925), Southampton (1933), Singapore (1937), and Genoa (1940). Floating docks, too, necessarily had to be larger. The first big floating dock built for the US Navy in 1899 had an over-all length of 525 ft and was capable of taking a 10 000-ton vessel. In 1948 a large floating dock at Portsmouth had a length of 860 ft and could lift a 54 000-ton vessel. Another 50 000-ton floating dock was completed at Bombay in 1947 and subsequently towed to Malta.

Closing the two arms of the Zuiderzee barrier-dam (1932).

VI. LAND RECLAMATION

Although not directly related to the needs of transport, land reclamation
may be mentioned here since the civil engineering techniques are much the
same. By far the biggest undertaking of this kind in the twentieth century
was the Dutch project for reclaiming the greater part of the land covered by
the Zuider Zee. This huge scheme began in 1920; it was interrupted by the
Second World War – during which the Germans destroyed part of the works
and flooded 20 000 hectares of regained land; and is still not quite complete.
When finished, some 200 000 hectares of agricultural land will have been
regained. The main work, started in 1920 and completed in 1932, was the
building of embankments, some 48 km long, joining North Holland and
Friesland via the island of Wieringen.

BIBLIOGRAPHY

Berridge, P. S. A. *The girder bridge after Brunel and others*. Robert Maxwell, London
    (1969).
Chevrier, Lionel. *The St. Lawrence Seaway*, Macmillan, Toronto (1959).
De Glopper, R. J. and Segeren, W. A. The Lake Ijssel reclamation project.
    Endeavour, **30**, 62 (1971).
Hammond, R. *Tunnel engineering*, Heywood, London (1959).
Hills, T. L. *The St Lawrence Seaway*, Methuen, London (1959).
O'Flaherty, C. A. *Highways*. Edward Arnold, London (1974).
Pequignot, C. A. *Tunnels and tunnelling*. Hutchinson, London (1963).
Schreiber, H. *The history of roads*. Barrie and Rockcliffe, London (1961).
Shirley–Smith, Hubert. *The world's great bridges*. Phoenix House, London (1953).
Steinman, D. B. and Watson, S. R. *Bridges and their builders*. Dover Publications,
    New York (1957).
Van Veen, J. *Dredge drain reclaim*. Martinus Nijhoff, The Hague (1952).

20

LAND TRANSPORT

In a sense, the most important developments in land transport were social and economic rather than technological. The world's railway systems expanded relatively slowly in the first half of this century and the main developments in locomotives and rolling-stock had been made. Similarly, the motor car of 1950 displayed few features that would have surprised a mechanical engineer of 1900. While technological developments were by no means negligible, they were far outweighed by the social consequences of greatly increased mobility for the population at large.

I. MOTOR VEHICLES

In general, motor vehicles in the present context can be taken to mean those driven by internal combustion engines. Nevertheless, we must briefly consider steam and electrically propelled vehicles, the latter including delivery vans, tram cars, and trolley buses.

The motor vehicle is essentially a collection of systems each carrying out a specific function. The most important, of course, is the power unit, mounted in a chassis/frame. Power has to be supplied to the road wheels via a transmission system including gearbox and clutch. In the absence of tracks, provision must be made for steering. A braking system has to be fitted to slow and, if necessary, halt the vehicle; the irregularity of the road has to be cushioned by tyres and springs. These systems, and their own individual components, were usually supplied by separate firms, and the motor industry was, and is, largely an assembly industry. As such, it lent itself to the adoption – not the invention – of the conveyor-belt system which was a key factor in making motor cars, from the famous Ford Model T onward, cheap enough for the mass market.

The engine. The development of the internal combustion engine as such has been discussed separately (Ch. 13), but not its specific role in road vehicles. In the very earliest vehicles there was great variety in the mounting of the engine, but by 1905 it was almost invariably mounted at the front. Rear-mounted engines – eliminating a propeller shaft – were re-introduced in the 1930s but the tail-heavy cars were difficult to control until design had been appropriately improved. Pioneers were Tatras in Czechoslovakia and Mercedes-Benz, but the best known is Ferdinande Porsche's Volkswagen, the first models of which appeared in 1936. They were not, however, to be a

Model T Ford Tourer, 1914.

familiar part of the motoring scene until after the Second World War, when Renault and Fiat began to mass-produce rear-engined cars in 1948 and 1955 respectively.

Transmission. An alternative method of eliminating the propeller shaft was to have a front-mounted engine driving the front wheels. The pioneer of this development was Citroen, who made three-quarters of a million such cars between 1934 and 1957. Some other European manufacturers followed suit, but in the USA this system was not used until after the Second World War (apart from some specialist vehicles), with the exception of a short-lived experiment by Cord in 1929.

If the engine is to continue to run while the vehicle is stationary some form of clutch is essential. During the first half of this century the dry friction clutch, in which two spring-loaded plates are disengaged by a pedal, was widely used. The main improvement was in the materials of construction, giving a much longer life before relining was necessary.

The chassis/frame. The earliest cars, and many later ones, consisted of a chassis to which the body – made of a combination of metal and wood, suitably upholstered – was attached. This design lent itself to the making of special bodies to meet the demands of individual customers. Later, however, chassis and body began to be combined as a single unit. The pioneer was the Budd Company of Philadelphia (1931), which used a welded structure, but manufacturers in the USA did not adopt it at that time and the main

Oldsmobile tiller-steered runabout, 1903.

development was in Europe. Citroen adopted it for his front-wheel-drive car of 1934, and, rather surprisingly, it was also taken up at an early date by European subsidiaries of General Motors of America (Opel, 1935; Vauxhall, 1938). It is a form of construction that lends itself to mass production because it is cheap, but by 1966 even Rolls Royce had adopted it.

Transmission and steering. In order to control speed, to climb gradients, and to reverse, a gear box must be incorporated in the transmission system. Some early American cars used planetary or epicyclic systems, but the standard soon became the manually operated sliding-pinion gearbox, commonly giving three or four forward speeds, and one reverse. Smooth meshing of the gears while on the move was not easy, and one of the techniques that novice drivers had to master was that of double-declutching. This was avoided with the introduction of the synchromesh gearbox (General Motors, 1929), though for some years it was a luxury confined to the more expensive cars, and even then not incorporated for all gears. A later development was the automatic gearbox based on a fluid coupling or torque conversion. The first to introduce fluid coupling was Oldsmobile in 1940.

In the earliest road vehicles power was transmitted to the road wheels by a belt drive or, later, a chain drive. By 1910 the latter was already old fashioned (except for motor bicycles) though it continued in use until after the First World War as, for example, on Trojan cars in Britain. The standard system was the propeller shaft running beneath the car and driving the rear axle through bevel gears. A problem here is that when rounding bends the two wheels must turn at different speeds; this was solved by the introduction of the rear-axle differential gear. Flexibility was conferred on the propeller shaft by a universal joint.

At the very beginning of the century a few cars steered by tillers or some form of handle bar, were still being made but for practical purposes the steering-wheel was universal throughout our period. The front wheels were mounted on stub-axles connected by an Ackerman linkage. This ingenious mechanism, patented by Rudolph Ackerman a century earlier for use on carriages but probably invented even earlier, ensured that on curves both steerable wheels turn about a common centre; without this, skidding occurs. Rotary movement of the steering-wheel was converted to translational movement of the steering linkage by various devices such as rack and pinion, worm and roller, or screw and nut.

Brakes. Dual braking systems were introduced at an early date. The handbrake, designed to hold the vehicle when stationary, normally acted on the transmission system, clamping the propeller shaft. When the vehicle was in motion, pedal-operated braking was applied to the wheels by shoes acting on the inner surfaces of drums. Normally only the rear wheels carried brakes, but as early as 1919 Hispano-Suiza introduced four-wheel brakes; generally speaking, however, four-wheel brakes were then limited to the more expensive models. By the end of the 1920s four-wheel brakes were fairly general. The movement of the pedal was at first transmitted to the wheels by wires, but, because they stretched, it was difficult to keep them in adjustment to ensure even braking. From the 1920s the more expensive cars began to have hydraulic systems and by mid-century this too was general.

Suspension. On early motor cars suspension was primitive and similar to that used on horse-drawn carriages. It was usual to have four axle-mounted leaf springs, one for each wheel, though the Model T Ford and some others had a single transverse spring across the front. In such systems the up-and-down movement of one wheel tipped the vehicle as a whole; to obviate this, independent suspension was introduced in the 1930s, initially on the front wheels only. Suspensions were usually based on coiled springs, though both Citroen and Volkswagen used torsion bars before the Second World War.

Tyres can, for practical purposes, be regarded as part of the suspension, in that they helped to cushion the shock of irregularities in the roads. First used for bicycles, they were applied to cars as early as 1895, though some early models had solid rubber tyres. The latter remained in favour as late as 1930, however, for heavy commercial vehicles, especially steam wagons. Initially, punctures had to be mended at the roadside, but by 1910 detachable wheels, making it feasible to carry a spare, had appeared. Two important developments were the use of reinforcing chemicals such as carbon black (*c*.1905) and the introduction of the corded tyre (*c*.1910). The corded tyre was improved by Goodyear in 1922 with the introduction of the highly extensible 'Supertwist'; in the 1930s rayon began to be used in Germany.

Although much of what has been said above relates primarily to motor cars, the development of goods vehicles followed a similar course. At first these were restricted to relatively short runs, for their low speeds and limited load-capacity made them uncompetitive with the railways. The railways, however, were quick to adopt them as an extension of their goods delivery service. Up to the First World War few lorries exceeded five tons, but by the late 1930s 15-tonners were in common use. To spread such heavy loads additional axles were introduced, initially two instead of one at the rear. At first the additional axle free-wheeled, but by the 1930s both were often power driven. At the same time two additional steerable wheels were introduced at the front of some very heavy vehicles.

A major development, not paralleled in private cars, was the articulated lorry in which a relatively light trailer was drawn by a three-or four-wheeled tractor. The latter could be disengaged and used with other trailers as required.

Thorneycroft steam omnibus, 1902.

Scammel frameless articulated tanker, *c.*1927. Vehicles of this type continued to be made until the 1960s.

Power unit of Stanley Steamer, 1920.

Foden overtype steam wagon, 1916.

Fageol Safety Coach (c.1930).

Bersey electric cab, 1897.

Among the most important developments of commercial vehicles, in the social sense, was the motor bus, which transformed life in country districts and speeded up public transport in urban areas. They were originally built on lorry chassis but very soon were custom-built. By 1910 London had over a thousand motor buses; by 1914 the number had doubled and horse-drawn buses had disappeared. By then, double-deckers were commonplace. In the 1930s the use of a chassis declined and, as in motor cars, chassis and frame were integrated. In road tankers this development had, in effect, already occurred in the 1920s, for the tank was rigid enough to serve also as the frame.

II. STEAM AND ELECTRIC VEHICLES

For road vehicles, steam had a longer and more successful life than is generally recognized. In Britain, though not so much elsewhere, steam lorries were still common in the 1930s, especially for very heavy loads, such as quarry stone. Traction engines, too, were a familiar and magnificent sight on the roads for pulling heavy trailers, often several in a row.

The steam car, however, was largely an American development. One advantage was the fact that no gearbox, and thus no difficult gear-changing, was required. A decided disadvantage was the difficulty and delay in getting started. Nevertheless, the White steam car continued in production until 1910 and the Stanley up to 1927. In 1923 the highly sophisticated Doble with a high-pressure boiler had appeared. It had an impressive performance, with a top speed of over 100 m.p.h., but it was far too expensive ($ 11 000) to compete with petrol. No more than forty-five were ever sold, and production ceased in 1932.

By contrast, electrically propelled road vehicles have continued in use up to the present day. They fall into two distinct categories: those that draw their electricity supply from an outside source and those that have a self-contained source of supply.

The first category comprises tramcars and trolley-cars. The first trams appeared well before the beginning of this century. They were mostly single-deck vehicles, drawing their power from an overhead cable by means of a trolley-arm. This supply system is cheap to install and maintain, because it is easily accessible, but is unsightly: in some cities – such as London, Washington, and Berlin – considerations of amenity prevailed and an underground conduit was used for the cable. Britain was almost unique in using double-decker trams.

Trolleybuses differ from trams in running at will on the road – subject to maintaining contact with their electrical supply – instead of on rails. The absence of the latter to return the current requires double overhead cables. Experimental models appeared at the turn of the century in both the USA and Europe but they did not become popular until after the First World War. At mid-century some 6000 were in use throughout the world, about 4000 in North America. Thereafter, they declined rather rapidly, partly because of the increasing difficulty of operating them in modern traffic conditions.

Battery-operated vehicles have been in use since the 1890s. The main obstacle to their wide use was, and remains, the difficulty of manufacturing batteries with appropriate weight: capacity ratio. Their speed and range between battery recharges were both low. Up to the First World War electric cars were quite popular, especially in the USA, as taxis and local runabouts. Thereafter their use was largely limited to small commercial vehicles, especially delivery vans making short runs and frequent stops. This limited market they have continued to hold, especially in Britain.

III. RAILWAY LOCOMOTIVES AND STOCK

Although an underground electric railway system was built in London in the 1890s and a stretch of the main-line track of the Baltimore and Ohio Railroad was electrified in 1895, the steam locomotive reigned supreme in

London and North Western Railway's locomotive, *Coronation*, built in 1911.

1900. In power and appearance and in precision of manufacture it had come a long way from Stephenson's *Rocket* of 1829 but in fact there had been few major innovations save the redesign of the fire-box to make possible the use of coal, rather than coke, as a fuel. The first decade of this century saw a major advance, however, with the introduction of superheated steam by Wilhelm Schmidt. This cut fuel consumption by over a quarter. At about the same time piston valves began to replace slide valves and the efficiency of boilers was much improved.

With higher speeds and heavier locomotives, more attention was paid to balancing moving parts, especially the driving wheels. The balance weights proved destructive to the track, however, as in effect they delivered a hammer blow to it with every revolution. To counteract this, locomotives were introduced with four cylinders instead of two. Originally these all drove one axle, but this tended to develop excessive stresses and in British practice this was overcome by driving two axles. In North America, two-cylinder locomotives were generally preferred.

Increases in speed and size were spectacular. In 1938 the streamlined British locomotive *Mallard* set a world record for steam of 126 m.p.h. In the USA – with its huge distances, wide-gauge track, and heavy loads – articulated locomotives appeared early in the century and by the 1940s the largest weighed 350 tons. Despite their size, their construction enabled them to negotiate quite sharp bends. The ultimate triumph of the steam locomotive was the Union Pacific 4000-class of 1941. With sixteen 68-inch driving wheels, these had a gross weight in excess of 600 tons and were capable of a speed of 70 m.p.h. While such developments were taking place, however, other forms of traction – electric and diesel – were appearing which were ultimately to eclipse steam.

Electric traction got off to an uneasy start, with several quite disparate systems finding their advocates. The most important of these were the low-

voltage direct current systems with third-rail pick-up – as developed by the
Southern Railway in Britain and some installations in the USA – and high-
voltage alternating current systems with overhead· pick-up. Within these
categories there were considerable variations. Thus direct current voltages
varied from 600 V (British Southern Railway) to 1500 V (Paris – Orleans)
and 3000 V (Belgium, Italy). In Hungary three-phase 3000 V alternating
current was used, but the highly successful Lotschberg line, opened in 1913,
employed single-phase 15 000 V alternating current. At mid-century,
however (1951), French railway engineers demonstrated that 25 000 V
single-phase alternating current could be used at the normal commercial
frequency of 50 cycles per second. This set the pattern for future
developments in France, Britain, and other European countries, as well as in
Japan and India.

Diesel traction was a relatively late development on the railways, dating
from the 1920s. Although, as we have seen in an earlier chapter on internal
combustion engines, Diesel engines lent themselves to the construction of
powerful units, the transmission of power to the driving wheels of
locomotives presented problems. As with motor cars, direct drive involved
the incorporation of a gearbox, unnecessary with steam or electricity, and
the construction and operation of this for the large power involved was
difficult. Experiments with the use of compressed air were made in Germany
in the late 1920s, but eventually electrical transmission was generally
adopted. In effect, the locomotive incorporates its own power-station. This
makes for a heavy unit and entails power losses on conversion, but in
practice it proved very successful.

Pioneers in this form of diesel traction unit were General Motors in the
USA, who introduced their famous 567 engine in 1935. The resulting

Silver Link, first of Sir Nigel Gresley's streamlined 'A4' Pacifics (1935).

Ancient and modern meet in the Sudan. Diesel-electric locomotive built by the English Electric Company for the Sudan Government Railway, *c*.1950.

locomotive had a capacity of 1800 h.p., which was far less than those of the large steam locomotives we have just described. These units were so designed, however, that they could easily be coupled together to give multiples of 1800 h.p. The ability to couple units was very important, for the Diesel engine, unlike steam, is at its highest efficiency when working at maximum load.

The Diesel has other significant differences from the steam-engine. On the debit side, initial cost, and the cost of fuel, is relatively high. On the credit side, the locomotive requires little preparation before starting; consumes fuel only when running; does not have to be raked out and serviced every time it goes off duty; and requires no stoker (though unions might still insist on a two-man crew). With these considerations in mind, railway operators in North America, Europe, and Asia widely adopted diesel-electric locomotives of the General Motors type after the Second World War.

While the general sequence of events in the development of steam, electric, and Diesel locomotives can be quite clearly discerned, local practice varied considerably. In countries such as Sweden and Switzerland, with plenty of cheap electricity, a policy of electrification was followed between the wars. In the USA steam traction predominated up to 1945, but dieselization was almost universal by 1960.

In rolling-stock a major development concerned the materials of construction: wood was largely replaced by steel, first in the USA and later in Europe. Open saloon-type passenger carriages replaced those with compartments opening on to a corridor, though some railways with heavy commuter traffic – such as the Southern in England – continued to use carriages having compartments with separate doors to the platform so that passengers could embark and alight more quickly. Standards of comfort varied considerably and reflected not only technological progress but also social attitudes. Thus the differences between the various classes of carriage were more pronounced in Britain then elsewhere. Only there, for example, could first-class passengers enjoy single-berth sleeper compartments; four-berth compartments, with a lack of privacy long deemed acceptable in the USA and on the Continent, were not introduced until 1929. In about 1900 oil and gas lighting, a considerable fire hazard, began to give way to electricity. Braking systems were improved in the inter-war years. One problem with both air and vacuum brakes was the delay which occurred along the length of the train: the brakes on the rear carriages of a long train came on some seconds after those on the front. To overcome this, the electro-pneumatic brake was developed.

Rolling-stock for goods traffic became much more specialized; the simple alternative of open trucks or closed wagons no longer sufficed. Tankers were introduced for liquids such as petrol and milk; hopper wagons for cement and minerals; bolster wagons for timber; and so on. Again progress was uneven. While vacuum brakes were widely used for express freight trains in the USA before the Second World War, Britain and much of Europe still used plain-coupled goods trucks, with only manually operated brakes on each, well after that war.

IV. TRACK AND SIGNALS

Heavier locomotives and rolling-stock, and higher speeds, made greater demands on the track and there were important changes in practice. In Britain, the cradle of railways, long-standing practice was to use a bullhead rail mounted in cast-iron chairs attached to wooden sleepers. In most other parts of the world, however, a flat-based rail was used, secured directly to the sleepers with spikes. This mode of construction gives a greater lateral stiffness and – an important point as labour became more expensive – is easier to maintain. Nevertheless, British practice was not altered until after the Second World War.

There were changes, too, in the type of sleeper. The traditional wooden sleeper began to be replaced in the 1940s by concrete ones: French practice was to use a concrete block under each rail, joined by a steel tie-rod. Steel sleepers were introduced on a limited scale in Britain in the 1930s. After the

Second World War continuous welded rail replaced the short sections; passengers were thereby deprived of the pleasure of estimating the speed of their train by timing the interval between the characteristic thud which marked the transition from one standard length to another. Tough steels, incorporating manganese or nickel, were introduced in the 1920s for sections of track – such as those leading up to points or cross-overs – subject to particularly heavy wear.

Increasing speeds and denser traffic focused attention on the improvement of the signalling system, which was mainly designed to eliminate the consequences of human error. A variety of mechanical, and later electrical, devices, were introduced – stemming largely from the work of John Saxby in Britain at the end of the nineteenth century – so that potentially dangerous combinations of signals and point-setting were impossible. Mechanical signalling by semaphores, worked by rods running alongside the track, gradually gave way to colour-light electrical signalling, though this was not general in Europe until after the Second World War. All such systems depend heavily on their observance by the driver and his prompt and correct response. In Britain, the Great Western Railway in 1906 introduced the system of automatically sounding an alarm in the driver's cab as each signal was passed, the sound varying according to the setting of the signal. If

Control panel in modern signal box at Bologna (Italian State Railways).

the driver failed to acknowledge the audiosignal the brakes were applied automatically. Such systems of Automatic Train Control became widely adopted in the 1920s and reached their logical conclusion in the USA during the Second World War, when trackside signals were dispensed with altogether on one all-freight system, the driver relying entirely on audiosignals in his cab. Another very important invention, made in 1872 but not widely applied until after the beginning of this century, was the track circuit. This provided the signalman with a display panel in his cabin showing the position of all trains for which he was responsible. It was one of the most important of all safety devices.

BIBLIOGRAPHY

Ahrons, E. L. *British Steam railway locomotives 1825–1925*. Locomotive Publishing Co., London (1927).
—— *British railways track*. Permanent Way Institution, London (1943).
Automobile Quarterly. *The American car since 1775*. Dalton, New York (1971).
Bruce, A. W. *Steam locomotives in America*. Norton, New York (1952).
Chapelon, A. *La Locomotive à vapeur*. J. B. Baillière et Fils, Paris (1938).
Cornwell, E. L. *Commercial road vehicles*. Batsford, London (1960).
Gairns, J. F. *Locomotive compounding and superheating*. Griffin, London (1907).
Gerogano, E. N. (ed.). *The complete encyclopedia of motorcars 1885 to date*. Ebury Press, London (1973).
Hinde, D. W. and Hinde, M. *Electric and diesel-electric locomotives*. Macmillan, London (1948).
Johnson, R. P. *The steam locomotive*. Simmons-Boardman, New York (1942).
Meeks, C. L. V. *The railway station*. Architectural Press, London (1957).
Montagu of Beaulieu, Lord, and Bird, Anthony *Steam cars 1770 to 1970*. Cassell, London (1971).
Nock, O. S. *British railway signalling*. Allen and Unwin, London (1969).
—— *Railways of Australia*, A. & C. Black, London (1971).
—— *Railways of Canada*. A. & C. Black, London (1973).
—— *Railways of Southern Africa*. A. & C. Black, London (1971).
—— *Railways of Western Europe*. A. & C. Black, London (1971).
Phillipson, E. A. *Steam locomotive design: data and formulae*. Locomotive Publishing Co., London (1936).
Sedgwick, Michael *Cars of the 1930s*. Batsford, London (1970).
Seth-Smith, Michael. *The long haul, a social history of the British commercial vehicle industry*. Hutchinson Benham, London (1975).
Shields, T. H. Evolution of locomotive valve gears. *J. Instn. Loco. Engrs*. **33**, 368–460, 1943.
Westinghouse Brake and Signal Company. *A century of signalling: John Saxby 1821–1913 and his part in the developing of interlocking*. Westinghouse, London (1956).
Williams, A. *A Life in a railway factory*. Duckworth, London (1915).

21

MARINE TRANSPORT

In many respects the development of marine transport resembles that of the railways. While the first half of this century saw much progress, there was relatively little innovation; certainly nothing to compare with the transition from sail to steam and from wood to iron that occurred largely within the second half of the ninteenth century. Ships were larger; methods of construction changed, steel replacing iron; many more specialized vessels appeared, especially tankers for oil and refrigerated ships for the meat trade; and, with the growth of commercial airlines, relatively fewer passengers travelled by sea. A very significant change, already alluded to, was the increasing use of oil as a fuel in place of coal. As in so many other fields, a steady technological progression can be seen but this did not take place everywhere at the same rate. In 1900 there were still some sailing trawlers with wooden hulls and plenty of iron-built tramp steamers. Even today, there are plenty of Arab dhows and Chinese junks. For special purposes, large sailing-ships continued to be built well into this century. German regulations, for example, demanded that all officers must be trained in sail; one consequence of this was the 5000-ton *Padua*, built as late as 1926. Although screw propulsion was almost universal, quite a few paddle-steamers were still in use at least until the 1930s, mainly pleasure vessels plying in shallow waters between tides.

I. CONSTRUCTION

Until the nineteenth century all ships were built of wood. This had many advantages but – apart from the difficulty of getting very large timbers – its properties were such that it was impracticable to build from it vessels much more than 300 ft in length. Brunel's *Great Eastern* (1858), the first of the great iron ships, was more than twice this length. By the 1880s, however, iron had largely given way to mild steel, thanks in large measure to the availability for the first time of steel that would meet the exacting safety and cost requirements of shipbuilders. By 1890 virtually no iron ships were being built on the Clyde. Both iron and steel ships were of riveted construction, a practice which remained general until the 1930s. Riveting then gave place to welding, again thanks in large measure to improved technology. As we noted in the context of bridge construction (Ch. 19), the early welding of large structures was not entirely satisfactory: the welding process itself set up

Wooden sailing ships and iron-hulled paddle steamers in Dover Harbour, 1904.

metallurgical changes and strains which sometimes proved the cause of catastrophic failure. By 1939, however, improved techniques had largely, though not entirely, eliminated this hazard. The outbreak of the Second World War, and the concomitant heavy loss of shipping, obliged the Allies to embark on a massive shipbuilding programme in which speed was of the essence. This gave a further impetus to the use of welding, for not only were riveters scarce by then but welders could be trained much more quickly. Moreover, welded ships were more efficient because of the smoothness of their hulls. By mid-century welding was almost universally practised for new ships. Nevertheless, riveted ships proved their worth; apart from not being liable to sudden general failure, small leaks quickly sealed themselves with rust.

In the Second World War losses of Allied and neutral ships through enemy action, primarily submarines, amounted to over 21 million tons, nearly twice the figure ($12\frac{1}{2}$ million tons) for the First World War. The losses in 1914–18, however, were such as to provide a powerful stimulus to shipbuilding both during and immediately after hostilities. To speed up production, building was limited as far as possible to a few standard types. This made possible a considerable degree of prefabrication, some of which was carried out in collaboration with the civil engineering industry, instead of assembly plate by plate on the skeleton of the hull. By the Second World War prefabricated units weighing up to 200 tons were not uncommon. Such

Gotaverken's shipyard, Arendal, Norway (1970).

large hull sections were not easily assembled and manœuvred in the cramped surroundings of old-fashioned shipyards; for this, and other reasons, Britain lost ground to rivals such as Japan and Scandinavia which possessed more modern yards.

Among the factors that encouraged the use of welding were the London naval convention of 1936, which put a premium on weight reduction. This was later achieved also by the increasing use of light alloys and the various forms of plastic then coming into general use. These, however, were not employed for the main structure but restricted to the interior and superstructure.

II. PROPULSION

At the beginning of this century, and well into it, the triple-expansion reciprocating steam-engine, introduced in 1880, was the most important propulsion unit. Some quadruple-expansion engines were also constructed. Working with superheated steam at pressures up to 400 lb per square inch, and in its later form with a reheating system, this was the ultimate development of the reciprocating steam-engine for marine use. Nevertheless, formidable rivals appeared at an early date, notably the steam turbine (Ch. 14) and the diesel engine (Ch. 13).

Parsons's steam turbine made a dramatic appearance at the Spithead naval review of 1897 and was quickly adopted for warships, coming into general use by 1910. Its main attraction was that it was relatively lighter and more compact than a reciprocating engine of equal power, smoother in action, and more flexible, but at the same time it was more expensive both to construct and to run. The turbine was also attractive to the operators of large merchant vessels, and as early as 1904 Cunard constructed for purposes of comparison two similar 20 000 ton liners, one equipped for turbine propulsion and one for conventional steam propulsion. The experience persuaded them to use turbines for later transatlantic liners such as the 32 000-ton *Mauretania* (1907) and the 45 000-ton *Aquitania* (1914).

For smaller vessels, however, the turbine was initially less popular, except to a limited extent in the USA. Up to the Second World War few cargo vessels were driven by turbines, but afterwards turbines were adopted for smaller ships and large container vessels and tankers.

The other rival to the steam-engine was the diesel. The first important vessel to use a diesel engine was the 5000-ton *Selandia* commissioned in 1912 by the Danish East Asiatic Company. Nevertheless, the diesel engine was not widely used for ship propulsion until after the First World War, when it was adopted for both cargo and passenger vessels. Among major successes in this class of vessel were the 18 000-ton *Gripsholm* built in 1925 for the Swedish American line and the 27 000-ton *Dominion Monarch* completed in 1939; the latter was then the largest and most powerful British passenger ship afloat except for the giant transatlantic liners.

During the First World War experiments were carried out to prevent the rolling of ships by installing massive gyroscopes weighing up to 600 tons. In 1925, however, the Japanese introduced the modern type of stabilizer in which under-water fins are controlled by small gyroscopes responding to the ship's motion.

III. SHIP TYPES

From very early days some degree of specialization in the design and construction of ships was manifest – largely, but not wholly depending on the average length of their voyages and whether they primarily carried passengers or cargo – but the twentieth century saw considerably more diversification, especially in freighters.

In passenger vessels there was, as we have noted, a great increase in size and speed. The splendour and extreme comfort of the very large liners – the most luxurious form of travel ever developed, unmatched by anything available today – should not obscure the fact that millions of passengers travelled on much less pretentious vessels. Indeed, at mid-century the decline of the ocean giants was already in sight; increasingly, long-distance

The *Queen Elizabeth*, launched 1938.

passengers were turning to the air, and liners found their market in the cruise business for tourists.

In 1900 the general-purpose tramp steamer epitomized the world's merchant ships: carrying capacity was around 10 000 tons and included bulk cargo such as coal, ore, and grain. Basically, they were of 'three-island' design – poop, bridge, and forecastle. Increasingly, the deck space between these was covered in, so that deck cargo could be protected from the weather. If only light covering was used and cargo openings fitted, the extra capacity so created was exempt from tonnage assessment. Later, this light superstructure was made more substantial, fitted with permanent hatchways, and made an integral part of the cargo space. From this evolved the modern cargo carrier, with deeper draught but fewer decks. To increase stability, the mid-section became broader and squarer, a trend we have already noted in connection with the design of dock entrances (p. 232).

As their name implies, bulk carriers were designed to carry a single commodity unpackaged. Important among this class of vessel were ore carriers built in the first decade of the century to carry iron ore on the Great Lakes and from Narvik. Their disadvantage was that they normally had to return home in ballast; this was overcome by developing a design that permitted carriage of either ore or oil.

As the world's transport became more and more dependent on oil,

T2 tanker converted for transporting aircraft during Second World War.

however, it was the tanker that showed the most striking development. Up to the First World War tankers remained comparatively small, few exceeding 8000 tons, but by mid-century they had more than doubled in size. The Second World War, with its heavy shipping losses, demanded mass production of standardized tankers. This demand was met by the American T-2 tankers, of 16 000 tons, of which 480 were built.

By mid-century larger tankers were being built, but of fairly uniform design. Internally they were divided into a large number of separate compartments – say thirty for a tanker designed to carry 25 000 tons of oil – by bulkheads running transversely and longitudinally. These compartments were separated from the rest of the ship by coffer-dams. Tankers designed to carry heavy oil, which becomes too viscous to pump when cold, had heating coils in these compartments. It had become universal practice to place the engine-room and machinery aft, which had two advantages. Firstly, it was easier to trim the ships by filling some of the compartments with water when running light; secondly, this arrangement reduced the risk of fire, always a particular hazard with vessels of this kind. In this respect they differed from most other ocean-going ships, in which the engine-room was located amidships for convenience in loading and unloading.

We can only briefly mention here the great variety of other specialized craft: fishing and whaling vessels were considered in Chapter 8, and

refrigerator ships in the context of food technology (Ch. 17). Submarines
will be considered separately in the context of military technology (Ch. 30).
Train-ferries were established in the nineteenth century – the *Leviathan*,
built in 1849 for the North British Railway Company, plied the Firth of
Forth between Granton and Burntisland – but many were subsequently
replaced by bridges. The first major train-ferry in this century in which the
rolling-stock ran straight aboard – as opposed to being transferred by
cranes – was the 4500-ton *Contra Costa* built in 1914 for use in California. At
mid-century there were regular cross-Channel ferries between Dover and
Dunkirk – reopened in 1948 after destruction during the Second World
War – providing a direct rail connection from London to Paris, and between
Copenhagen and Malmö. By that time drive-on car ferries were in
widespread use, though some of the earlier ones – such as that across the
Severn at Aust – were replaced by bridges. The demand for car ferries was
greatly stimulated by the post-war enthusiasm for holidays abroad, and
regular crossings of the Channel and the North Sea were established. These
very large ferries, with capacious holds, carried not only motor cars but also
long-distance lorries.

IV. DESIGN AND SAFETY

Even today total catastrophe at sea cannot be avoided, and large, well-
found vessels still disappear without trace. Nevertheless, this century has
seen a very considerable improvement in safety at sea: at the beginning of
the century the mortality rate among seamen from accident was 0.14 per
cent, but at mid-century it was only 0.03 per cent. Similarly, the percentage
loss of ships (in relation to tonnage) fell from 2.0 per cent to 0.23 per cent
over the same period.

The major factors in this improvement were design and legislation. The
work done in the nineteenth century by such pioneers as William Fairbairn
and William Rankine led to great progress in naval architecture, and in
particular to a better understanding of the interaction between the hull and
the wave action to which it is subjected. The importance of compartmental
design – already discussed in the context of tankers – was widely recognized
as a means of limiting the consequences of serious holing of the hull. The
high freeboard inherent in later designs also gave a greater reserve of
buoyancy.

These changes were partly the consequences of a natural process of
evolution, but partly also the result of legislation obliging less scrupulous
owners to do what the more conscientious did as a matter of course. Samuel
Plimsoll's Merchant Shipping Act of 1876, setting a limit to the depth to
which a seagoing vessel might be laden, was only one of a succession of

internationally agreed regulations designed to improve safety at sea. Even so, progress was slow; not until 1906 was international agreement reached on the way in which the position of the Plimsoll line should be calculated. In 1913 agreement was reached on the need to provide sufficient lifeboats and life rafts for all passengers and crew; the strengthening of these regulations in 1930 included a requirement that some motor-driven lifeboats should be provided. Regulations were also introduced to ensure that adequate fire-fighting equipment was carried.

Although the distinction between passenger-carrying ships and freighters is fairly sharp, it is not absolute. From time immemorial many cargo vessels have carried some passengers, but not until the shipping convention of 1948 was the position clearly defined. This limited the number of passengers carried by cargo ships to twelve; if there was accommodation for more passengers, the ships were classified as passenger vessels and came under more stringent safety regulations.

A major hazard at sea, apart from fire and tempest, has always been that of shipwreck. Improvements both in cartography, which resulted in the provision of more accurate charts, and in navigation, particularly the invention of the gyro-compass and of radar, greatly reduced the danger of ships running around through faulty navigation. These developments will be considered in a later chapter (Ch. 23) on navigation.

BIBLIOGRAPHY

Baxter, B. Comparison of welded and riveted ship construction. *Engineering*, 22 July 1955.
Eyres, D. *Ship construction*. Heinemann, London (1972).
Mitchell, W. and Sawyer, L. Empire ships of World War II. *Journal of Commerce*, Liverpool (1965).
—— British standard ships of World War I. *Journal of Commerce*, Liverpool (1968).
Murray, J. M. *Merchant ships 1860–1960*. Royal Institution of Naval Architects, London (1960).
Parkinson, J. R. *The economics of shipbuilding*. Cambridge University Press (1960).
Walton, T. *Steel ships*. Griffin, London (1901).
—— Baxter, B. *Know your own ship*. Griffin, London (1970).

22

AERONAUTICS

Many technological developments in the twentieth century were essentially extrapolations of existing ones. Aeronautics is very different, however, in that for practical purposes it is entirely a product of the twentieth century; moreover, by mid-century the main characteristics of modern flying had already been established. This is not to ignore the fact, of course, that lighter-than-air craft and gliders were developed before 1900; nor that space flight, a development from military rockets developed in the Second World War, was not achieved until just after the middle of the century.

Although relatively unimportant, airships – stemming from the first ascent of a man-carrying Montgolfier balloon in 1783 – can appropriately be considered before powered fixed-wing aircraft, even though Sir George Cayley had been successful with a man-carrying glider in 1809 and thousands of glider flights had been made by the end of the century.

I. AIRSHIPS

Balloons filled with hot air or, later, with hydrogen or coal gas had – and still have – the disadvantage that the speed and direction of flight are largely uncontrollable: there is nothing corresponding to the degree of manœuvrability achieved by sailing craft. Controlled flight, which distinguishes the airship or dirigible from the balloon, had to await the availability of a suitable power unit; in the event this was, as with aeroplanes, the internal combustion engine. Nevertheless, H. Giffard carried out near Paris a seventeen-mile flight in an airship driven by a three-horsepower steam-engine in 1852. In 1884 C. Renard and A. C. Krebs, also in France, achieved a speed of 14 m.p.h. on a five-mile circular course in an airship powered by a nine-horsepower electric motor. Not until 1903, however, the year in which the Wright brothers achieved their success at Kitty Hawk, was a controlled cross-country flight made in a true dirigible. In that year the *Lebaudy* flew from Moisson to Paris, a distance of some forty miles. It was powered by a forty-horsepower Daimler petrol engine.

The *Lebaudy* was a semi-rigid airship. The hydrogen gas which lifted it was maintained at a slight pressure in an internal ballonet which sufficed to maintain the shape of the external envelope. The engine and passenger accommodation were attached to a long spar suspended beneath the envelope. Some 800 airships of this kind were built between 1902 and 1960 but most were relatively small and were used on short-range work, such as military patrols. A notable exception was Umberto Nobile's *Norge* in which he flew over the North Pole in 1928. He was wrecked on the return journey, and Roald Amundsen – who had flown over the Pole in an aeroplane in 1926 – lost his life at sea in a subsequent rescue attempt. After the Second World War the US Navy built a few much larger airships of this kind.

The best-known airships, however, were the rigid types associated with Count Ferdinand von Zeppelin. In these the gas is contained in one or more cells within a rigid streamlined hull. Between 1900 and the outbreak of the Second World War about 160 Zeppelin-type airships were built; the first really successful one was the LZ4 of 1908. In the First World War the Germans devoted considerable effort to the building of large Zeppelins – about fifty in all – for military purposes, including bombing raids against the Allies. However, the huge size, low speed, and lack of manoeuvrability of these leviathans made them no match for the kind of aeroplane appearing at the end of the war.

In 1917 the L59 made a successful return flight from Germany to Africa and in 1932-7 the *Graf Zeppelin* made regular commercial flights to South America. Britain, too, was interested in commercial flights (as early as 1919 the R34 made the first return flight across the Atlantic) and in the 1920s had plans to establish regular Empire flights. In the USA the *Los Angeles*, a Zeppelin acquired as part of German reparations, continued flying until 1939.

These developments made only a trifling contribution to commercial flying because of the rapid development of the aeroplane. In any case, airships had serious inherent disadvantages, including their vulnerability to storms when moored. Most serious, however, was the highly inflammable

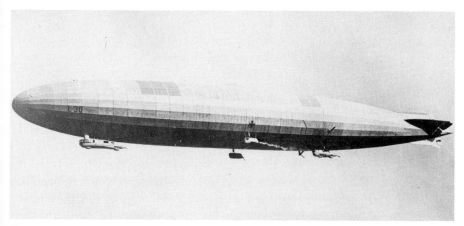

Zeppelin L.30, a rigid airship developed for the German Navy (1916).

nature of the hydrogen with which they were filled; some major disasters, such as the total loss of the R101 (1930) and of the *Hindenburg* (1937), caused much public apprehension. Although helium offered a safe and economically feasible alternative, and was first used in the USA in 1921, it was not universally available and its lifting power was very much less than that of hydrogen. Airships still have their advocates, but there has been little serious interest in them since the last war.

To conclude this section, mention may be made of the role of barrage balloons in anti-aircraft defence during the Second World War. They were used to protect large cities and also during the Allied invasion of Europe. At its peak, Balloon Command in Britain had a flying strength of some 2400 balloons.

II. EARLY AEROPLANES

In the nineteenth century there was much interest in heavier-than-air gliders and a number of attempts were made to sustain flight by means of engines. Aerodynamics developed considerably as a science, and by the middle of the century Sir George Cayley (d. 1857) had established the main design features necessary for stability. He calculated, for example, the power required to achieve a given load and speed, and it was apparent that no engine then available had a sufficiently high power: weight ratio to make controlled flight possible. Such a unit became available, however, with the advent of the petrol-fuelled internal combustion engine, patented by Gottlieb Daimler in 1889. Once these two separate lines of development met, powered flight became not only feasible but inevitable, though it was, in fact, not until 17 December 1903 that the Wright brothers, Wilbur and Orville, made their first historic flight in North Carolina. Even then the event attracted little public attention. Their pusher-type biplane incorporated various important design principles, such as wing-warping to effect lateral control, but not all those that Cayley had laid down years earlier; in particular, there was no vertical tailplane for directional control.

In the summer of 1908 Wilbur Wright went to France and gave a series of public demonstrations that for the first time established the Wright brothers' claim – which some had questioned, giving special credit to the Frenchman Henri Farman, who actually first flew only in January of that year – to be the true fathers of modern aviation. Temporarily, the scene of action was transferred to France. Some thirty Wright-type tail-less biplanes of the pusher type were built, but Louis Blériot pioneered a tractor-type monoplane which became internationally famous when he made the first flight across the English Channel on 25 July 1909. Although the monoplane was later to become dominant, it was not widely adopted at that time,

The Wright Flyer. This machine achieved the first controlled and sustained flight under power on 17 December 1903.

because, weight for weight, it was easier to build a strong and rigid biplane. For many years the tractor-engined biplane was the standard type of aeroplane.

An important innovation in design was the use of ailerons, initially to control roll but subsequently also to assist directional control. The first to use them was probably Robert Esnault-Pelterie in 1904, but it was Farman who first introduced large ailerons as substantial additions to the control system in 1908. F. W. Lanchester's *Aerial Flight* (1907-8) was a major contribution to the science of aerodynamics but made little immediate impact on design, which remained largely empirical.

Although the petrol engine had solved the power: weight problem, it had not done so by such a margin that minimizing weight became unimportant, and all these early aircraft were constructed basically of wood, wire, and fabric. Even then, the use of fabric was restricted largely to the aerodynamically important surfaces, and protection for the pilot was minimal. The earliest engines (Ch. 13) were essentially the same as those developed for motor cars: the Wright brothers used a four-cylinder water-cooled engine rated at about 6 kg per h.p. While water-cooled engines continued in use, air-cooled alternatives of different design were developed. In these the cylinders were either arranged radially about a central crankshaft or the whole engine rotated about a stationary crankshaft. One of the most successful early engines of this type was the French Gnome (1907).

III. INFLUENCE OF THE FIRST WORLD WAR

Military aircraft will be considered in a separate chapter on military technology (Ch. 30), but the influence of the First World War on the aircraft industry generally was so great that they must be mentioned also at this point. In 1914 there were probably in existence in the whole world no more than 5000 aeroplanes, of very mixed design and performance, but their potential military importance – then thought of primarily in terms of observation – was already recognized by military strategists, if not by field officers. Government moneys therefore became available for their development and, in general, research and development in the aircraft industry have ever since always been provided in one way or another through defence contracts. By 1918 some 200 000 aeroplanes had been built and experience in construction and flying correspondingly enlarged. At first these wartime aircraft were mostly two-seaters, carrying pilot and observer, but quite soon the latter became also an air-gunner, even though equipped at first only with small arms. Later the role of pilot and gunner was combined, and single-seater fighters fitted with machine-guns appeared; they were used both for aerial combat and for attacking troops on the ground. This involved firing forward through the revolving blades of the propeller, a technical problem solved first by fitting the propeller with deflector plates and later, more successfully, by an interrupter mechanism that allowed the gun to fire only when the propeller blades were not in line with it. This device was first used, with great success, in the German Fokkers, and gave them a considerable temporary superiority.

Bomber aircraft were developed by both sides, though the Germans also made extensive use of their Zeppelins. Bombers included not only single-engined aeroplanes for short-term daylight flights but also multi-engined types designed primarily for night attacks on relatively distant targets. Such aircraft were the forerunners of post-war commercial transports; one of the most successful was the British de Havilland D.H.4, introduced in 1917. While these early aeroplanes were generally biplanes with open cockpits, they embodied most of the features of modern aircraft and resembled them in general appearance. The open fuselages of pre-war aircraft had by then given way to wholly enclosed ones, though the cladding was still fabric.

The end of the war in November 1918 posed major problems for industry on both sides as production for military purposes abruptly came almost to a stop. The aircraft industry, which from primitive beginnings had expanded enormously in response to almost exclusively military demands, was particularly severely hit. There were many thousands of serviceable aircraft, a large and well-organized production capacity, but virtually no immediate demand. It was, therefore, inevitable that attention should be directed to the possibilities of civil aviation.

The de Havilland D.H.4 light bomber (1917) later adapted for civil use.

IV. CIVIL AVIATION IN THE INTER-WAR YEARS

Even before the First World War attempts were made to establish regular air services to carry both goods, especially mail, and passengers, but circumstances were then quite unfavourable to their success. Aeroplanes were unreliable, noisy, and uncomfortable. Moreover, they were still demonstrably unsafe and few civilian passengers would have entrusted themselves to them. At the end of the war the situation was quite different: reliability, safety, and – to some extent – standards of comfort had all improved. Surplus military aircraft which could be adapted for civilian use were available in large numbers. Not until the mid-1920s did aircraft specifically designed for civilian use make their appearance, and even then they were greatly influenced by current military programmes.

Rather surprisingly, in view of the obvious potential of the American domestic market, civil aviation initially developed most strongly in Europe. The first regular international flight, between London and Paris, was inaugurated in 1919, and continued for more than a year. In that same year John Alcock and Arthur Whitten Brown gave a glimpse of future developments when they made the first crossing of the Atlantic by air, and R. and K. Smith made the first flight from England to Australia. In 1924 a US Army Air Force team made the first round-the-world flight. Private operators found their economic problems at least as great as their technological ones: with the aircraft available it was virtually impossible to compete with surface transport. The only satisfactory answer was govern-

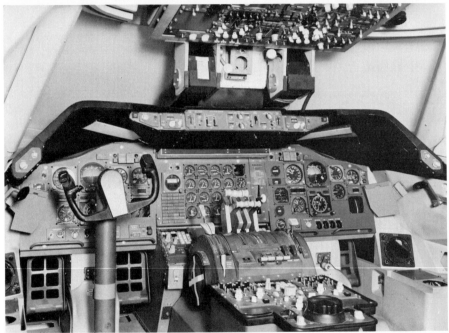

(*Top*) Cockpit and instrument panel in Voisin biplane in which Alcock and Brown made
their first trans-Atlantic flight in 1919. It contrasts sharply with that of a Boeing 747 (*bottom*)
c.1971.

ment subsidies, and in the early 1920s the principal European countries set
up their own national airlines – KLM, SABENA, Imperial Airways,
Lufthansa, etc. There was quickly established not only a European but also
a worldwide organization. In the United States, air transport was at first
used largely for mail; after 1925, however, this government business was

transferred to subsidized private operators, who soon began to carry passengers as well. By the end of the decade the USA had established a commanding lead over its nearest rival, Lufthansa, and Pan American had begun to operate internationally.

The inter-war years showed two important trends in fuselage design and construction. The first of these paralleled a development in the motor-car industry where, as we have noted, the separate chassis and body were giving way to a unit combining both. In aircraft this took the form of stressed-skin construction in which the skin contributed substantially to the over-all strength of the structure. Limited use was made of this before the war and much more during it, especially in Germany. At the same time materials of construction changed, welded steel tube beginning to replace wood for the framework. It was, however, the advent of duralumin (p. 127) that marked the major step forward in this respect. This occurred in 1917 when Junkers began to make aircraft frames from duralumin and covered them with duralumin sheet, though at first the latter was not stressed. This alloy became the most important structural material for aircraft, and from the early 1920s the sheeting was stressed and thus contributed to the strength of the fuselage. The Junkers F13 (1919) was the first civil aircraft to be built with a metal frame and cladding, but the USA was the main pioneer of this important new development: the Lockheed Vega of 1927 was an outstanding early example.

Fokker D.VIII, a highly successful German fighter aircraft with welded steel-tube fuselage. The wing frame was of wood.

In the 1930s twin-engined all-metal civilian aircraft became usual; an early example was the Lockheed Electra, seating ten passengers, but perhaps the most famous was the Douglas DC3. Because of its reliability and economical performance it was widely exported and was built in thousands for the US government during the Second World War; about 500 were still operational twenty years after the end of hostilities. The DC3 could carry 20–30 passengers; it was followed by the DC4 with twice the capacity.

By 1939 the size, speed, and comfort of aircraft had all greatly increased and long-haul scheduled flights were firmly established. To avoid bad weather the refuge of higher altitudes was sought; in order to achieve this, and still maintain the comfort of passengers, the cabin had to be pressurized. In 1939 Boeing introduced the pressurized Stratoliner, designed to cruise at 250 m.p.h. On military aircraft individual air supply to the crew sufficed at high altitudes; thus the DC4, designed as a pressurized civilian aircraft, began its career as an unpressurized military one, as did the Lockheed Constellation.

In 1945 the standard long-distance airliner was driven by four air-cooled engines, with the interior pressurized and air-conditioned. It flew at around 20 000 ft. and cruising speeds were 250–300 m.p.h. The development costs for these sophisticated aeroplanes were very high, in the region of $30 million, and it was clear to designers and manufacturers that the road of evolution would inevitably lead to bigger and more powerful aircraft. To reduce development costs a policy of 'stretching' was devised. Current

Douglas DC3, one of the most successful of all transport aeroplanes, developed in the 1930s.

models were so designed that without basic change in the airframe design the fuselage and wings could be expanded to increase capacity.

For most citizens flying meant occasional travel as a paying passenger by scheduled flights, and up to the Second World War only a small minority of people had ever flown at all. There was, however, also another very important kind of flier – the private aviator. The demand here was for light single- or two-seater aircraft, though at the upper end of the market there was eventually a demand for small private aircraft with cabins for 6–8 passengers. One of the earliest and most successful of light private aeroplanes was the British Moth (1925) – a two-seater biplane with a 60-h.p. Cirrus engine. Some 2000 were built, followed by over 9000 of an improved version, the Tiger Moth. A little later (1927) Hans Klemm introduced the L.25 low-wing monoplane in Germany and this was followed by a number of popular light aircraft in the USA. There the Aeronca appeared in 1928 and the E2 Cub in 1930; both were high-wing monoplanes. The latter was followed by improved versions, the Piper Cub and the Super Cub; in all, some 40 000 light aircraft of this series alone were produced. In the inter-war years most aircraft of this type had fabric-covered fuselages made of wood or tubular steel. D. A. Luscombe (1934) in the USA was one of the first to manufacture a light aircraft made entirely of metal.

Gliding, to which much attention had been directed in the nineteenth century, again became of special interest from the 1920s. At that time

The de Havilland Moth (1925), the first widely used light aeroplane.

Germany was restricted as to powered flight by the Treaty of Versailles, and gliding provided an opportunity for training potential pilots. Gliding also became a popular sport, and by mid-century improved designs and techniques had led to some remarkable achievements. Distances of 500 miles and heights of over 40 000 ft. had been achieved. During the Second World War large numbers of gliders were used for carrying assault troops.

V. HELICOPTERS

As aircraft grew faster and bigger, and stalling speeds became higher, longer runways for landing and take-off became essential. This set important limitations to their use, for it was impossible for them to operate in congested areas or areas in which basic facilities were unavailable. Moreover, it is the motion of an aeroplane relative to the air that keeps it up, and thus in the nature of things it cannot remain stationary. None of these objections applies to the helicopter, by which is meant an aircraft that sustains itself by a powered horizontal rotor or rotors that accelerate the air downwards and thus provide a reactive lift. By varying the pitch of the rotor blades a forward force for propulsion can be generated. A small vertical rotor serves to prevent the fuselage rotating in the same sense as the horizontal rotors. In the event of engine failure, the windmill effect of the rotors enables an emergency landing to be made. With so much in their favour, it may be surprising that so few helicopters – probably no more than 5000–6000 up to 1960 – were built. The main reasons are their low forward speed and their heavy fuel consumption, which is uneconomic and limits range.

Nevertheless, there are situations in which these disadvantages are relatively unimportant and for special purposes helicopters have been uniquely useful. Perhaps their most important single use has been military, as transports, mobile observation posts, and, most particularly, for the speedy and comfortable evacuation of wounded. This last use was forcefully demonstrated in the Korean War (1950). Other important uses have been developed in airport link services, agricultural spraying, servicing oil rigs, and emergency rescue work of all kinds.

Helicopters are entirely a twentieth-century development. A very few experimental machines were built in France before the First World War, but even in the 1920s serious difficulties were encountered in maintaining stability, and it was not until the 1930s that the helicopter became a practical proposition. In 1935 the French Breguet-Dorand 314 underwent successful trials and continued in use until 1939, but it never went into production. In 1936 the German FW61 was introduced and a few were used by the German forces in the Second World War.

The most important development, however, was the helicopter designed

Sikorsky V.S.300 A helicopter (1941), with single main rotor and small anti-torque rotor at rear.

by Igor Sikorsky in the USA. The prototype flew in September 1939, but the production model (VS 300A) was not available until two years later. Some 400 Sikorsky helicopters were used by the Allied forces during the war. These machines were powered by 75-h.p. petrol engines, but after the war gas turbines were introduced.

VI. JET PROPULSION

At mid-century civilian aircraft of all kinds were driven by traditional piston engines, but for military aircraft quite a new form of power had by then been adopted. This new form was jet propulsion, so named because propulsion is effected by discharging a powerful jet of air, much as certain marine creatures move by ejecting jets of water. Three main lines of development were followed. In the turbo-jet engine, the one most widely used, the propulsion jet is generated by gas turbine and centrifugal compressor. In the ram-jet, used primarily in certain guided missiles, air is compressed in the inlet chamber by the forward velocity of the unit; used to burn fuel in a combustion chamber; and the hot exhaust gas is discharged through a rear jet. In rocket propulsion the engine requires no atmospheric air to burn the fuel: the fuel is ready-mixed with an oxygen-yielding chemical.

Messerschmitt Me 262, the first operational jet fighter (1944).

The development of jet propulsion is particularly associated with A. A. Griffith and F. Whittle in Britain and H. von Ohain and M. Mueller in Germany. They were supported by their respective governments, who were conscious of the potential advantages for high-speed military aircraft. The German and the British engines both ran on test beds by about 1937 but the Germans were in the air first. The Heinkel He 178 was airborne in August 1939 – a week before the outbreak of the Second World War – and the Gloster E28/39 in May 1941. In the event, however, jet aircraft became operational only in the last year of the war: the Messerschmitt Me 262 and the Gloster Meteor both entered service in 1944. In the civilian field the British de Havilland Comet was put into service in 1952 but had to be withdrawn two years later following disasters which proved to be due to metal fatigue in the fuselage. However, it was back in service in 1958, and by a margin of only a few days British Overseas Airways beat Panam, with the Boeing 707, in the race to offer the first commercial jet flights across the Atlantic. Meanwhile the Russians had put their first jet transport, the Tupolev Tu-104, into service in 1956.

From 1950 the future of air transport – as distinct from light aircraft – lay with the jet engine, though many traditional aircraft with piston engines were operational long after that, and indeed still are. For a time turbo-prop engines – in which gas turbines drove conventional propellers – had a considerable vogue, as in the highly successful Vickers Viscount (1953).

The middle of the twentieth century saw two highly significant events in

The de Havilland Comet (1952) the world's first jet transport.

the history of air transport. In 1948/9 the Russian blockade of Berlin was broken by the Allied airlift in which 2½ million tons of supplies were flown into the beleaguered city within the space of fifteen months. The second has already been remarked on: in 1957 more passengers flew across the Atlantic than travelled by ship.

VII. SPACE FLIGHT

In a sense, space flight has no place in a history concerning itself with the first half of this century, for the first manned space flight – a single orbit of the earth by Yuri Gagarin in *Vostok I*–was not made until 12 April 1961. Nevertheless, the technological developments which made this possible go back to the 1920s, with very considerable contributions from military technology during the Second World War. It is, therefore, inappropriate to omit this aspect of aeronautics altogether.

During the period of our immediate concern, the main centres of interest in rocketry were in Russia, the United States, and Germany. As early as 1903 K. E. Tsiolkovsky, professor of physics at Kaluga, had published his *Investigation of outer space by means of a reaction apparatus*, in which he set out what are now recognized as basic concepts of extra-terrestrial rocket flight. Among these were the need both for a very high exhaust velocity and for carrier rockets – successive stages of a rocket being discarded as the fuel in them was exhausted. In 1929 he successfully launched a multi-stage rocket. In the USA, R. H. Goddard carried out a long series of experiments while professor of physics at Clark University 1919–43. Even before this, he had extended Tsiolkovsky's theory of rocket flight and in 1914 built a two-stage rocket powered with solid propellent. Realizing, however, that much greater power could be generated with liquid fuels, he turned his attention to mixtures of petrol and liquid oxygen, and in 1926 launched the first rocket of this type. His objective was to make meteorological observations at high

Model of Vostok I, the first
manned spacecraft, 1961.

Robert Goddard (USA) with his
first liquid oxygen/petrol rocket,
1926.

altitudes, and in 1929 he launched his first rocket carrying instruments – camera, barometer, and thermometer. During the next ten years, working mostly at Roswell in New Mexico, he made important innovations, including self-cooling combustion chambers, automatic steering systems embodying gymbals and gyroscopes, and centrifugal pumps for petrol and liquid oxygen. By 1935 his A-series rockets had attained a height of 2.3 km.

In Germany, Hermann Oberth developed the theory of rocket propulsion in the 1920s and designed a spacecraft to be propelled by liquid fuel. His enthusiasm led to the formation of the German Rocket Society in 1927. In the early 1930s trial flights were made in Berlin under the directions of Klaus Riedel and Willy Ley, again using liquid fuels. How far the Germans were familiar with Goddard's work has been questioned, though as he published over 200 patents it is hard to believe that they were altogether ignorant of it. In 1932 the German Army became interested, through its Weapons Development department. They enlisted the services of Wernher von Braun, most famous of all rocket specialists, who in 1937 was appointed Director of a new experimental rocket station at Peenemunde, on the Baltic. Here a huge establishment was quickly built up, which by 1939 had a staff of 3500; by 1945 this had risen to 20 000. In addition to the direct effort there, much related work was contracted out to German universities.

By 1935 Robert Goddard had developed quite a sophisticated launching platform for his rockets.

German V–2 (A–4) rocket: this particular example was one captured by the Allies in 1945 and fired by a German crew.

The immediate objectives at Peenemunde were purely military and are thus most appropriately considered in a later chapter (Ch. 29) on military technology. A family of potential long-range rocket weapons was developed, the first major success being achieved with the V-2 in October 1942; this achieved a range of 190 km, a height of 96 km, and a speed of over 5000 km per hour. Its fuel was a mixture of alcohol and liquid oxygen. It is significant that after the first successful trial flight the Commandant of Peenemunde remarked that the practicality of space flight had been demonstrated.

After further technical improvements, the large-scale manufacture of V-2 rockets was initiated in Germany; from September 1944 more than a thousand were launched against southern England, killing over 2500 people and seriously injuring 6000. The menace was finally ended when the launching sites were overrun by the Allies. After the war von Braun went to the USA and eventually became (1960) Director of the George C. Marshall Space Flight Center. There he played a major part in the Apollo programme which led to the first lunar landing in 1969.

The Apollo programme was America's response to the Russian space project that in 1961 shocked the American public as much as the explosion of the first Russian atom bomb in 1953. It provides a striking example of the far-reaching consequences of technological progress: the ten-year Apollo programme designed to re-establish American technological superiority cost the USA $24 billion. Despite its initial success, the Russian programme ran into difficulties. The moon-landing programme was halted after fatal disasters in 1967 and 1971; nevertheless, Russia had considerable successes with unmanned flights.

BIBLIOGRAPHY

Beaubois, H. *Airships: an illustrated history*. Macdonald and Janes, London (1973).
Brooks, P. W. *The modern airliner: its origins and development*. Putnam, London (1961).
— *Historic airships*. Hugh Evelyn, London (1973).
Davies, R. E. G. *A history of the world's airlines*. Oxford University Press, London (1964).
Davy, M. J. B. *Aeronautics: heavier-than-air aircraft*. HMSO, London (1929).
— *Aeronautics: lighter-than-air aircraft*. HMSO, London (1934).
— *Aeronautics: propulsion of aircraft*. HMSO, London (1930).
— *Interpretative history of flight*. HMSO, London (1937).
Dollfus, C. and Bouche, H. *Histoire de l'aéronautique*. L'illustration, Paris (1932).
Gibbs-Smith, C. H. *A history of flying*. Batsford, London (1953).
— *Aviation: an historical survey*. HMSO, London (1970).
— *The re-birth of European aviation*. HMSO, London (1974).

— and Brooks, P. W. *Flight through the ages*. Hart-Davis MacGibbon, London (1974).

Hodgson, J. E. *The history of aeronautics in Great Britain*. Oxford University Press, London (1924).

Joint Dod-Nasa-Dot study. *Research and development contributions to aviation progress* (RADCAP). Department of the Air Force, Washington, DC (1972).

Kelley, F. C. *The Wright brothers*. Harrap, London (1944).

Miller, R. and Sawers, D. *The technical development of modern aviation*. Routledge and Kegan Paul, London (1968).

Petit, E. *Histoire mondiale de l'aviation*. Hachette, Paris (1967).

Simonson, G. R. *The history of the American aircraft industry*. MIT Press, Cambridge, Mass. (1968).

Taylor, J. W. R. *A picture history of flight*. Hulton Press, London (1955).

Vivian, E. C. and Marsh, W. C. *A history of aeronautics*. Collins, London (1921).

Welch, A. and Welch, L. *The story of gliding*. John Murray, London (1965).

Ward, B. H. (ed.). *Flight – a pictorial history of aviation*. Year – the Annual Picture History, Los Angeles (1953).

23

NAVIGATION

In 1900 navigation meant, for all practical purposes, navigation at sea, and this had been brought to a high degree of precision. The basic requirement was to be able to measure latitude and longitude, since these are the parameters which determine a ship's position on the globe. Latitude could be measured by making astronomical observations with a sextant. The measurement of longitude depended, in effect, on being able to measure the time relative to a fixed meridian (usually that of Greenwich) at which the Sun passed through the meridian and this in turn depended on the availability of chronometers capable of keeping precise time for very long periods at sea, for only in port could they be precisely set. At the equator, an error of only four seconds in a chronometer corresponds with an error in the calculated position of one mile. In the absence of a chronometer – which was not standard equipment at the beginning of this century – longitude could be computed by the method of lunar distances, which involves observing the movement of the Moon relative to the stars and then making elaborate computations on the basis of tables given in the *Nautical Almanac*, first published in 1767.

With latitude and longitude known, the ship's position could be precisely located on the appropriate chart and the future course plotted. This course was steered by means of the magnetic compass. The use of iron and steel for constructing ships seriously affected the traditional compass, but in 1876 Sir William Thomson (Lord Kelvin) patented a greatly improved version. The compass bearing, however, indicated only the direction of the axis of the ship: the actual course would be affected by drift resulting from wind, currents, and tides, for which the navigator had to make allowance. It depended also on the ship's speed, traditionally measured by the taffrail log. This was essentially a board tied to a knotted cord in such a way that when thrown overboard it strongly resisted movement through the water. The ship's speed was measured by the number of knots that slipped through the observer's hand in a given period of time (usually 30 seconds). By 1900 this had been largely supplanted by propeller-actuated devices trailed astern. Steamships could make a fair estimate of their speed through the water by noting the engine revolutions.

Finally, it was necessary to measure the depth of water, especially when navigating in the presence of shoals. This was commonly done with a lead

4-inch compass and binnacle made to
the design of Lord Kelvin, 1876.

Sperry Mk. II marine gyro-compass.

and line worked from the bows. The bottom of the lead was usually concave and a little tallow was put in the depression; this brought up a small sample of the sea bottom, itself often a useful indication of position.

By 1900 these aids to navigation were so well developed that under favourable conditions any capable navigator knew his ship's position very accurately all the time. In practice, however, conditions were not always favourable; heavy cloud or fog, for example, might for long periods preclude the possibility of making the necessary observations. Moreover, the word 'capable' is advisedly used with reference to the navigator: precise observations and complicated computations were essential for success.

From the turn of the century, however, new methods began to supersede those which had, for the most part, been traditional for centuries. Some were mechanical, but most were based on some form of electrical phenomena.

Of mechanical devices, the most important was the gyro-compass, which depends on the fact that a spinning gyro-wheel will react to disturbance by precession and the reassumption of its original direction. This phenomenon had been used in the 1890s to maintain torpedoes on course, but in 1910 Hermann Anschütz-Kaempfe built a gyro-compass that was installed in a German battleship. In the same year, Elmer Sperry – an American engineer who had been experimenting with large gyro-stabilizers (p. 253) for ships – introduced an experimental gyro-compass in the USS *Delaware*. Within twenty years the gyro-compass had largely replaced the magnetic compass except in small vessels. By 1916 Anschütz-Kaempfe had successfully coupled the gyro-compass to the steering mechanism of a Danish ship and thus opened the way to automatic course-steering.

Early aircraft used magnetic compasses but, because of the dip of magnetic lines of force, these were reliable only when the aircraft was on a straight, level course. From the mid-1920s larger aircraft began to use gyro-compasses; magnetic compasses were also carried to enable the gyro-compasses to be re-set when conditions were favourable.

I. ELECTRICAL SYSTEMS

The greatest advance, however, stemmed from the advent of effective long-distance radio communication, marked by Guglielmo Marconi's trans-atlantic transmission in 1901. So far as ships at sea were concerned this had two major advantages. Firstly, it became possible to maintain constant and instant communication with shore bases and with other ships; this was particularly valuable in case of emergency and led to the introduction of the internationally recognized SOS distress signal (p. 311). More significant, however, was the fact that the Greenwich time-signals could be received anywhere, making it possible to correct ships' chronometers daily. From

1914 the *Nautical Almanac* appeared as the *Nautical Almanac Abridged for the Use of Seamen*; there was no longer any need to include the old tables of lunar distances used for computing longitude. In 1948 an international convention on safety at sea made it mandatory for all vessels of more than 500 tons to be fitted with radio, but this was really no more than a recognition of general practice.

It soon became apparent, however, that radio had much more to offer in the field of navigation. In particular, it was discovered that the bearing of a radio transmitter could be determined by using a loop-shaped aerial with the receiver. The intensity of the signal received varied according to whether the axis of the loop was in line with the transmitter or at right angles to it. In practice, as the earliest receiving sets had no amplification system, secondary coils were used to intensify the signals and a third rotating (gonio) coil to indicate the bearing of the transmitter. It was thus possible to set up radio beacons – akin to the old lighthouses – along the coasts to guide ships at sea. Equally, the position of ships sending out distress signals could be determined by taking cross-bearings from suitably placed receivers. By feeding the signals to a cathode-ray tube the compass bearing of a transmitter could be indicated by a 'needle' of light displayed on the screen.

Radio direction-finding solved the problem of both latitude and longitude. A ship's position could be pin-pointed without astronomical observations, and was as easily obtained in bad weather as in good, except in severe electrical storms which produced excessive 'atmospherics'.

About 1925 an alternative to the lead line was introduced; this was the echo-sounder. It is a sophisticated instrument, but simple in principle. A pulsed sound signal is generated beneath the hull of a ship and the signal reflected from the sea bed is detected by a receiver on board the ship. As the interval between the signal and reception of its echo depends on the depth of water, it is possible to obtain a continuous plot of the surface of the ocean bed beneath the vessel. In practice, the trace recorded is usually a little blurred, as the sea bed is rarely hard and clean; from the appearance of the trace an experienced operator can form an idea of the nature of the bottom. The echo-sounder derived from ASDIC (Anti-Submarine Detection Investigation Committee) devices developed in the latter part of the First World War to locate enemy submarines.

Navigation in ports and their approaches, where waters are narrow and traffic dense, presented particular problems. About 1930 experiments were carried out with what was called the leader system. This depended on the use of a submarine cable laid on the bottom of the channel: signals sent through it caused induced signals in receivers on board ships which enabled them to follow its course to their destination. In some cases two cables were

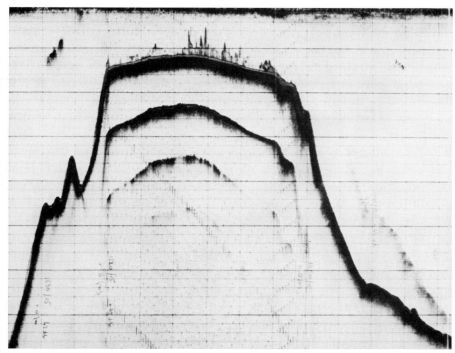

Echo-sounding trace; this example outlines the Gettysburg seamount in the Atlantic.

laid, for ingoing and outgoing shipping respectively. Although effective, this system was never widely adopted, perhaps because of its cost.

Quite early in the use of radio beacons it was noticed that observations were liable to distortion by flocks of birds or passing aircraft, and radar makes practical use of this interference. In 1924 Edward Appleton, using reflected radio waves, demonstrated the existence of a layer of ionized gas high in the atmosphere. In 1927 radiolocation methods had been used to locate thunderstorms, a hazard to the growing number of aviators. In principle, radar closely resembles sonar: a pulsed beam is emitted and reflections from a distant object are displayed on a cathode-ray tube, which can indicate both the bearing and the distance of the object. In practice, the development of this simple principle into effective instruments, especially ones light enough to be used in aircraft, presented great difficulties. In particular, only a minute fraction of the transmitted energy is reflected back and, because of the very high speed of light, 500 ft. of there-and-back distance corresponds to a time lapse of only a millionth of a second between signal and echo. Two related, but separate, lines were followed. On the one hand, radar was used to locate distant targets, such as enemy ships and aircraft; on the other, it was used for navigational purposes.

A first requirement for effective radar is an intense short-wave emission. Sharp images can be obtained only if the wavelength is relatively short in comparison with the target. An important development, therefore, was the magnetron, invented by Albert Hall of General Electric in 1921, which generated adequate power at a wavelength of less than 50 cm. It was developed in Britain, where subsequently the cavity magnetron was invented in 1939; this made possible high-power transmission in the centimetre/millimetre wavelength band. A simple antenna transmits equally in all directions and consequently the signal-strength in any given direction is relatively weak. A single aircraft at a distance of 150 miles from the transmitter reflects back to it less than a thousandth of a billionth of the output energy. An important factor in the utilization of very short wavelengths was, therefore, the development of focusing devices – parabolic or dished aerials – which concentrated the signal within a narrow beam. These aerials could be rotated to scan any given area, up to a full circle, much as a submarine commander scans the horizon by rotating his narrow-angle periscope.

However, this is getting a little ahead of events. In the 1930s, with war once again in prospect and the likelihood that aircraft would play a dominant role in any war, radio detection devices excited widespread interest. In the USA this interest arose from the radio exploration of the upper atmosphere by G. Breit and M. A. Tuve; in Germany Rudolf Kuhn demonstrated a

Prototype cavity magnetron, requiring to be made to very exact tolerance ($\pm 25 \mu$m).

simple radar device as early as 1933. Although the Germans had developed an effective early-warning system by 1939, their effort in this field was not strong, partly because of inadequate organization of the necessary co-operation between specialists and partly because they persisted in using wavelengths that were too long. This in turn may have reflected the German belief, held at top governmental level, that the war would be short and sharp: it would be over before radar was needed for military purposes.

In the event, Britain proved the pioneers, and radar was a vital factor in the ultimate success of the Allies in what turned out to be not a blitzkrieg but a long war of attrition. By 1935 Britain had five operational radar stations and within a few years fifteen more had been added. The name of Robert Watson-Watt is particularly identified with the development of radar, and the importance of his contribution was recognized after the war with an award of £50 000 by the Commission on Awards to Inventors, the largest it made.

The origin of the word radar is interesting. In February 1935, Watson-Watt submitted to the UK Radio Research Board a remarkable memorandum in which he set out very clearly the basic principles of detecting aircraft by means of radio. It included the wavelength that should be used; the importance of cathode-ray oscilloscopes as detectors; the need for a central control room at which the course of an aircraft could be plotted; and the need for means of discriminating between hostile and friendly aircraft. During that year, at a remote airfield at Orfordness, detection was demonstrated at distances up to $19\frac{1}{2}$ miles and further development was authorized. For reasons of secrecy the work was said to relate to ionospheric research and was referred to as RDF (radio direction finding). Ironically, this deceit became truth. At first it was thought that a single station could determine only the distance of an aircraft and not its direction; quite soon, however, RDF meant what it said: the range, direction, and height could all be measured from a single station. Nevertheless, the term RDF continued in use up to 1944, when it was agreed that the American term radar (RAdio Detection And Ranging) should be adopted internationally.

Early radar equipment was heavy and could be used only at ground stations or on ships at sea. The advent of the cavity magnetron made airborne radar a possibility. In 1942 it was used for antisubmarine work by Allied coastal patrol aircraft and a year later was introduced into night fighters and bombers. For the latter it could give an indication not only of enemy aircraft but also of the nature of territory beneath the aircraft (H_2S system): water was non-reflecting, and so appeared dark, whereas buildings reflect light and so built-up areas appear as bright patches. As equipment and techniques improved, ground features were revealed in considerable

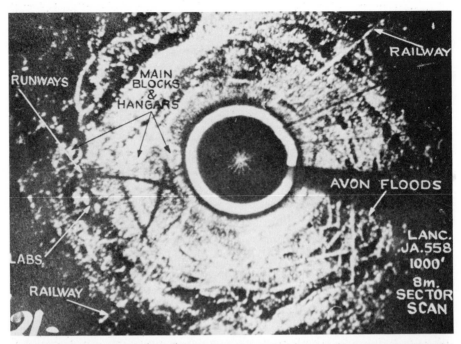

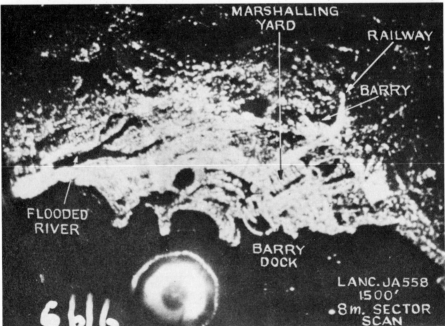

Early results obtained with K-band H$_2$S radar equipment (December 1944): (*Top left*) Defford Aerodrome with adjacent railway lines and flood water. (*Top right*) Part of north coast of Bristol Channel. (*Bottom*) Docks and marshalling yard at Barry, near Cardiff.

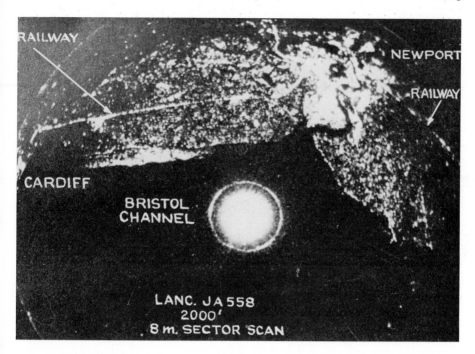

RAILWAY

NEWPORT

RAILWAY

CARDIFF

BRISTOL
CHANNEL

LANC. JA 558
2000'
8 m. SECTOR SCAN

detail. This made it very much easier to identify targets at night, even through thick cloud.

Radar was a valuable aid to aerial navigation, but alternative methods – based on the radio beacons already described – were developed. In the USA a whole network of radio beacons was set up in 1928 which was rather more sophisticated than its predecessors. At each station two transmitting loops were installed, one of which transmitted A in Morse code (dot/dash) and the other N (dash/dot). An aircraft approaching on correct course would pick up a steady note resulting from an equal combination of these two; any deviation would cause the note to change. Each station had its own call-sign and frequency, so that they could be unequivocally identified. This radio range system was adopted also in Australia and Canada but was less popular in Europe.

Three important wartime navigational developments were Gee, Loran, and Oboe. The first of these worked on the 3–15 metre waveband and depended on measuring the time intervals between the reception of signals from three ground stations. This enabled a high-flying aircraft to fix its position to within one mile at ranges up to 300 miles. Loran, an American development, was based on co-ordinating signals from two fixed stations using different pulse-rates. As its name implies, it was used for long-range work, mostly by ships, up to 750 miles. The range could be doubled, though with reduced accuracy, by using waves reflected from the lower layers of the ionosphere. Oboe was used to guide bombers on to their targets by means of signals sent from the aircraft to two fixed ground stations. Observers there could locate the aircraft, and give the signal for the release of bombs, with an accuracy of 50 yards at ranges of 200 miles.

III. POST-WAR DEVELOPMENTS

The great military importance of radar resulted in intensive research during the war years, and in the five years between the end of hostilities and mid-century the main development was the application of existing techniques to civil needs.

At sea, radar became an invaluable aid in the prevention of collision with other vessels, icebergs, exposed rocks, and similar obstacles; in negotiating harbour entrances; and in fixing position precisely.

In civil aviation radar continued to be invaluable in navigation during flight and in traffic control at airports, which became increasingly complex as the density of traffic increased. The bigger airports began to use ground-control approach systems by which pilots could, even in bad visibility, be talked down until the runway lights became visible to them. As early as 1928 the Royal Aircraft Establishment, Farnborough, experimented with an automatic pilot, linking the gyro-compass to the controls. They achieved an accuracy of 1 mile at a range of 200 miles. During the Second World War the Germans were successful with automatic navigation in their V-1 rockets, and in 1947 a US aircraft flew the Atlantic under automatic control throughout, including take-off and landing.

BIBLIOGRAPHY

Anderson, E. W. *Principles of air navigation*. Methuen, London (1951).

Hitchens, H. C. and May, W. E. *From lodestone to gyro-compass*. Hutchinson, London (1952).

Hughes, T. P. *Elmer Sperry: inventor and engineer*. Johns Hopkins Press, Baltimore (1971).

International Hydrographic Review. *Radio aids to marine navigation*. Monaco (1952).

Ratcliffe, J. A. Robert Alexander Watson-Watt. In *Biographical Memoirs of Fellows of the Royal Society*, vol. 21 (1975).

Richardson, K. I. *The gyroscope applied*. Hutchinson, London (1954).

Sonnenburg, G. J. *Radar and electronic navigation*. Newnes, London (1952).

Sperry, E. Aerial navigation over water. *Transactions of the Society of Automotive Engineers*, **12**, 153, 1917.

Watson-Watt, R. A. *Three steps to victory*. Odhams, London (1957).

Weems, P. H. V. *Air navigation*. Van Nostrand, New York (1942).

— and Lee, C. V. *Marine navigation*. Van Nostrand, New York (1952).

24

PRINTING AND PAPER

The second half of this century saw revolutionary changes in printing methods, but up to 1950 techniques were very largely developments of ones used in the latter part of the nineteenth century. Photographic methods were widely used for the reproduction of illustrations, but text was still almost wholly derived from metal type. This is not to say, however, that there were not substantial improvements in typesetting methods and in printing. In general, the main emphasis was on quantity rather than quality. A growing, and more literate, population stimulated demand for popular newspapers and magazines and cheap books. These last included not only reprints of the classics but also a wealth of new titles specifically designed for a large and relatively undiscriminating readership. There was, too, a substantial increase in demand for educational books of all kinds. This development was dependent on the availability of large supplies of cheap paper, an aspect which we will consider later.

I. TYPESETTING AND PRINTING

In 1900 three principal methods of printing were firmly established. Letterpress printing from raised inked surfaces, using cast metal type set by hand, was a direct descendant of the original method used by such pioneers as Gutenberg and Caxton. Intaglio printing is the reverse of this; the matter to be printed is cut into a plate – originally copper, later steel – which is inked and then wiped clean, leaving ink only in the depressions, from which it can be printed by applying it under pressure to paper. The 1890s saw an important development known as the half-tone process. In effect, this was a means of converting photographs – by a further photographic process – into closely packed dots on a metal plate: the darker the part of the photograph being reproduced, the denser the dots and the more ink they carried. The half-tone plate was made by interpolating a rectangularly ruled glass screen: the finer the screen, the more faithful the reproduction. The importance of the half-tone process was that it was purely mechanical, needing no artistic skill. The *Daily Mirror* (London, 1904) was the first newspaper to be illustrated entirely by this photo-mechanical process.

Lithographic processes involved printing from stone, as the name implies. They depend upon the fact that if a design is drawn on a smooth stone – usually a finely grained limestone – with a greasy pencil, the areas so treated

Making photographs for half-tone printing process, *c*.1905.

repel water. If the stone is thoroughly damped, only the untreated parts absorb water. Consequently, if a thick ink is applied by roller to the stone it will adhere only to those parts corresponding to the original design, and thus a replica of the latter can be transferred from the stone to paper. If a succession of stones were prepared, and treated with differently coloured inks, fine colour reproductions could be made (chromolithography). In its original form lithography was a slow manual process, but in the 1870s flatbed printing machines based on the manual process were introduced. In the next ten years zinc (later aluminium) plates, which could be curved to fit a roller, were introduced. Before the end of the century offset lithography was in use. In this the design is transferred from the plate to the paper by intermediary (offset) impression on a rubber roller. The elasticity of this roller permits fine reproduction on relatively coarse papers that are unsuitable for direct printing from a rigid metal surface. In the early years of this century rotary offset litho printing machines were introduced, notably by the firms of Harris in the USA, George Mann in Britain, and Vomag in Germany. In 1912 Vomag introduced an offset litho press, fed from a continuous roll of paper, capable of printing 7500 sheets per hour. Although the term lithography was retained, printing stones were by then no longer used except for artistic work.

In 1923 the Rotaprint offset litho machine was introduced in Germany, followed by the Multilith in the USA, primarily for use as an office duplicator and for other small jobs. After the Second World War, however, these machines were developed for more substantial jobs, using basic texts produced by special typewriters, and began to rival conventional jobbing machines.

In the 1920s, also, the diazo (dye-line) printing process was introduced for making multiple copies of small originals. Translucent photographic copies were first made and then printed, using ultraviolet light, on to paper coated with an emulsion containing a diazo compound (a chemical intermediary much used in the dyestuffs industry) and a sensitizer. The print was developed by treatment with ammonia.

Most printed matter – that is to say, words rather than pictures – was printed from metal type, and there were important developments in the way in which the type was set. Up to the First World War – and to a diminishing extent after it – some setting was done directly by hand, virtually as in Gutenberg's day; that is to say, by selecting individual letters from a subdivided tray and setting them in lines in a frame. By the beginning of this century, however, this traditional method was being replaced by faster mechanical ones. Two main kinds of machine were in use, the Linotype and the Monotype. Both cast type as required from molten metal – a mixture of lead, tin, antimony, and copper.

Monotype keyboards, early 1920s.

As its name implies, the Linotype set a complete 'line of type' in one operation. It did so by drawing the required letters from a magazine, using a keyboard, and then presenting them to the orifice of a mould into which molten type metal was injected. The result was a bar of metal, with the letters imprinted on it in reverse, ready to be put into the frame for printing. Arrangements had to be made, of course, to adjust the spacing so that the line was 'justified'; that is to say, there was no empty space at either end.

The name Monotype, too, indicates the mode of operation. By this machine letters are cast singly and assembled into lines. The caster is controlled by jets of compressed air regulated by a perforated ribbon – rather like that of a pianola – generated by the keyboard.

The Linotype machine was introduced in 1886, for the *New York Tribune*, but in 1900 an improved version appeared. The Monotype first came on the market in 1894; an improved model was introduced in London in 1900 by Cassell & Co., the publishers. The Intertype machine of 1912 was essentially a variant of the Linotype; it was first used for the New York *Journal of Commerce*. In theory, the Linotype or Intertype machines are the fastest, capable of setting 12–14 lines per minute, though in practice operators achieved little more than half this. Monotype machines set about 5 lines per minute. The Ludlow typesetter resembled the Linotype in producing a line of type but the individual characters were assembled by hand for casting. It was normally used for casting large sizes of display type.

As the casting process was controlled by a keyboard there was in principle no reason why keyboard and caster should form part of the same machine. Teletypesetters were proposed as early as 1905 but not until 1928 was a satisfactory machine, the Fairchild, demonstrated in New York. Teletypesetters were installed in the House of Commons to send Parliamentary reports to *The Times* newspaper. The telegraphic signals were controlled by punched tape which could be prepared in advance of transmission and, if desired, filed for future use.

Although there had long been an interest in photographic methods of composition, real success was not achieved until 1949, when the Fotosetter was introduced by the Intertype Company. In this, the desired succession of letters is transferred, via a camera, from a master collection. The print size can be adjusted by varying the photographic magnification. Printing plates are then prepared from the film for the press. The British Monophoto used perforated tape from Monotype machines to bring the sheet of film into the correct position for the desired character to be photographed on to it.

Two principal forms of press remained in use: flat-bed and rotary. In the flat-bed press the prepared frames, each corresponding to one page of the finished product, are positioned in a chase on a flat sliding bed. Plates for

Intertype Fotosetter, first installed
commercially in 1949.

providing illustrations are locked in position with the type. For convenience
in the subsequent folding and cutting of the printed sheet for binding, the
frames are normally laid in multiples of four. The prepared type is inked,
the paper is positioned on top of it; the paper is gripped by the roller, which
carries the sheet round with it and brings it in contact with the inked type as
the frame moves underneath; and finally the printed sheets are drawn away
from the machine for folding, trimming, binding, etc. Normally, presses of
this type are horizontal but in 1920 the Miehle Company of Milwaukee
introduced a compact vertical form.

Flat-bed machines were widely used for relatively small jobs, and had
much to commend them. They were, for example, very flexible and lent
themselves to the use of various sizes of paper; they were suitable for long and
short runs; and they were cheap to install. They were, however, relatively
slow, printing around 2000 sheets per minute, though later versions of the
Miehle vertical press could print up to 5000.

Such speeds were not adequate for long runs of popular newspapers and
periodicals which had to be completed quickly. For these, rotary presses
were used in which the inked type was carried on the circumference of a
rotating cylinder. They thus differed from flat-bed machines in which the

cylinder served to press the paper against the type. The initial stages of preparing this cylindrical sheet of type were similar to those for flat-bed work. When the type is firmly locked in the chase, however, a flat matrix (mat) of papier mâché is laid on top of it and a mould of the type is taken by applying pressure. From this a stereotype plate is cast in type metal to fit the curvature of the cylinder of the machine. At the beginning of the century a wet paste was used to make the mould, and this had then to be dried. Shortly afterwards, however, a dry sheet made from paper pulp – invented by H. W. Wood in 1900 – began to be used.

Apart from purely mechanical improvements allowing faster working, the speed of such presses was increased by incorporating several printing units in one machine. One of the leading manufacturers was R. Hoe and Company, founded in the USA early in the nineteenth century. In the 1890s they produced in succession double, quadruple, and sextuple machines. In 1902 the Hoe Octuple could print an eight-page newspaper at the rate of nearly 100 000 per hour. In 1909 there were 2600 daily newspapers in the USA alone. By 1925 a 24-cylinder machine was available. Machines with such outputs required a continuous supply of paper rather than the single large sheets which were used for most – but not all – flat-bed machines. They were fed from large rolls of newsprint (p. 296) and provision was made for splicing a fresh roll on to the end of an exhausted one so that the machines need not be stopped.

II. THE TYPEWRITER

Although printing from movable type dates – at least in Europe – from the fifteenth century, the nature of the process was such that it was suitable only when a considerable number of copies were required. Up to the third quarter of the nineteenth century, correspondence, legal documents, etc. were still written by hand; a copy could be made by using damp tissue and a handpress. Although mechanical writing machines were proposed at least as early as 1714, the first typewriter in the modern sense was that invented by an American printer, C. L. Sholes, in 1867 and manufactured at the Remington factory – already making firearms and sewing-machines – in New York from 1873. After a slow start, the innovation, which came at a time when the increasing size of businesses called for more systematic correspondence and record-keeping, was widely adopted. By 1900 at least forty makes of typewriter were on the market, and they opened up a completely new field of employment for women. Thus, in 1881 there were in England and Wales only 7000 women clerks; thirty years later there were 146 000.

By 1900 the main features of the modern typewriter, including the inked

Remington No. 10 typewriter, 1907.

Specialist typewriter for musical work – the 'Musik'.

IBM Executive typewriter 1959.

ribbon, had appeared, but there was still much scope for improvement. Two major disadvantages needed to be overcome. Firstly, only one typeface could be used and every letter had to be assigned the same space; unless the ribbon was kept in good condition, and the keys were evenly struck, the appearance of the finished page was poor. Typewritten pages could be reproduced by lithography but the quality did not compare with conventional printing. Secondly, the lines could not be 'justified'; this again adversely affected the appearance of the page.

In W. A. Burt's typographer, patented in 1829, the type characters were mounted on a semicircular band of metal, and some of its immediate successors used wheels or sleeves. This system permitted the type-face to be changed easily, but by 1900 it had been replaced almost universally by type-bar machines in which the characters are individually mounted at the end of bars, which strike the paper when the appropriate key is struck. The familiar QWERTY keyboard was designed to minimize the consecutive striking of adjacent characters, with consequent risk of the bars sticking. As already mentioned, this had the disadvantage of ruling out an easy change of type, but in 1933 the VariTyper went back to the old system of mounting the type on a semicircular metal plate that could be easily changed. Four years later a line-justifying device was incorporated, but not until 1947 was the problem of differential spacing for individual characters solved. This made it possible to prepare high-quality typewritten copy acceptable for offset printing.

Electric typewriters were proposed in the 1870s but not until 1923 were they first marketed, by the North Eastern Appliances Co. This business eventually merged with International Business Machines (IBM) which systematically attacked the market. In 1946 they produced the IBM Executive, with proportional spacing of letters. This led to the Selectric model of 1961, incorporating the now familiar 'golf ball' head, an ingenious and effective solution to the problem of changing the type-face quickly.

III. PAPER

As we have already noted, the output of the printing industry expanded enormously during the first half of this century, particularly at the cheaper end of the mass market. This was possible only by the introduction of bigger and faster machines, but it depended also on the availability of correspondingly larger quantities of paper, and here, too, the emphasis was on quantity and cheapness rather than quality.

Traditionally, paper was made from rags or textile waste, as a certain amount of high-quality paper still is. The sorted rag is macerated and boiled to a pulp with pressurized steam. The bleached and washed pulp is then spread evenly on rectangular sieves, allowed to drain, and finally pressed and dried.

By the middle of the nineteenth century, however, the supply of rags was proving inadequate and increasing use was being made of esparto grass, native to the Mediterranean region; the Spanish product was generally of the highest quality. This yielded a high-quality paper entirely satisfactory for the printing processes of the day. By 1890 between two and three hundred thousand tons of esparto were being imported into Britain alone. It was introduced by George Routledge, who pioneered cheap book production with his shilling Railway Library (1848) which eventually included 1000 titles, of which *Uncle Tom's Cabin* alone sold half a million copies.

Although esparto was still in use in the 1950s it had by then long been insufficient to meet the needs of the popular press, and by the end of the nineteenth century extensive use was being made of wood-pulp as a raw material; by 1950 80–90 per cent of all paper was made from this. Initially the wood was simply ground with water in a mill to separate the fibres, and paper was made from it in much the same way as from rag pulp. The product is, however, of poor quality as regards appearance, strength, and durability.

Better results were obtained with the use of chemical pulp in which wood chips are boiled with soda, sulphite, or sulphate. This process removes nearly all the lignin, in contrast to mechanical pulp which may contain as much as 95 per cent of the original wood supplied to the mills. These processes were in use in 1900, and the next important development was the introduction of various bleaching processes to give a good white product. Another important development, effected by Kamyr AB in Sweden (1938–44), was to transform pulp manufacture from a batch process into a continuous one. At that time a batch digester could convert 1000 cubic feet of wood into 10 tons of pulp within twenty-four hours; the continuous digester nearly doubled this output.

Pulp production demands good supplies of suitable timber and plenty of water. It demands also facilities for dealing with large quantities of noxious waste, a problem that became more serious as the protection of the environment became a subject of increasing concern. In the twentieth century the main producers were Canada, Sweden, and Finland, with the newspaper industry by far their biggest customer. In 1946 these countries together produced approximately 5 million tons of chemical pulp and 4 million tons of mechanical pulp. The USA was the biggest producer of chemical pulp (7 million tons), and the paper industry was among the ten largest in the country, but much of the paper was made from imported pulp wood. Apart from its use in printing, much paper – mostly coarser grades – is absorbed by the packing industry. The division is by no means sharp, however, as much packaging paper and card is printed with brand names, directions for use, and so on.

In the twentieth century paper was still made by a process developed by the Fourdriniers, members of a Huguenot family which came to England, via Holland, after the revocation of the Edict of Nantes in 1685. Early in the nineteenth century Henry and Sealy Fourdrinier became associated with Bryan Donkin, an engineer who had already built a successful paper-making machine and was interested also in printing machinery. After heavy initial losses, they developed the continuous paper-making machine that was still widely used a century later. In this, dilute pulp is fed on to an endless moving belt of wire mesh, on which water is drained and sucked out. Further water is then pressed out by rollers, and the resulting felted web is finally dried on heated rollers and wound on to a reel.

Although the principle of the Fourdrinier machine remained unchanged, there were important practical improvements, mainly designed to speed up the process and provide wider and longer rolls of paper. In 1950 the largest machines of this kind could produce a roll 25 feet wide at the rate of 14 miles per hour; in 1900 the standard width was 8 foot. The main factor making such speeds possible was the introduction of anti-friction bearings, in place of sleeve bearings, in 1922.

The chief rival to the Fourdrinier machine was the cylinder machine developed by John Dickinson in 1812. In this the early stages were similar,

Continuous paper-making machine, c.1900.

but the web was built up on a cylinder mould covered with wire mesh. This type of machine was used mainly for the production of board, rather than paper, and by the end of the century multi-cylinder machines were being made in the USA. Such machines made possible not only variations in thickness and quality, but also provision of board differently coloured on the two sides.

Paper is commonly subjected to various finishing processes. As it comes from the machine it is very absorbent, like blotting-paper, and must be sized to make it ink-proof. This is done by incorporating resin or gelatine. The former can be incorporated in the original pulp but this treatment is not appropriate for gelatine. Gelatine sizing, used mostly for good quality hand-made papers, is effected by passing the finished paper through a suitable bath (tub-sizing). For certain printing processes, especially colour printing, a smooth watery paste of china clay (up to 30 per cent) is applied to the surface of the finished paper.

Good surface finish is achieved by calendering, which involves passing the paper through stacks of chilled iron rollers at the end of the machine.

BIBLIOGRAPHY

Beeching, W. A. *Century of the typewriter*. Heinemann, London (1974).
Clapperton, R. H. *The paper machine: its invention, evolution, and development*.
    Pergamon, London (1967).

Lilien, O. M. *History of industrial gravure printing up to 1920*. Lund Humphries, London (1972).

Moran, James. *Printing processes*. Faber and Faber, London (1973).

—*Printing in the 20th century*. Northwood Publications, London (1974).

Shorter, A. H. *Paper making in the British Isles. An historic and geographical survey*. David and Charles, Newton Abbot (1971).

25

ELECTRICAL COMMUNICATION

In 1900 the application of electricity to communication was no novelty. Not only had the industrialized nations developed extensive internal telegraph networks but reliable international systems also existed. The first successful submarine telegraph cable, linking Britain and France, was laid in 1851; only seven years later the Atlantic, too, had been bridged. Various forms of automatic telegraph, notably that of Charles Wheatstone (1858), had greatly increased the rate of transmission. As early as 1880, W. H. Preece, chief engineer of the Post Office in Britain, reported that 5000 Wheatstone instruments were operating, at the rate of 180–190 words per minute; the annual volume of traffic was equivalent to 15 million columns of *The Times*. Overall, cheap and rapid telegraphic communication had been established between and within the principal countries of the world.

The position with regard to the telephone was rather different. Although earlier work had been done, the first effective telephone was that of Alexander Graham Bell, patented in the USA – only a few hours before a similar application by Elisha Gray – and Britain in 1876. By that time, as we have noted, the telegraph was firmly established and represented heavy capital investment and an important source of revenue. Understandably, there were doubts about the viability of this new and so far untested rival. There were, too, legal complications. In many countries the telegraph service was a government monopoly and the telephone was generally regarded as being in effect an extension of the telegraph. Consequently, private telephone operators could normally operate only under licence. However, licences were generally issued fairly readily to suitable applicants, though some countries, such as Germany, inaugurated the telephone service as a state monopoly (1877). As long as the role of the telephone was uncertain, many governments were happy to leave the business to private concerns; in the twentieth century these were normally absorbed into the state systems as their licences expired. The early private companies established in the latter years of the nineteenth century were mostly small and local: it is estimated that in 1880 there were no more than 30 000 telephones in the USA and only 5000 in Europe. Although the telephone had then arrived in the technological sense, it differed from the telegraph in having yet to make its social impact. By 1900 there were over a million telephones in use in the USA, but this represented less than two for every 100 people. After America, the Scandinavian countries were among the earliest to greet the telephone with enthusiasm.

Meanwhile, however, a rival to both was on the point of emerging. In the last quarter of the nineteenth century one of the great problems in physics was whether forces act at a distance without need for any intermediate matter, or whether action between bodies depends on changes in an all-pervasive aethereal medium. On the whole, Continental physicists favoured the former view while those of Britain accepted the latter, which was developed mathematically by James Clerk Maxwell in his classic *Electricity and Magnetism* of 1873. A corollary of Maxwell's theory was that there should exist electric waves having properties similar to those of light. In the 1880s Heinrich Hertz, working at Kiel, began to seek experimental proof of the existence of such electric waves. He achieved this and succeeded in both generating electric waves from the sparks of an induction coil – with wavelengths of about 24 cm – and detecting them at a distance of some 60 feet. He showed, too, that they travelled with a velocity similar to that of light (186 000 miles per second) and, like light, could be reflected and refracted. Roughly concurrently with Hertz, Oliver Lodge – then professor of physics in Liverpool – also began experimenting with the generation of electromagnetic waves. Hertz's interest was scientific and philosophical, but Lodge perceived that these waves could be put to practical use in telegraphy. This possibility he demonstrated at a meeting of the British Association for the Advancement of Science in 1894. To detect the signals, Lodge used a coherer, a device depending on the fact that under the influence of electromagnetic radiation the electrical resistance of a metallic powder (originally iron, later nickel and silver) changes. However, the French physicist E. Branly can also lay claim to the invention of this important device, about 1890/1. In Russia, A. S. Popov was working on similar lines and in 1895 demonstrated the generation and reception of electromagnetic waves over a distance of 80 metres.

Thus at the turn of the century the experimental verification of purely theoretical hypotheses had led to several practical demonstrations of the feasibility of wireless communication. In the event, however, the development of radio as a practical method of long-distance communication was not effected by these pioneers but by a newcomer, Guglielmo Marconi. Son of a wealthy Italian landowner and of an Irish mother, he had little formal education save some special instruction in physics by A. Righi at Bologna. In 1894, at the age of twenty, he read an account of Hertz's experiments and began to repeat and extend them at his family estate. Within a year he had sent and received signals over a distance of more than a mile and conceived the idea of pulsing them so that messages could be transmitted in the Morse code used for telegraphy. As he was unable to interest the Italian government in his work, he moved to London, where he could get advice on procedure from some of his mother's family. He also got collaboration from Captain (later Admiral) H. B. Jackson, who later introduced wireless telegraphy into the Royal Navy. In June 1896 he took out his first UK patent for a system of wireless telegraphy based on Hertzian principles but using much longer – 300–3000 metres – wavelengths. Once launched he made rapid progress. By 1899 he had realized the importance of resonance and began to use a simple tuning device which he called a jigger. His inventive genius was matched by his business acumen. In 1897 The Wireless Telegraph and Signal Company was formed in London; in 1900 this became the Marconi's Wireless Telegraph Co., with worldwide affiliations. The range of reception was rapidly increased – mainly by the use of larger antennae set at a height – and on 12 December 1901 he scored a spectacular success with the first transatlantic transmission from Cornwall to Massachusetts. This was a sequel to bridging the English Channel in March 1899. From this time on he devoted his time increasingly to the management of his business affairs, recognizing that this must include the ability to attract first-class technical men. In the first decade of the century he established a commanding position in Britain and Canada. In 1909, aged only thirty-five, he shared the Nobel Prize for physics with Ferdinand Braun, another important pioneer of wireless telegraphy.

In the field of equipment manufacture, Marconi soon encountered foreign opposition, as, for example, from the German firm Telefunken (1903). He therefore adopted a policy of not selling equipment to all comers but of building up his own system of land-based transmitters; installing his own equipment and operators on board ships; and forbidding communication with any operators except those within his own system. As we shall see, however, he was not able to maintain this position.

Thus in 1900 the principal forms of electrical communication had been established but were in very difficult states of development. The electric telegraph was already commonplace and an accepted amenity. In urban areas the telephone was beginning to be used on the basis of exchanges – including some automatic ones serving a number of subscribers and the beginnings of an international network – but it was yet to make a major social impact. Radio, except to the more imaginative of those directly concerned, was no more than an untried novelty. Electronic engineering was in its infancy, and television, its most vigorous offspring, was as yet undreamed of.

I. THE TELEPHONE

The principle of the telephone is very simple. If sound waves fall on a thin diaphragm the latter vibrates in tune with them. These vibrations can be translated into electric impulses which set up corresponding vibrations in the diaphragm of a receiver at the other end of the line, and these in turn set up sound waves matching those falling on the transmitter. There are, of course, major practical difficulties. In particular, the wave pattern of the human voice is very complex and the ear is very sensitive; for a verbal message to be even comprehensible, let alone acceptable as a public utility service, the correspondence between transmitter and receiver must be very close. Moreover, even the best electrical conductors have some resistance, which causes the signal strength to fall as distance increases. For long-distance work periodic boosting is necessary; this also is a potential source of cumulative signal distortion. Again, currents flowing in telephone cables are susceptible to outside influences – such as currents flowing in adjacent telegraph cables – and these, too, may affect the quality of reproduction.

In the earliest Bell telephones the transmitter and the receiver were in fact identical. Speech vibrations caused electrical variations in a coil wound round a magnet set close behind the diaphragm, and this process was reversed at the other end. While this remained the pattern for receivers, Edison developed an improved form of transmitter in 1878 which was still widely used in the 1950s. In this, the microphone is packed with carbon granules and the vibration of the diaphragm affects the packing of these and thus the electrical conductivity of the carbon as a whole. Here the voice does not generate the electrical transmission but modulates current supplied to the microphone. This permits a more powerful signal.

The basis of a telephone system is that any subscriber can call any other subscriber. This requires some signal – usually a bell or a light – at the receiving end so that the subscriber there knows that he is being called. For a very small number of subscribers using a shared line some simple code could indicate who was being called, but if the number increased beyond a quite low threshold some sort of exchange was necessary, at which an operator could be notified of a subscriber's requirement and make the necessary connection to the circuit of the subscriber being called. The first such exchange to be worked commercially was probably that opened in January 1878 at New Haven, Connecticut; it enabled any pair from eight subscribers to be connected. The first exchange in Britain – at Coleman Street, in London – opened a little over a year later. Most early exchanges were based on the cross-bar system (1880) employing a board with an array of vertical and horizontal conductor bars; one vertical bar was assigned to each subscriber. The connection was made by the operator inserting a plug

Instrument supplied by
National Telephone Co.,
c.1910.

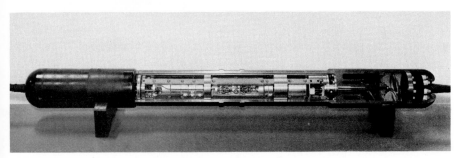

Submarine repeater for Atlantic telephone cable, 1956.

to complete the circuit at the appropriate intersection. At that time, and for a long time to come, there was no problem in engaging operators at economic rates. Manual exchanges were, therefore, widely used until well into the second half of the twentieth century. Nevertheless, as early as the 1890s A. B. Strowger, in the USA, developed an automatic step-by-step switching system in which two-motion switches are activated by dialled-in signals. Dialling of a series of numbers, introduced in 1896, actuated a series of switches in turn until the desired circuit was completed – hence the name step-by-step. The switchgear is first driven vertically to the appropriate height and then a feeler arm is rotated horizontally to engage with an appropriate terminal. In America, the first exchange operated on the Strowger system was opened in 1897, and Amsterdam had a similar

Strowger automatic telephone switch 1897/8.

Strowger automatic telephone, c.1904.

Multiple-Type switchboard (U.K. Post Office Type CB1) 1925.

exchange a year later. However, it was 1912 before the first automatic exchange was opened in Britain (Epsom), and such exchanges were not widely used anywhere until after the First World War.

By this time a new source of revenue and an extension of the service had appeared. Coin-operated public call-boxes enabled subscribers to use the service when away from their homes or places of business. They also made the service available to those who were unable or unwilling to afford a permanent installation.

The telegraph had already demonstrated that information could be carried for long distances by means of electrical impulses in conducting wires, and in theory there was no reason why long-distance telephony should not be equally feasible. There were, however, considerable practical difficulties, of which telegraphy itself was one. As a matter of convenience, because they were usually in common ownership, telephone and telegraph wires were commonly carried on the same poles or in the same conduits. In such circumstances electromagnetic induction between the two systems effectively prevented telephony over more than about thirty miles. This could be overcome by using twisted twin-cable conductors for the telephone, instead of the single wire with earth return. By 1900 this device was generally used, though in Britain the Post Office had used it from the very beginning of its service in 1881. It was found, too, that copper wires were greatly superior to iron. In Belgium, F. von Rysselberghe, introduced in 1882 'chokes' in telegraph circuits to reduce the induction effects to an acceptable level; this led to the development of means of carrying telephone messages on telegraph wires instead of having a dual system. Since an extensive international telegraphic system already existed, this development immediately opened up the possibility of long-distance telephone networks extending to nearly 20 000 miles. In the USA separate systems developed in the east and the west of the country, based on heavy-duty copper conductors to improve transmission strength and quality. Not until 1915 was a transcontinental line inaugurated, from New York to San Francisco. Rather surprisingly at so late a date, for valve amplifiers (p. 315) had been available for several years, signals on this line were at first boosted by electro-mechanical 'repeaters'. Soon, however, the service was much improved with the replacement of these by valve amplifiers.

The big expansion of telephone services, particularly cheap long-distance calls, came in the 1930s and resulted mainly from the application of new techniques developed primarily for radio. In particular, the fdm (frequency division multiplex) system, by which incoming signals were allocated to a range of frequency bands, made it possible for a single coaxial cable to carry a number of different conversations simultaneously. The foundation of this system had, in fact, been laid by M. Leblanc and M. Hutin in France around the turn of the century, but it was some thirty years before practical use was made of it. A major contribution was the invention of negative feedback by H. S. Black in America in 1927. By feeding back some of the output signal to the input, interaction between different channels is reduced and higher amplification is possible without undue fluctuation in signal strength. Initially, long-distance telephony had been dependent on the incorporation of loading coils to increase the inductance, but by 1950 these had virtually disappeared in favour of valve amplifiers.

As we have already noted, automatic exchanges actuated by alphabetical/numerical codes dialled by the subscriber were introduced at an early date. Nevertheless, they were rather slowly adopted for general use, partly because the Strowger switch was a somewhat elaborate and expensive device. The line switch, in which a rotating arm searches a bank of subscriber-line terminals, was much cheaper.

By mid-century the world's telephone system was very large and the volume of traffic had led to widespread automation. Two main systems were in use. In the satellite system a large central exchange was linked with a number of smaller local ones which handled all calls in their area without involving the central exchange at all. The latter was called into operation only when one subscriber needed to contact another on a different satellite exchange. In densely populated areas, such as New York or London, the so-called director system was employed. The first three digits of a seven-digit number were indicative of a particular local exchange and, when dialled, directed the subscriber to that exchange; dialling the remaining four digits triggered selector switches which searched out the line of the subscriber being called.

This left the problem of subscribers wishing to call another area or to make an international call. Until mid-century this still involved the personal intervention of an operator who – with the increase in the number of channels made possible by fdm – could usually dial the main exchange desired without delay. In the 1950s, however, subscriber trunk dialling (STD) began to be introduced, enabling a caller to dial direct to any other subscriber in his own country or, in due course, abroad.

The growth of all public utilities – gas, electricity, water—is naturally limited by the extent of the distribution system, and the availability of the commodity offered. The three examples mentioned above differ from the telephone service in that new customers have merely to be connected to the nearest point on the supply system. New telephone subscribers, by contrast, have to be connected by twin cables all the way to their local exchange, where there has to be sufficient switchboard gear to deal with the additional load. This requirement restricted the rate of growth, especially under wartime conditions when there was a shortage of electrical equipment of all kinds. Nevertheless, the over-all growth of the world telephone system in the first half of this century was remarkable. Thus the USA had 17 million telephones in 1934 and 32 million in 1947; the corresponding figures for Britain are 2 million and 4 million, showing a comparable rate of growth. The world total in 1934 was around 33 million; by 1976 it was to be 380 million. In terms of social impact, however, the number of telephones per 100 of population is more instructive than the total. Table 25.1 gives statistics for some of the principal countries.

Table 25.1.   Telephones per 100 population

Country	1934	1947
United States	13	23
Canada	11	16
New Zealand	10	16
Denmark	10	14
Sweden	9	18
Switzerland	9	16
Australia	7	11
Norway	7	11
United Kingdom	5	9
Spain	1	2
Russia	0.5	1

Within countries there were substantial variations. Thus in 1947, in the United Kingdom, the figure was 19 for the greater London area but only 7.6 in Glasgow. Similarly, American figures ranged from 36 (Los Angeles, Chicago) to 24 (Baltimore).

II. WIRELESS

The early history of radio communication up to Marconi's trans-atlantic transmission in 1901 has already been briefly told. From that point on, progress was very rapid, but two separate though related threads have to be followed. The most important, in the present context, is the technological thread – improvements in transmission and reception. This cannot be separated, however, from consideration of the uses to which wireless was put and of the methods introduced to regulate it in the interests of all concerned.

On the technological side the most important advance was the thermionic valve, with the development of which the names of J. A. Fleming and Lee de Forest are particularly associated. Though neither can lay claim to the discovery of the underlying principles, both belonged to that relatively small group of inventors who can perceive the immense practical importance of what others see only as an academic advance.

Fleming thermionic diode, 1905.

De Forest 'audion' valve.

Fleming, the first professor of electrical engineering at University College London, was appointed consultant to the Edison Electric Light Company in 1882, and in this capacity visited Edison in the USA in 1884 and there had demonstrated to him what is called the Edison effect: if a conducting plate is placed in an electric lamp bulb and connected to the positive side of the filament, a current flows from the filament to the plate. No current flows, however, when the plate is connected to the negative side. On this basis Fleming constructed the first diode valve in 1904, so called because it restricts the flow of electricity to one direction, just as a valve controls the flow of water or air in a pipe.

In 1906 Lee de Forest, a pioneer of radio broadcasting in the USA, applied for a patent for a three-electrode (triode) valve which could be used to amplify weak electrode currents and (according to an extension of the patent made a few months later) also act as a detector. This was an important innovation but unfortunately gave rise to prolonged litigation. Fleming claimed that the triode was dependent on his diode; de Forest claimed to have been ignorant of Fleming's patent when applying for his own. The courts originally upheld Fleming, but more than forty years later found in favour of de Forest. For de Forest the triode was essentially a working device; interpretation of the physical principles owes much to Irving Langmuir of General Electric and E. H. Armstrong of Columbia University.

Whatever the facts may be as regards priority, the triode represented a major advance, as measured by the fact that at mid-century some 200 million thermionic valves – many of a highly sophisticated nature – were being made annually. For amplification, the triode valve (known briefly as an audion) could be used in series to magnify the effect by a cascade system. Moreover, when combined with an oscillator (Alexander Meissner, 1913) it could be used to generate powerful radio signals. In 1919 W. G. Housekeeper patented a new method of sealing base metals through glass. This allowed the anodes of large valves to be water-cooled and opened the way to very high-power transmitters.

At about this time the heterodyne circuit, invented by R. A. Fessenden, in the USA, came into use. Up to 1912 the main function of a receiver was to switch a direct current on or off in response to a received signal. The heterodyne system combined the received signal with a local wave generated at the receiving station. In 1920 the more sophisticated super-heterodyne technique was patented by Armstrong.

Until the end of the First World War it was generally considered that the lower wavelength limit for commercial transmission was 200 metres, and shorter wavelengths were assigned to amateurs in 1918. They quite quickly

Burncleft's 'Ethyodyne' superheterodyne receiver and horn loudspeaker, 1925.

showed that short wavelengths had unsuspected possibilities and in December 1921 they achieved transatlantic transmission and reception at 200 metres. More strikingly, a young Dutch engineer, L. J. W. van Boetzelaar – working under very primitive conditions in Hilversum – succeeded in establishing loud and clear communication with Malabar, India, on 26 metres.

Temporarily, however, we must go back on our tracks, for the assignment of wavelength is an important matter to which we have so far made only passing reference. The spark-gap transmitters used in the earliest days of wireless radiated waves covering a large bandwidth, and a single transmitter could virtually monopolize a large area. Nevertheless, they were very easily operated and were not formally prohibited by international agreement until 1930. With the advent of tuning devices, however, it was possible to assign to any given transmitter a restricted bandwidth, leaving other bands free for other operators. Such an assignment was possible only by international agreement, and this was difficult to reach in the face of the restrictive policy of Marconi's company. The first international radio-telegraph conference met in Berlin 1903; it comprised seven nations including the USA, Germany, the United Kingdom, and Italy. It achieved little beyond drafting a convention, which Britain and Italy refused to sign, declaring that all land stations should be bound to receive and dispatch telegrams for ships at sea – the application with which radio-telegraphy was then most concerned – regardless of what radio system they were using. More important, it paved the way for the First International Radiotelegraph Conference (Berlin, 1906), which twenty-nine nations attended. The main issue again was to break 'Marconism' and make communication independent of the source of the equipment used. Some progress was made: coastal stations were required to conform to the rules of the international telegraph service, distress signals were given priority, and interference – sometimes indulged in maliciously – was to be avoided as far as possible. Specific wavebands were assigned to various services, and the Bureau of the International Telegraph Union (1865) at Berne was authorized to undertake administrative responsibility. Incidentally, this Conference adopted SOS as the universal telegraphic distress signal – May day (from the French *m'aidez*) for spoken calls – though it was some years before this wholly displaced the earlier CQD.

The next Conference was convened in London in 1912, and met in a different atmosphere from its predecessor. The sinking of the *Titanic* on her maiden voyage only three months earlier, with the loss of 1513 lives, was in everybody's mind. It stressed the importance of all sea-going vessels being equipped with radio. Some ships within radio range were not so equipped and knew nothing of the disaster; others, notably the *California*, had tried to give warning of ice but had been warned off the air by the *Titanic* herself. Also in the minds of those present was the collision in thick fog between the American *Republic* and the Italian *Florida* in 1909. Wireless signals from the *Republic* brought a rescue vessel to the scene within half an hour and the entire ships' companies – 1700 in all – were saved. The dramatic arrest in 1909 of the notorious murderer Crippen and his mistress, as a result of a radio message from the captain of the transatlantic liner in which they were fleeing to Canada, had attracted enormous publicity. Thanks to wireless, almost the entire world, save for the two fugitives themselves, knew the reception awaiting them. It was in this atmosphere that agreement was reached that radio communication should not be restricted by the nature of the equipment used; Marconi, the main offender, had in fact abandoned the practice some months earlier. This was a major step forward; radio technology was at last in tune with an international code of practice.

The First World War gave a powerful stimulus to the development of wireless because of its military significance (Ch. 29). This led, in the inter-war years, to the development of radar (Ch. 23) and the application of radio in aeronautics (Ch. 22), including its use for air-traffic control. These were all important lines of application, but, so far as the general public was concerned, the most striking development in wireless communication was the advent of public broadcasting services. Beginning modestly in the 1920s, they grew within a decade into an international service, providing entertainment, news and weather bulletins, and educational programmes for a rapidly increasing number of homes in the western world. By mid-century radio was virtually universal – with the transistor an important factor in reducing the cost and size of receivers – and it had been joined by a powerful newcomer, television. In 1950 the production and transmission of radio and television programmes, and the manufacture of receivers, was an enormous and expanding industry and one which did not exist, and was not even foreseen, at the beginning of the century.

The growth of radio amateurs, and the allocation of short-wave bands to them after the First World War, has already been mentioned. Their numbers were swelled by the many who had become proficient in wireless telegraphy through their training with the armed forces. By that time they were increasingly transmitting and receiving in sound rather than in Morse code, using headphones to pick up the signals. Very soon, however, more powerful transmitters and amplifiers made it possible to dispense with headphones and use loudspeakers. The earliest receivers were hand-built by enthusiasts – who often made some of the simpler components themselves – but gradually manufacturers entered the field, offering sets that could be

Two-valve amplifier, Mk. III, for military use, 1917.

operated by those with no technical knowledge. At first they were decidedly utilitarian in appearance, but by the 1930s the working parts were enclosed in decorative cabinets leaving only the controls outside. By then they mostly operated from mains electricity, instead of the batteries of early sets, but a good outdoor aerial was desirable for satisfactory reception.

Public broadcasting began almost simultaneously in the USA and Europe, but developed on rather different lines. In the USA transmitters had to be licensed although there was no power to refuse a licence. In Europe there was official opposition to broadcasting because of fears that it would interfere with vital communications; when this resistance was broken down, it remained policy in many countries to make broadcasting a government monopoly in some form or other. Probably the first broadcast as part of an advertised regular service was that given on 2 November 1920 by the Westinghouse Station KDKA in Pittsburgh. This was followed by a literal rash of new broadcasting stations in the USA, many of them very short-lived. In the last quarter of 1921, 21 were licensed; in the first half of 1922 the figure was 264. The number of receivers rose correspondingly: 50 000 in March 1921, 750 000 in May 1922. This movement was strongly promoted by the manufacturers of radio receiving sets.

In Europe the pattern was similar. In France, broadcasts from the Eiffel Tower began in February 1922; the first programme of the British

Broadcasting Corporation was made in November 1922; German broadcasting began in October 1923. Growth was rapid: over a million radio licences were operative in Britain in 1925 and half a million in Germany.

Broadcasting ran at once into the same difficulties as wireless telegraphy: in the absence of agreement about the allocation of wavelengths confusion ruled the air and there was serious interference between stations. Worse, broadcast programmes could interfere with important radio-telegraphic services affecting safety. This problem led to the formation of the International Broadcasting Union in 1925 which by 1939 had obtained a fair measure of agreement with regard to long and medium waves, the bands most directly concerned. This agreement was generally observed by the belligerents in the Second World War, and discussions were resumed shortly after its conclusion. By that time a new factor – frequency modulation (FM) – had appeared, which greatly enlarged the scope for allocation. In this the signal is conveyed by altering the frequency rather than the amplitude of the carrier wave. FM transmission has various advantages, including less liability to interference between stations and from static and lower power for a given signal strength. Although FM broadcasting lies outside the period of our present concern, its theoretical possibility had been demonstrated as early as 1922 by J. R. Carson, and Armstrong had shown its feasibility in 1933.

If the first quarter of the twentieth century can be generally designated the age of wireless telegraphy, the second can be regarded as the age of broadcasting. The growth of broadcasting and the advancement of radio engineering went hand in hand. The huge sums of money involved, and the fierce competition between manufacturers led to greatly increased interest in research and development both within the industry and in the universities. From this emerged a new profession, that of radio engineer.

The origin and importance of the thermionic valve in both transmission and reception have already been noted. Up to the end of the First World War the triode was the only valve in general use, but in the inter-war years more sophisticated types were developed, mainly to reduce coupling between electrodes. As early as 1915, Walter Schottky in Germany had suggested a four-electrode tube (tetrode) to improve amplification, and A. W. Hull of General Electric suggested that coupling could be reduced if a screen was interposed between the anode and the control grid. The first practical version of the screen-grid valve was due, however, to H. J. Round (1926) in Britain, and it was subsequently developed in a variety of forms. Suppression of secondary emission in a tetrode is difficult, especially at high voltages, but a solution to this was found in the pentode, developed by Philips in Holland in 1928. In this a suppressor grid is introduced between

'Pam' radio receiver, 1956. This was the first transistor model
to be made in Britain.

the screen grid and the anode, and maintained at the filament potential. During
this period valves varied enormously in size and complexity, ranging from
the giant valves used in powerful broadcasting stations to the miniature
valves used for portable radios.

In very early receivers crystal devices were used as valves. If a fine wire
(cat's whisker) is pressed lightly on the surface of semi-conducting material
(such as the mineral galena, lead sulphide), current will flow in only one
direction; that is, it acts as a valve. Such devices were abandoned with the
advent of thermionic valves but they were taken up again, in a more
sophisticated form, to meet the demands of radar and computers. The new
devices were based on silicon crystals and tungsten or molybdenum wire.

It seemed clear that, if a third electrode could be added to such solid-state
devices, a new type of amplifier would be possible, and J. Lillienfeld, for
example, patented such a device in 1928. Practical development did not
occur until after the Second World War, however, when a research team at
the Bell Telephone Laboratories studied the problem systematically; its
members included J. Bardeen, W. H. Brattain, and W. Shockley. Their
work led in 1950 to the development of the transistor, a miniature device
which was to have enormous consequences for electronic engineering in the
second half of this century.

Another important innovation, particularly in respect of the introduction of miniature components, was the printed circuit. Up to the Second World War circuits were built up from insulated wire and soldered joints. During the war, printed circuits began to be used, which adopted the techniques of the graphic arts to electronics. In effect, the dried ink was an electrical conductor and the circuit was printed on to an insulator. This facilitated mass production, saved weight and space, and increased reliability by eliminating or reducing the number of soldered joints. One of the first applications was in the proximity fuse (p. 377); this involved a circuit printed on ceramic and was developed in 1940 by Globe-Union for the National Bureau of Standards. Printed circuits were also developed in Britain by P. Eisler at about the same time. In 1947 the ECME (electronic circuit making equipment) process was introduced by J. A. Sargrove in Britain. This was designed for the mass production of radio receivers. It involved stencilling all the basic elements of a circuit on to a plastic base by spraying with zinc.

The small complex circuits that were developed in the 1940s were vulnerable to shock. This was overcome by the introduction of so-called potted circuits in which a whole unit was encased in a rigid block of cold-polymerized resin.

III. TELEVISION

In a sense, television lies largely outside the scope of the present work, for its social impact was minimal before 1950; its growth in the last thirty years has been phenomenal. Nevertheless, it cannot be ignored because not only were there some important practical achievements in the inter-war years but some relevant work was done even before 1900. Two distinct lines of development were pursued, the photo-mechanical and the electronic. In the event, the former petered out in the 1930s and the latter proved to be the main stream of evolution.

As long ago as 1884 the German physicist Paul Nipkow proposed an electronic television system based on the rotation of polarized light in a magnetic field. He also devised the Nipkow disc, a mechanical scanning device with small perforations to partition the picture. In effect, the picture was broken down into a pattern of spots of varying intensity and reassembled by the receiver; there is an analogy here with the half-tone printing plate. In 1907 Boris Rosing, in St. Petersburg, suggested that the cathode-ray oscillograph – invented by Braun in 1897 – could be used to convert electrical signals into visual patterns. A year later A. A. Campbell Swinton in Britain proposed an electron-beam scanner. Thus it is not unreasonable to say that the history of modern television began a little before the start of this century.

Nevertheless, the first to achieve practical television, and organize a regular public transmission service, was John Logie Baird. In 1924 he succeeded in transmitting a picture over a distance of a few feet using a Nipkow disc with a photo-electric cell placed behind it. In 1926 he gave a public demonstration in London and a year later he televised pictures between Glasgow and London – a distance of about 700 km – using a telephone cable. He then founded the Baird Television Development Company, which had a considerable, but short-lived, success. In 1928 Baird spanned the Atlantic, and a year later the newly founded British Broadcasting Corporation began an experimental programme. In 1932 viewers in London were able to see the finish of the Derby horse-race run at Epsom, Surrey. He gave demonstrations in Berlin, Paris, and Stockholm.

Up to 1932 the BBC programme was provided by Baird's company, but from then until 1937 it was the responsibility of the BBC. In 1935 a commission set up to consider the future of television in Britain recommended adoption of the rival all-electronic system. For a further two years the BBC transmitted programmes using both systems – which did not encourage viewers to invest in either – but finally abandoned Baird's. Sets cost half the price of a small car and no more than 2000 were sold by the end of 1936.

Baird has been criticized for seeking to develop a photomechanical system that was outmoded before it was launched. With the benefit of hindsight this

Baird's experimental television transmitter 1925/6.

is true, but we must view his work – for which he never got any recognition – in the context of its day. His system was launched in the 1930s, the heyday of the cinema. The latter, too, was based on a photomechanical system, but there was no doubt about its success and the immense fortunes being made out of it. Nor was his approach unique: other workers – notably C. F. Jenkins in the USA and D. von Mihaly in Hungary – were working with mechanical scanners in the 1920s.

In 1908 Campbell Swinton made proposals for a television system in which cathode-ray tubes were used for both transmission and reception and he elaborated these ideas over the next twelve years. He envisaged the image being scanned in $\frac{1}{25}$ th second and resolved into 400 000 points of light which would then be reproduced on the receiving screen. However, the first to test these ideas experimentally seems to have been Vladimir Zworykin, a Russian electrical engineer who joined Westinghouse in 1920. At that time Westinghouse were interested primarily in radio and Zworykin had little opportunity to develop his television work. He resigned, but shortly returned under terms which allowed him to retain rights to inventions he had made up to 1919, with Westinghouse having the option to purchase them later. In 1929 he became director of research (and later vice-president) in the Radio Corporation of America (RCA). Here he received greater encouragement and by 1938 had developed the iconoscope, the first effective television camera. This contained a screen coated with tiny droplets of an alloy of silver and caesium. As a light beam systematically scans the screen, each droplet emits electrons in proportion to the intensity of the light.

At about the same time Philo Farnsworth, an independent inventor working in California, developed (1927) a television system based on what he called an image dissector tube. Difficulties arose through patent conflict but were eventually resolved.

Electronic television became a practical possibility shortly before the Second World War. The BBC began a regular service – two hours a day, six days a week – on 2 November 1936. RCA began experimental transmission from the Empire State Building in the same year; their first public service was 30 April 1939, when Roosevelt was shown in the opening ceremony of the World Fair. The outbreak of war delayed development of the service on both sides of the Atlantic; even in 1942 there were no more than 10 000 sets in the USA and perhaps twice this number in Britain. After the war there was very rapid progress, but television for mass audiences is essentially a feature of the second half of this century. In 1947 there were still only 34 000 television sets in use in Britain; by 1953 the number had jumped to $2\frac{1}{2}$ million and an estimated 20 million people watched the Coronation service.

Bush 9-inch television receiver, 1950.

IV. SOUND REPRODUCTION

Mechanical sound recording was well established in 1900 and continued in general use up to the Second World War; it is still in vogue among a few enthusiasts. Long before that, however, electronic devices were in use for both recording and reproduction. In considering these, two main lines of development must be followed: sound recording for playback to individuals or groups, as in dance-halls; and sound recording as an accompaniment to motion pictures, which became general from the 1920s. This second development is most conveniently considered later, in the context of cinematography (Ch. 26).

As has constantly been apparent in earlier chapters, few inventions have clear-cut origins, and Edison's phonograph of 1877 was not quite the first step in sound recording, though it may be regarded as such for present purposes. The phonograph embodied a stylus attached to a diaphragm. The stylus cut a continuous helical groove in a rotating cylinder, the depth of the groove corresponding to the vibration of the diaphragm; the same instrument was used to play back the recording. This so-called hill-and-dale system was short-lived at that time though it was reverted to later. In 1887 E. Berliner in the USA (who coined the word gramophone) introduced the disc system in which a spiral groove is cut from the circumference inwards (in a few early ones, from the centre outwards) and the undulations were horizontal. This only slowly displaced the original cylinder-type recordings. As late as 1908 the Sears Roebuck catalogue, for example, listed both cylinder and disc records and machines to play them on.

Parmeko disc recording machine, *c*.1932.

Edison's phonograph, 1877.

At this stage, recording and reproduction became separated. Recording was first done on a wax disc which was then copper-plated. The copper plate was attached to a heavy backing to form the master matrix, from which any number of copies could be made with a thermoplastic resin. These flat records, easily stored, were then played back on the purchasers' own machines.

The electrification of the gramophone took two forms. Firstly, electric motors were used in place of clockwork ones, which required frequent winding and were liable to slow down disconcertingly before the record was finished. Secondly, valve amplifiers – introduced about 1926 – made it unncessary to depend on sheer volume of sound to make a satisfactory recording; at the same time greater sensitivity to frequency was achieved. Similar developments took place in reproduction. In the mechanical system, a steel (later fibre) needle set up vibration in a diaphragm in tune with the undulations of the track on the record. Commonly a horn was used to amplify the sound. Later, electrical pick-ups converted the vibrations of the stylus into signals which were amplified and used to activate a moving-coil loudspeaker. The weight of the stylus, which causes wear of the record and restricts response to variations in the groove, was steadily reduced, from about 3 ounces in the 1920s to 1 ounce in the late 1930s. By 1950 the crystal head commonly weighed only a few grams.

An inherent disadvantage of the gramophone record is that the playing time is relatively short. There are two ways of extending it: by making the grooves closer together or by reducing the speed of rotation. Immediately after the Second World War the long-playing (LP) record was launched; it had a speed of $33\frac{1}{3}$ revolutions per minute compared with the traditional 78, and a playing time of about 20 minutes. Records of 45 rpm were also introduced, with a playing time, on a 7-inch disc, of about 8 minutes.

Finally, we must mention a quite different system of recording and reproduction. As long ago as 1898 the Danish electrical engineer Valdemar Poulsen demonstrated that sound could be recorded by variations in the magnetization of a wire, but it was not until after the Second World War that magnetic recording became of commercial importance. During that war, German technologists developed $\frac{1}{4}$-inch recording tapes consisting of a flexible base coated with a fine-grain magnetic oxide. This became generally available about 1948 and was particularly important for sound recording in the motion-picture industry. It was also widely used in office tape-recorders, which had become a normal feature of commercial life, because of the ease with which the message could be erased, enabling the same tape to be used repeatedly. Later, tape cassettes began to replace gramophone discs.

German military tape recorder used in the Second World War.

A defect inherent in all the sound reproduction systems considered here is that natural hearing integrates signals received by our two ears, whereas reproduced sound falls on a single receiving microphone. While experiments on stereophonic reproduction systems were carried out in the early 1930s by A. D. Blumlein in Britain and in the Bell Telephone Laboratories in the USA, stereophonic reproduction was not commercially successful until after 1950.

BIBLIOGRAPHY

Appleyard, R. *Pioneers of electrical communications*. Macmillan, London (1930).

Baker, W. J. *A history of the Marconi Company*. Methuen, London (1970).

Bowers, Brian. *Sir Charles Wheatstone*. HMSO, London (1975).

Braun, Ernest and MacDonald, Stuart. *Revolution in miniature; the history and impact of semiconductor electronics*. Cambridge University Press (1978).

Briggs, Asa. *The History of Broadcasting in the United Kingdom*. 3 vols. Oxford University Press, London (1961–70).

Dibner, B. *The Atlantic cable*. Burndy Library, Norwalk, Conn. (1959).

Dummer, G. W. A. *Electronic inventions and discoveries 1745–1976*. (2nd edn.) Pergamon, Oxford (1978).

Dunlap, O. E., Jr. *Radio's 100 men of science: biographical narratives of pathfinders in electronics and television*. Harper, New York and London (1944).

— *Marconi, the man and his wireless*. Macmillan, London (1937).

Fahie, J. J. *A history of wireless telegraphy 1888–1899*. Blackwood, Edinburgh and London (1899).

Fessenden, H. M. *Fessenden – builder of tomorrow*. Coward-McCann, New York (1940).

Fleming, A. *Memories of a scientific life*. Marshall, London (1934).
— *Fifty years of electricity*. Wireless Press, London (1921).
Geddes, K. *Guglielmo Marconi, 1874–1937*. HMSO, London (1974).
Hancock, H. E. *Wireless at sea: the first fifty years*. Marconi International Marine Company, London (1950).
Harlow, A. F. *Old wires and new waves: the history of the telegraph, telephone, and wireless*. D. Appleton-Century, New York and London (1936).
Hilliard, J. K. The history of stereophonic sound reproduction. *Proceedings of the Institute of Electrical Engineers*, May 1962, p. 776.
Hunt, F. V. *Electroacoustics*. Wiley, New York (1954).
Institute of Radio Engineers. 50th anniversary number of *Proceedings*. London (1962).
Institution of Electrical Engineers. *Thermionic valves 1904–1954*. London (1955).
Jolly, W. P. *Marconi: a biography*. Constable, London (1972).
MacLaurin, W. R. *Invention and innovation in the radio industry*. Macmillan, New York (1949).
Mance, O. *International telecommunications*. Oxford University Press, London (1943).
Marconi, Degna. *My father Marconi*. McGraw-Hill, New York (1962).
Marland, E. A. *Early electrical communication*. Abelard-Schuman, London (1964).
Michaelis, A. R. *From semaphore to satellite* (published on centenary of the International Telecommunication Union). ITU, Geneva (1965).
Moore, C. E. and Spencer, K. J. *Electronics – a bibliographical guide*. MacDonald, London (1965).
Pawley, E. L. *BBC Engineering, 1922–1972*. BBC, London (1972).
Pocock, R. F. and Garratt, G. R. M. *The origins of maritime radio*. HMSO, London (1972).
Rhodes, F. L. *Beginnings of telephony*. Harper, New York (1929).
Robertson, J. *The story of the telephone: a history of the telecommunications industry of Britain*. Pitman, London (1947).
Robinson, H. *The British Post Office: a history*. Princeton University Press (1948).
Rolo, C. F. *Radio goes to war*. Faber and Faber, London (1943).
Ruhmer, E. *Wireless telephony* (trans. from German). Crosby Lockwood, London (1908).
Shiers, G. *Bibliography of the history of electronics*. Scarecrow Press, Metuchen, New Jersey (1972).
Shridharani, K. *Story of the Indian telegraph: a century of progress*. Post and Telegraphs Dept., New Delhi (1953).
Snel, D. A. *Magnetic sound recording*. Philips, Eindhoven (1958).
Sturmey, S. G. *The economic development of radio*. Duckworth, London (1958).
Tiltman, R. F. *Baird of television – the life story of John Logie Baird*. Selley Service, London (1933).
Transistors – the first 25 years. Special volume of *Radio and Electronic Engineer* (1973).
Tricker, R. A. R. *Early electrodynamics*. Pergamon, Oxford (1965).
Tyne, Gerald F. J. *Saga of the vacuum tube*. Indianapolis, (1977).

26

PHOTOGRAPHY AND CINEMATOGRAPHY

The latter part of the nineteenth century saw not only the basic principles of photography firmly established but also the process itself so greatly simplified that amateurs with no technical knowledge, and not much money to spend, could take photographs of outdoor subjects with a high probability of success. The turning-point in this respect was George Eastman's Kodak camera of 1888. This was a simple mass-produced hand-camera loaded with a length of roll-film sufficient to take 100 pictures. When all the film had been exposed the camera was sent back to the firm and returned reloaded, together with prints from the original film. 'You press the button – we do the rest' was Eastman's slogan. In essence, this is what the majority of amateur photographers do today, save that the film alone is returned and not the whole camera. By 1900 one person in ten in the USA and Britain owned a camera, though photography was less popular on the Continent. Professionals used more elaborate equipment and normally did their own processing, but the processes employed were basically the same. Cinematography, too, had arrived by 1900 though it was still very primitive; the first public showing of a motion picture to a paying audience took place in Paris in 1895, with an attendance of 33 people. From this insignificant beginning grew a vast new industry that reached its peak with the lavish films of the 1920s and 1930s, thereafter declining after the war with the growing popularity of television.

Though events in the first half of this century were thus an extrapolation of what had already been achieved, there were nevertheless some very important developments. The performance of cameras and films improved enormously. Colour processes for professional use appeared before the First World War and were greatly improved in the inter-war years; by mid-century colour processes had been so greatly simplified that they were poised to capture the amateur market. Finally, in 1948, Edwin H. Land launched his Polaroid camera that gave an instant picture.

Although still photography and cinematography have much in common, it is convenient to discuss them separately. One important bridge was the 35-millimetre film stock that is common to cinematography and the miniature camera that became popular with the launching of the Leica at the Leipzig Fair in 1925. We will, therefore, begin with the evolution of still photography.

I. THE CAMERA

Almost throughout the first half of this century the professional photographer relied heavily, both for studio and for some outdoor work, on the traditional camera with bellows extension; rising and cross front; ground-glass focusing screen; and interchangeable holders for glass photographic plates. Even up to the Second World War wood and brass remained the main constructional materials, as in the German Kühn-Stegemann camera produced from 1925 to 1940. The post-war Swiss Sinar camera was similar in design but used light alloy in place of wood. In view of the great progress that had been made in the design and construction of smaller hand-cameras this now seems decidedly old fashioned, but the fault lay with publishers rather than photographers. For reproduction in magazines and elsewhere very high quality prints were essential, and it is true that enlargements from 35-millimetre and other small stock sizes were not initially good enough. But the publishers insisted on 8 × 10 inch contact prints long after this technical difficulty had been overcome. *Life*, for example, insisted on 8 × 10 prints until they were forced to accept enlargements because of wartime problems in the 1940s; the fashion

Until the Second World War photographic plates remained popular even among amateurs. The manufacture of wooden dark slides – here seen at Houghton's Ensign Works – was an important industry (1907).

Kühn-Stegemann camera, 1925.

Sinar camera, embodying optical bench system, 1948.

The Leica I, first of the miniature cameras, 1925.

Carlton twin-lens reflex camera, 1885. Such cameras did not become popular until the 1930s.

magazine *Vogue* held out for ten years longer. As a rigid policy this was folly, for the smaller camera greatly increased the opportunities of the press photographer and made it possible to obtain pictures unattainable with more cumbersome equipment.

For the amateur, the situation was quite otherwise. While he was by no means indifferent to quality, he rated convenience very highly and favoured the small camera that he could easily carry around with him. As long as contact prints were the rule rather than the exception, the size of camera was largely dictated by the size of the picture desired. As early as 1903 a so-called vest-pocket camera was marketed in France taking a $4\frac{1}{2} \times 6$ cm picture, about the smallest thought tolerable. Up to 1930, however, most cameras took bigger pictures than this, say 6×9 cm. Box cameras were simple and therefore cheap; after the First World War most took roll-film, but plenty – like the Primus – were still in use which took a magazine of glass plates that had to be loaded and unloaded in the dark. Compactness was achieved with the advent of the folding camera in which the front dropped down, taking the lens and bellows with it. The first of the true miniature cameras, made with the precision of a scientific instrument, was the Leica (1925). This was the beginning of a long line, of which one of the best known is the Zeiss Ikon, first marketed in 1932. Germany led the world in the manufacture of these high-precision miniatures, but after the war lost much of the market to Japan. For those requiring a sub-miniature camera the Minox was marketed in 1937; it took fifty 8×11 mm pictures on special film.

All cameras must have a viewfinder to show the photographer what image is falling on his film. In the cheaper cameras this was done by reflecting the picture on to a tiny ground-glass screen by means of a lens and a mirror set at $45°$, or just by two simple frames arranged something like the sights of a gun. These were not very accurate, especially for close-up work when the difference between the optical axis of the viewfinder and that of the camera lenses was important because of parallax. For more expensive cameras more sophisticated optical viewfinders were developed. The reflex camera – which was experimented with, but not widely adopted, in the 1880s – represented a different approach to this problem. The twin-lens reflex (such as the Rolleiflex, 1929) was essentially a double camera, in which the upper lens reflected the picture by a fixed mirror on to a ground-glass screen which the photographer could watch up to the instant of making the exposure; this reflected picture was identical with that projected on to the film. In the single-lens reflex, which is more compact, the mirror is hinged and moved out of the way when the exposure is made. Among the best known of early models were the Kine Exacta (1936) and the Reflex Korelle (1936).

The lens is, of course, a principal feature of the camera. For cheap cameras a simple meniscus lens, giving a clear image from a few feet to infinity, sufficed; the aperture was so arranged that the exposure was about right for open-air work on a bright day using a popular brand of film. Simple refinements were an extra lens that could be pushed into place for portrait work and an extra stop that could be used to adjust the exposure for dull days.

For more ambitious work, much more elaborate lens systems were required. For example, if work is to be done in dull light, the light-gathering capacity of the lens must be increased; that is to say, its maximum aperture must be large relative to its focal length. The design and manufacture of such lenses is exceedingly complex, as two major defects must be avoided. Firstly, spherical aberration must be avoided; that is, straight lines in the object being photographed must be rendered as straight lines in the image, even at the edges. Secondly, chromatic aberration must be avoided. The refractive index of light depends on its colour, and with a simple lens the blue image will not be found at exactly the same distance from the lens as the red one, resulting in slight – but perceptible – blurring in the photograph. To obviate these defects a lens must be built up from a number of very precisely ground components, made of glass having different optical properties. This in itself raises problems, for light is lost by reflection at the interfaces between these components. From about 1950 this source of loss, which may be as high as 50 per cent, was much reduced by 'blooming' lenses with a very thin film of magnesium fluoride.

Just before the beginning of the century, good anastigmatic lenses worked at an aperture of around f 6.8, but the Tessar lens of 1902 was designed for a maximum aperture of 4.5, and the Leica Elmax (1925) operated at f 3.5. By the 1930s operating apertures of f 2.8 were not uncommon, with the Ernostar down to f 1.8. These maximum apertures were, of course, not always required, and an iris diaphragm was incorporated so that smaller apertures could be used as desired. For large apertures very precise focusing was necessary; this was achieved by screwing the lens mount in and out against a graduated scale. Depth of focus was difficult to achieve at large apertures – though advantage was sometimes taken of this for artistic purposes – and the smaller apertures were, therefore, in frequent use. As the light-gathering power of a lens depends on the square of its aperture, a f 2.8 lens is 16 times faster than an f 11 one.

With miniature cameras different lenses could be fitted to suit the work in hand. In the 1930s the zoom lens had been developed for cine cameras so that close-up and distant views could be taken without changing lenses, and in 1958 Voigtlander developed this for hand-cameras.

Correct exposure depends on the total quantum of light falling on the sensitive photographic film. Given a fixed intensity of light, a large aperture calls for a short exposure, and vice versa. It is, therefore, necessary to incorporate in the camera some device which will strictly control the exposure time. In the early days of photography, with relatively insensitive film and small-aperture lenses, the exposure was measured in seconds and the photographer simply removed a cap from the lens and then replaced it. By the beginning of the century, however, much shorter exposures were called for, down to as little as $\frac{1}{100}$th second, and quite complex mechanisms were necessary to achieve this. For consistent results there were two basic requirements. Firstly, the timing had to be accurate and remain so after repeated use. Secondly, the opening and closing of the shutter had to be so contrived that the exposure time was the same for the whole of the film – otherwise the photograph would be bright at the centre and dull at the edges, or vice versa.

At first, mechanical shutters were normally placed in front of the lens – perhaps because this is where the lens cap used to be – but, with the growing use of compound lenses, it became usual to place them between the components. One of the best and most successful of early shutters was the Compur, made by Deckel of Munich, which appeared in 1912. Initially its fastest speed was $\frac{1}{250}$th second but by 1935 this had been reduced to $\frac{1}{500}$th second. For rather cheaper cameras the Prontor shutter ($\frac{1}{200}$th second) was commonly used.

If even shorter exposures were required – as in press work in which fast-moving objects had, as it were, to be frozen – cameras were fitted with focal-plane shutters. In these, a spring-loaded roller blind was passed directly in front of the film or plate; exposures as short as $\frac{1}{1000}$th second were attainable. It was, in fact, a type of shutter that had been experimented with as early as the 1860s, but it was not until about 1900, and then mainly with press photographers, that it became popular.

An alternative way of controlling exposure was to operate in a dull light keeping the shutter open, and to illuminate the subject with a flash of light of controlled duration and intensity. Electric light was used from a very early date, and flash powders – magnesium mixed with sodium chlorate – from the 1880s. Magnesium ribbon, too, was used and had the advantage that exposure could be measured in terms of the length of ribbon burnt. The electrically operated flash bulb dates from 1929.

II. FILMSTOCK AND PAPER

In principle, the photographic process is very simple. An emulsion containing silver salts is coated on a transparent base, originally glass, but increasingly, as we have noted, celluloid or similar flexible film that can be rolled up. Celluloid is dangerously inflammable and was a particular hazard in the cinema industry. In the 1930s it was replaced by the much safer cellulose acetate. When the camera lens projects a picture on to such a film a 'latent' image is formed and this can be 'developed' by means of certain chemical agents. Development consists in the reduction of colourless silver salts to opaque particles of silver. The more intensely the film has been illuminated, the more silver is formed; thus bright parts of the picture appear dark in the developed film and vice versa. To obtain a picture corresponding to the original the process must be reversed; this involves placing the transparent negative against paper coated with a light-sensitive silver film and exposing it to light. The great advantage of the negative is that an unlimited number of positive prints can be made from it.

The rendering of coloured objects or scenes satisfactorily in black-and-white presents difficulties, as every artist knows. In the case of photography the difficulties are aggravated by the fact that the ordinary silver bromide photographic film is far more sensitive to the blue end of the spectrum than to the red. Thus blue objects appear too bright and red ones too dull. Sometimes advantage could be taken of this to achieve artistic effects, but generally speaking the photographer sought strict tonal rendering in accordance with what the eye sees. The simple answer was to fit over the lens a filter that retarded the blue light but transmitted the red unimpaired; the penalty was that this extended the exposure time. In the first decade of this century the same effect was achieved, notably as a result of German research, by incorporating appropriate dyes in the emulsion itself.

A second problem is that the silver produced in the film by the development process is not uniformly dispersed, as a black dye might be; instead, it consists of minute separate particles. These are so small that with good film they are not apparent in contact prints. Difficulties could arise, however, when enlargements were made, as became mandatory with the advent of miniature cameras. They were compounded by the fact that the tendency towards grain, as it is called, was increased by forms of heat treatment used to increase the speed of the film. The problem was resolved in two ways, in the 1930s. Firstly, manufacturers produced fine-grain but slower films which were fairly generally acceptable in view of the shorter exposures made possible by wide-aperture lenses. Secondly, new chemicals (notably paraphenylenediamine) were introduced which – although slower acting – reduced grain very considerably.

Up to the 1930s much contact printing from negatives was done by amateurs on so-called daylight paper. This was based on a silver chloride emulsion which darkened to a pleasing sepia colour without need for a developer; all that was necessary was a fixing bath of hypo (sodium thiosulphate) to stop the process when the picture had reached the desired intensity. The making of such prints was a pleasant occupation for a sunny afternoon; if desired, the finely graded prints could be toned to black-and-white in a platinum-containing bath.

For many purposes, and particularly with the use of 35-mm film, enlargement was essential; for this the much more sensitive bromide paper was necessary. It required the use of a dark-room – or at least a darkened room – and the print had to be developed as well as fixed. Gaslight papers occupied an intermediate position. Based on a chloro-bromide emulsion they were – as their name implies – fast enough for the production of prints by artificial light. Introduced in the 1880s they continued in use, without change in name, long after electricity had almost totally replaced gas for lighting.

III. SENSITIVITY

To produce predictable and consistent results there had to be some measure of the sensitivity or speed of plates or film. Most speed ratings were based on the response of film under standard lighting conditions, and the first to be widely adopted was the H and D (F. Hurter and V. C. Driffield) system, developed in Britain in 1890. Abroad the system proposed by J. Scheiner of Berlin (1894) was favoured, until replaced in 1936 by the DIN (Deutsche Industrie Norm) scale, which in effect did no more than increase the Scheiner rating by an addition of ten. In the USA yet another system (ASA) was proposed by the American Standards Association. After the Second World War DIN and ASA were universally adopted to the exclusion of others.

Experienced photographers using graduated apertures, accurate shutter speeds, and film of standard speed rating, could judge the correct exposure pretty accurately whatever the circumstances; millions of amateurs did quite well with standard film in box cameras set permanently at f 11 and $\frac{1}{25}$ th second. Nevertheless, for good results it was desirable to be able actually to measure the intensity of light falling on the subject. Early exposure meters were ingenious but simple. Some were based on the darkening of a strip of sensitized paper in a fixed period of time. Others indicated the correct speed and shutter setting by adding together a series of factors related to aperture, shutter speed, season of year, time of day, bright sun or cloud, distance of subject, and so on. In the 1930s more sophisticated meters began to appear based on the photoelectric properties of selenium – this allowed light intensity to be measured by observing a needle traversing a scale. Usually they were separate accessories, but a few high-quality cameras had built-in photoelectric exposure meters in the mid-1930s. All these devices had the defect that they did not necessarily measure the intensity of light falling on the film at the moment of exposure, which is what really matters. In the 1960s, this was rectified with the introduction of exposure meters set within the camera behind the lens (Pentax, 1964).

IV. COLOUR PHOTOGRAPHY

Colour photography is based, like colour printing, on combining images in three primary colours – red, green, blue. Even before the beginning of this century the American photographer F. E. Ives (1893) produced his Kromscop viewer in which three colour-separation negatives could be viewed as a composite whole, and there were also techniques for overprinting images formed in the primary colours. However, these were all techniques requiring skill and patience and lay in the realm of the really enthusiastic amateur or the professional. For the tens of millions of ordinary amateurs something very much simpler, and compatible with existing equipment, was required.

The first major step in this direction was the Autochrome process of A. and L. Lumière (1907). They manufactured photographic plates in which the sensitive emulsion was underlaid by a transparent film containing a mixture of starch particles each dyed in the primary colours; each one thus served in effect as a colour filter. A standard camera could be used, but the plate had to be developed a second time after dissolving away the first silver image (reversal process) and then viewed as a transparency. The result was satisfactory but inclined to be rather dark because of the opacity of the starch granules. Shortly afterwards (1908) L. D. Dufay launched a similar process. Agfa entered the field in 1916 with a colour-plate process which was marketed as Agfacolor film in 1932.

The above were all additive processes, so called because the different coloured lights are added to each other: if light of all colours is transmitted fully, the result is white. Subtractive processes, by contrast, are ones depending on combining from three separation negatives three corresponding positives in colours complementary to the negatives. The three positives can be viewed together as a transparency or superimposed on white paper to form a print. Among early examples of subtractive processes were Uvachrom (1916) and the Autotype carbro process (1930).

In the 1930s two major colour-film projects were launched, by Kodak (1935) and Agfa (1936) respectively. They were subtractive processes involving three differently coloured layers of sensitive emulsion. The end result was a colour transparency which could be viewed by transmitted light or projected on to a screen. By their nature they gave bright pictures with a wide range of tones. Nevertheless, the popular demand was for conventional photographs – that is, ones printed on paper – even though these of necessity gave a smaller range between highlights and shadows. The Second World War restricted development along these lines, and in the event the popular colour print was a post-war development.

V. INSTANT PHOTOGRAPHS

As every photographers knows, an inherent disadvantage of the conventional process is the delay between taking the photograph and seeing the result. For the keen professional or amateur who does his own processing, the delay is a matter of hours even if he is working close to his own facilities; for the snapshotter who sends his film away for developing and printing only when a complete roll has been exposed, it may be weeks or even months before he sees results. Apart from being tiresome, this delay normally precludes the possibility of taking a second shot of the same subject should the first be unsatisfactory; the only safeguard is the wasteful one of taking a series of photographs over a range of exposures in the hope that one at least will be satisfactory.

There is, therefore, considerable appeal in a process that provides an instant picture. Such a process was perfected in 1947 by the American inventor Edwin H. Land. Essentially the negative and the positive film are combined in a single pack separated by a thin layer of jelly containing the

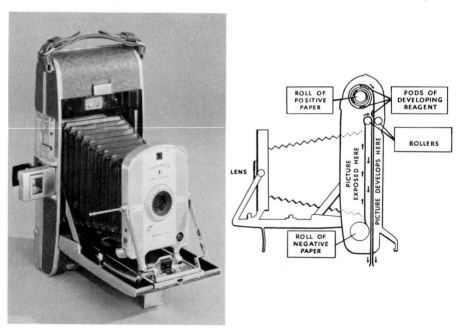

Polaroid Land camera Model 95 (1952) showing method of working.

developer. When the exposure has been made, the pack is passed between rollers, which release and spread the developer. After a minute, or less, the positive print is peeled off. Originally designed for black-and-white prints, it was extended to colour in 1963.

The great advantage of virtually instant prints is offset by three disadvantages: the process produces only one direct print, for the negative cannot be re-used; it is rather expensive in comparison with other processes; and, perhaps most serious, the development process is very sensitive to temperature. In ordinary photographic development the temperatures of the bath can be carefully controlled, but Polaroid prints have to be developed at the ambient temperature – or, more precisely, – the temperature of the developing jelly in the pack. Although the makers give instructions for varying the development time according to temperature, these are in practice not easy to observe, and it is difficult to get good results in very cold or very hot weather.

VI. CINEMATOGRAPHY

We have noted that the first public showing of a moving picture took place almost unnoticed in Paris in 1895. Britain had a public cinema in Bishopsgate in 1905 and Pittsburgh's famous Nickelodeon opened in the same year. Only ten years later cinemas were numbered in hundreds, and by 1914 there were 3500 in Britain alone. In 1946 the world total was 83 000. The great boom, however, was in the inter-war years, when a huge new industry, catering for a popular audience numbered in hundreds of millions, developed. In 1937 cinema takings in the USA totalled over $ 1000 million and world investment in the industry was around $ 2500 million. Its centre was unquestionably Hollywood – with 40 per cent of world output – but most countries developed some kind of film industry of their own. Up to the First World War the main European centres were Britain and France. While the emphasis was on commercial film production for mass cinema audiences, cinematography also became a popular amateur hobby.

Such an explosive development demanded standardization of film. At the turn of the century many sizes were in use, but the most popular was that used by Edison; this was 35-mm film with four perforations per inch for the sprocket drive on both camera and projector and 16 frames per foot. In 1909 this was internationally adopted and remained in use for some twenty years. For amateur work, designed for use in smaller cameras and projection on smaller screens, 16-mm film was introduced in 1923, but this was also soon used for much commercial work not intended for large-screen showing, such as educational films. Kodak introduced an 8-mm film for amateurs in 1932.

Film-drying racks, c.1910.

As in still photography, there were problems arising from the fact that cine film was much more sensitive to blue light than red, but panchromatic film was introduced in 1919, and became popular from 1925.

Film producers tackled the problem of colour at an early date but not until the 1930s was real success attained. Then Technicolor so transformed film-going for countless millions that the word became almost synonymous with anything that seemed larger than life. It depended on simultaneously making three separate films in red, blue, and green in a special beam-splitting camera. From these three colour negative films three positive films were prepared, and from these the colour was transferred to blank film. The end result was a positive transparency in colour that could be screened with ordinary projectors. By the 1940s, however, three-colour multiple-layer processes such as Kodacolor had been adopted which dispensed with the need for a special camera.

The Technicolor dye-transfer process was first used in 1932, for a Walt Disney cartoon film, but the first full-length film to use the process was *Becky Sharp* in 1935. The first colour film made by the multiple-layer process was *Thunderhead, son of Flicker* in 1948; this heralded the rapid demise of the dye-transfer process, with its need for a special camera.

Though the Technicolor process described above dominated colour cinematography from the mid-1930s, it was by no means the first attempt to introduce it. In the very earliest days, even before the turn of the century, colour films were made by laboriously colouring each frame by hand. An illusion of colour could also be produced by simply tinting the whole film appropriately: blue, for example, could suggest a moonlit scene. This was possible, however, only because films were short. With the advent of longer films, from about 1907, the process was speeded up by applying colour mechanically with the aid of stencils. This process was used for some twenty years.

The Kinemacolor process, developed by G. A. Smith in 1909, ingeniously utilized the persistence-of-vision phenomenon that is the basis of all cinematography: the image is retained on the retina until the next frame appears on the screen, thus giving the illusion of continuity. In Kinemacolor, black-and-white film was exposed at twice the normal rate, using red and green filters alternatively for successive frames. The processed film was projected on to the screen through similarly alternating filters. To the eye, the effect was the same as that of superimposing the red image on a green one. Although widely used for several years, Kinemacolor had three inherent defects: it required special cameras and projectors; good colour rendering requires a combination of three primary colours, not two; and moving objects tended to have coloured fringes because the red and green exposures were not made simultaneously, but in succession and thus the images were not strictly superimposed.

Like colour, sound was experimented with extensively before a satisfactory process was developed. Not surprisingly, for both were developing at about the same time, the first attempt was to associate motion pictures with Edison's phonograph, and limited commercial success was achieved in the 1890s. For various reasons, both commercial and technological, this development was not popular. Commercially, film producers were doing well enough with silent films enlivened by a cinema pianist providing background music. Technologically, it was difficult to synchronize vision and sound, and if they got even very slightly out of step the effect could be ludicrous. Further, it was difficult, in the days before electrical amplification, to generate enough sound to fill a cinema. Nevertheless, in 1925 the Vitaphone Company adopted a silent film (*Don Juan*) for use with gramophone records playing an appropriate musical accompaniment. This attracted considerable attention, and two years later they produced a film designed for sound from the outset; this is generally regarded as the first talkie. It was *The Jazz Singer* and included vocal sound as well as music. It was a considerable success, but this was limited by the scarcity of cinemas wired for electrical sound amplification.

In the event, the future of sound lay in a different direction. Much as

Baird stimulated interest in television and recognition of its possibilities, without inventing the process finally adopted, so Vitaphone demonstrated that talking pictures could command commercially viable audiences. The successful process was a photoelectric one, in which the sound was recorded on the film simultaneously with the picture. Basically, sound from a microphone was made to modulate a light signal falling on a strip along the edge of the film. When the film is screened a beam of light passes through this strip on to a photoelectric cell, giving a signal proportional to the intensity of the light transmitted; this signal is amplified electrically and activates a loudspeaker, which vibrates in phase with the original sound falling on the microphone. In brief, the sound is recorded in the form of a continuous ribbon of film of continuously varying density; a variant, not differing in principle, was for the light signal to produce on the strip a serrated track of variable width.

The pioneer of this form of on-film sound recording was E. A. Lauste, an associate of Edison who came to England in 1904. Although he quickly achieved technical success, his process was not commercially successful. After the First World War, Tri-Ergon in Germany took it up but they, too, were unsuccessful commercially. Success was finally achieved in the USA. There Lee de Forest's Phonofilm (1926) became the basis of Movietone. The RCA Photophone process was introduced in 1928 and in the same year Western Electric introduced a process in which the intensity of light falling on the soundtrack was controlled by the vibration of two metallic reeds activated by the microphone. Almost overnight talkies took over from silent films, the limit being set by the availability of equipment. During 1930 the number of American cinemas wired for sound rose from 8700 to 13 500, and only 5 per cent of major films made were silent.

The universal adoption of sound made necessary certain important changes in filming technique. For the making of silent films, noise in the studio did not matter, but for talkies a silent studio was essential. Even the noise of the cameras was obtrusive, and they had to be housed in sound-proof cabins. Later, cameras that were silent in action were developed. In addition, the speed of filming and projection had to be restandardized. Up to 1930, the notional standard for both filming and projection remained 16 frames per second, this being about the minimum that would avoid a flickering effect owing to one image on the retina fading before the next succeeded it. In practice, however, cinema owners often speeded the rate of projection up to 20 frames per second, partly to reduce flicker still further and partly – in a highly competitive business – to show their customers more film in a given time. With the advent of sound, however, strict control of filming and projection speeds was essential and this called for more

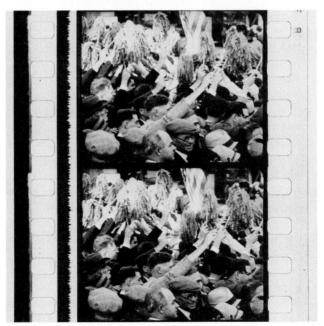

Alternative methods of recording sound on film. (*Top*) variable density, (*bottom*) variable area sound track.

sophisticated projectors. These too had to be silent in operation. To meet the new requirements, a new standard of 24 frames per second was adopted.

As has been noted in the context of sound recording (p. 321), magnetic tape was developed in Germany during the Second World War, but not adopted by the film industry until just after mid-century. The use of a dual sound-track opened the way to stereophonic effects.

BIBLIOGRAPHY

British Photographic Manufacturers Association. *Photographic industry of Great Britain*. London (1918).

Brown, B. *Talking pictures*. Pitman, London (1931).

Coissac, G. M. *Histoire du cinématographie de ses origines jusqu'à nos jours*. Gauthier-Villars, Paris (1925).

Eder, J. M. *The history of photography* (trans. E. Epstean). Columbia University Press, New York (1945).

Fielding, R. *A technological history of motion pictures and television: an anthology from the pages of the Society of Motion Picture and Television Engineers*. University of California Press, Berkeley (1967).

Gernsheim, H. and Gernsheim, A. *The history of photography*. Oxford University Press, London (1955).

—— *A concise history of photography*. Thames and Hudson, London (1965).

Holmes, C. *Colour photography and other recent developments of the art of the camera*. London (1908).

Hurter, F. and Driffield, V. C. *Photographic researches*. Ferguson, London (1920).

Lecuyer, R. *Histoire de la photographie*. Baschet, Paris (1945).

Mees, C. E. K. *From dry plates to Ektachrome film*. Eastman Kodak, New York (1961).

Moholy, Lucia. *A hundred years of photography 1839–1939*. London (1939).

Quigley, M., Jr. *Magic shadows: the story of the origin of motion pictures*. Georgetown University Press, Washington (1948).

Ramsaye, T. *A million and one nights: a history of the motion pictures*. Simon and Schuster, New York (1926).

Robinson, D. *The history of world cinema*. Stein and Day, New York (1973).

Stenger, Erich. *The history of photography: its relation to civilization and practice* (trans. E. Epstean). Easton (1939).

27

COMPUTERS

The treatment of computers presents difficulties that have already been encountered with other topics, such as space flight and atomic power: their social impact is a part of the history of the second, rather than the first, part of this century, although their roots can be traced back much earlier. In the case of computers this difficulty applies with full force to the electronic devices that now dominate the scene, for the first commercial electronic computer (UNIVAC I) was not made until 1951. We must recognize, however, that mechanical and electro-mechanical calculating devices have a far longer history and that some important innovations were made in this field in the first half of the century.

Today, and even in 1950, the term computer is a misnomer, for it implies a calculating device. The role of the modern computer, and some of its mechanical predecessors, is, however, much wider than this, for it is widely used for the processing, storage, and retrieval of information of all kinds. It was, indeed, the need to analyse the results of the US Census of 1890 that inspired the Hollerith punched-card system.

I. MECHANICAL DEVICES

Calculating devices are almost as old as mathematics; the Romans, Greeks, and Egyptians used a simple counting board known as the abacus. The Chinese and Japanese had similar devices in which counters were slid along a series of parallel rods or grooves. It is interesting to note that as late as 1946, almost at the end of our period, an American skilled in the use of the best available electric desk calculator failed in competition with a Japanese clerk using the traditional *soroban*.

The true ancestor of modern calculators is an adding machine invented by Blaise Pascal in 1642. Like many of its successors it depended on a series of interconnected dials numbered from 0 to 9 and corresponding to units, tens, hundreds, thousands, etc. A complete revolution of one dial effected one-tenth of a revolution of the next highest one and so on. The machine could be used for multiplication by repeatedly adding the multiplicand (number to be multiplied) to itself the necessary number of times $(17 \times 3 = 17 + 17 + 17)$. Thirty years later the mathematical philosopher Leibniz made a greatly improved machine, with a register to store the multiplicand, which was much more positive in action. The interaction

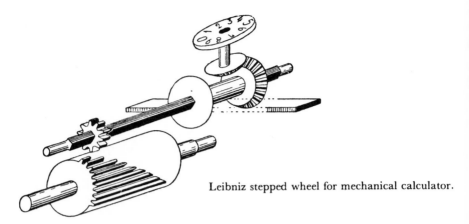

Leibniz stepped wheel for mechanical calculator.

between successive ten-digit dials was effected by means of cylindrical cog-wheels each carrying ten teeth: the first extended one-tenth of the width of the wheel, the second two-tenths, and so on. This 'stepped wheel' device was incorporated in almost all later mechanical calculators. He also included a gear train by which multiple additions could be effected by a single movement of a lever. Leibniz's principles and designs were sound, but beyond the general engineering skill of his day. Although a few models of his calculator were made, of which one survives, the manufacture in quantity of machines of this kind had to await the development of more advanced precision-engineering techniques some 150 years later.

Meanwhile a different approach had been made to the problem of laborious calculations, such as were necessary in commerce and in the preparation of navigational tables. In 1614 John Napier – remembered also for a mechanical multiplication system known as Napier's bones – published his invention of logarithms, which is the exponent indicating the power to which a number must be raised to produce a given number. Thus if $X^2 = Y$, 2 is the logarithm of Y to the base 2. From the calculating point of view, the importance of this is that numbers can be multiplied by adding logarithms and divided by subtracting them, thus reducing multiplication and division to a very simple operation. To make use of this mathematical device it was, however, necessary to produce tables of logarithms which was itself a very great labour; once prepared, however, the tables could be used for as many different calculations as desired. In the 1780s, for example, the French government authorized the preparation of a new table of logarithms (and trignometrical functions); the task occupied nearly a hundred human computers for two years.

Logarithms were mainly used for calculations on paper, but it was immediately clear that they lent themselves to mechanical computation. If

Mechanism of Leibniz calculator.

numbers were expressed on a linear scale in terms of their logarithms, and the length on one scale was added to that on another, the combined length represented the logarithm of the result of multiplying the two numbers together. As early as 1654 Robert Bissaker had devised the prototype of the modern slide-rule in which one logarithmically graduated scale slides against another; the only major development to this, made about the middle of the nineteenth century, was the transparent cursor, with a hair-line to set one scale precisely against the other and to read the result. The accuracy of a slide-rule depends primarily on the length of the scale, and much ingenuity was exercised in increasing this without making the instrument too unwieldy. The device most commonly used to achieve this was that of engraving the scale spirally or helically. In the twentieth century, a popular version of the slide-rule was the Otis-King calculator which, although less than 6 inches long, embodied a helical logarithmic scale 66 inches long. In the 1970s, the slide-rule was to be eclipsed by pocket-size electronic computers.

Not until the 1820s were machines of the type visualized by Leibniz manufactured in quantity. The first was marketed, initially in France and Germany, as the Arithmometer and continued to be manufactured up to the Second World War. It was claimed that it would multiply two eight-figure numbers in 18 seconds. Many machines of a similar kind were also manufactured. The Arithmometer embodied Liebniz's 'stepped wheel' device. In 1872, in the USA, F. S. Baldwin invented an alternative pin-

Interior of Colmar calculating machine, showing stepped wheels; calculators of this kind were made until about 1930.

wheel device. In this, the number of cogs on a calculating dial could be varied from 0 to 9 by a simple lever setting. These pin-wheels meshed with ten-toothed gear wheels: if one tooth on the pin-wheel was operative it moved the ten-toothed wheel one-tenth of a turn for each revolution, and so on. Baldwin's device is generally referred to as the Ohdner wheel, after W. F. Ohdner who was among the first to incorporate it in a calculating machine. One of the best-known mechanical calculators incorporating an Ohdner wheel was the Brunsviga, first manufactured in 1892; Brunsviga machines continued in use until the middle of this century.

Yet another popular type of calculator was the Millionaire, which was essentially a mechanized version of the multiplying device invented by Napier (Napier's bones). It was roughly three times as fast as the Arithmometer. By the outbreak of the First World War some 3000 had been sold, the later versions being driven by small electric motors.

The machines so far considered depended on the numbers involved being displayed and operated by dials and levers, but towards the end of the nineteenth century keyboard machines began to appear. These were perhaps suggested by the success of the early typewriters (p. 292). The highly successful Comptometer invented by D. E. Felt was first marketed in 1887, and improved two years later by the incorporation of a device by

Millionaire calculating machine: motorized versions of this machine were made up to the First World War.

which results could be recorded on a roll of paper. In the Comptometer the calculating mechanism was activated by the keys themselves. In the print-out calculator developed by W. S. Burroughs at about the same time, the keys served to set the mechanism but it was not put in motion until a lever was pulled. The advantage of this was that the operator could check his work step by step and correct it if necessary instead of having to start afresh if any error was made.

In the first half of the twentieth century, keyboard machines of the Felt/Burroughs type, recording results on a paper-roll, were standard equipment in banks and commercial offices, and even before the First World War they were being made at the rate of thousands per year. They continued to be made, in increasingly sophisticated forms, such as the Facit, up to 1960. Although many were electrically driven, and thus less fatiguing to the operator, the calculator itself was purely mechanical and this set a limit to their speed. The importance of the electronic calculator was that it had no moving parts and thus operation was vastly faster.

Before considering electronic devices, however, we must go back on our tracks and consider another line of mechanical development that was to produce devices widely used in the first half of this century. This line begins with the work of Charles Babbage, Lucasian Professor of Mathematics at

Early Comptometer, *c.*1890.

Cambridge (1828–39), who conceived the idea of a machine that could not only make the elaborate computations necessary for the compilation of astronomical and navigational tables, but could print the results as well. This could not only speed up the work but – no less important – would eliminate the possibility of human error in copying and printing. In Britain, for example, the important *Nautical Almanac* (p. 277) was being criticized for the number of its printing errors. By 1822 Babbage had demonstrated a 'difference engine' which showed the feasibility – if not the practicability – of his ideas. In the following year the British Government, concerned at mounting criticism of existing tables, gave him financial support which ultimately totalled £17 000. In the event the machine was a failure, despite some fourteen years' work, and for this there seem to have been two main reasons. Firstly, Babbage was far too ambitious; at a time when the manufacture of the relatively simple Arithmometer presented great engineering problems, it was optimistic to hope that the far more complex difference engine could succeed. Secondly, we must look to Babbage's temperament: he seems to have been constitutionally unable to apply himself to work in hand and was forever distracted by new ideas. Among these new ideas was his analytical engine, designed to undertake a wide range of arithmetical computations instead of being limited to 'differencing'. This machine, too, was never completed, but it is nevertheless

One of the 'mills' of Babbage's analytical engine (as built by his son H. P. Babbage about 1910).

historically important, and relevant to twentieth-century practice, because it introduced important new principles that were widely publicized and adopted later. Firstly, it contained a mill and a store. The mill comprised the calculating machinery; the store, as its name implies, registered the data supplied to it and could also store, or memorize, intermediate stages in a computation. Secondly, the engine was programmed by means of punched cards, an idea which Babbage may well have adopted from the Jacquard loom (1801). The punched cards were 'read' by rods passing through their perforations.

This punched-card principle was taken up by an American statistician, Herman Hollerith, of the US Census Bureau, founded in 1879, to analyse the results of the 1880 census. In the event, the analysis was completed and published only shortly before the 1890 census was due to take place. The slowness of this process, and the fact that the American population was growing rapidly, led Hollerith to turn his attention to some form of mechanization. After briefly experimenting with perforated rolls of paper he came up with the famous punched-card system that is named after him. Briefly, relevant information is recorded as a pattern of holes on a card – in Hollerith's case, measuring 3×5 inches. Each card was 'read' by a machine carrying an arrangement of spring-loaded pins corresponding to the potential pattern of holes. If the pin encountered a hole, indicating perhaps the profession or the age of the citizen being registered, it passed through the card into a pool of mercury below it, thereby completing an electrical circuit. Contrariwise, if a pin encountered solid card, it was stopped and no electrical contact was made. Electro-mechanical counters were used to count the number of cards having identical patterns. This simple technique was highly successful and, despite a 25 per cent increase in population, the 1890 census took only one-third as long to analyse as that of 1880.

Realizing the wide application of his device, Hollerith left the Census Bureau in 1896 and established the Tabulating Machine Company to exploit it. He developed a variety of machines and introduced the keyboard principle in 1901. Subsequently, metal brushes making contact with a metal plate through the holes in the cards replaced the pins and mercury. In 1911 the Hollerith system was used for the British census. This was the beginning of a considerable series of successes. Hollerith merged with two other companies – one making time-recording machines and the other scales – to form the Computing-Tabulating-Recording Company (1911). In 1924 this became the International Business Machines Corporation, which in due course was to be a world leader in electronic computers.

Meanwhile, history was repeating itself at the US Census Bureau, where another statistician, James Powers, had been appointed in Hollerith's

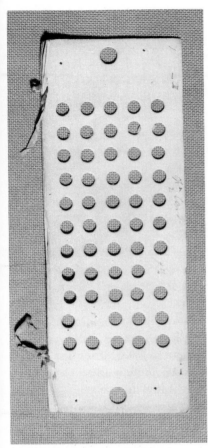

Programming card for Babbage's analytical engine

Powers tabulator, 1910.

place. He introduced a purely mechanical punch-card system in place of the existing electro-mechanical one. This worked well in the 1910 census, after which Powers in turn left to set up his own business, the Powers Accounting Machines Company. In 1927 this merged with Remington, a major typewriter manufacturing company, to form Remington Rand, later (1955) the Sperry Rand Corporation. This, too, was a major manufacturer of electronic computers.

The punched-card system of storing, retrieving, and analysing information was extremely flexible and widely adopted in industry and commerce, especially in the United States. The future, however, lay not with mechanical or electro-mechanical devices but with electronic ones, a change in the line of development we have already seen in the case of television.

Punched-card calculating unit, Great Northern Railway, Peterborough, 1920.

Before leaving the subject of mechanical calculators brief mention should be made of a special example of this type of device, namely the cash register. Designed originally to prevent dishonesty by shop assistants, it became widely used for automatically recording sales, issuing receipts, and, in the more sophisticated versions, provided much additional information for the use of management. The leading manufacturer was the National Cash Register Company, founded by J. H. Patterson in the USA in 1884. By 1950 it had made more than four million cash registers. Other leading manufacturers were Burroughs and Ohmer.

II. ELECTRONIC COMPUTERS

By the mid-1930s various developments had occurred which made possible the realization of Babbage's ideas without having to solve the mechanical problems which had defeated him. These were in particular the punch-card system just described and a variety of automatic selecting devices developed primarily for electrical communication systems. These possibilities were realized by Howard Aiken, of Harvard University, in 1937 and he invited International Business Machines to collaborate with him in their development. The result, in August 1944, was the IBM Automatic Sequence Controlled Calculator (ASCC). This is commonly regarded as the prototype of modern electronic computers and it is certainly true that it initiated an entirely new line of calculators and information processors. Nevertheless, it was not strictly an electronic computer, for it was based on electrically driven number wheels, controlled by an elaborate system of electro-magnetic clutches and programmed by punched tape. Although less ambitious than Babbage's analytical engine, it was 50 ft long, weighed 5 tons, and included nearly a million components and 500 miles of wire. Although it could multiply two eleven-digit numbers in three seconds, this was impossibly slow by the standards of only a few years later. Nevertheless, ASCC remained in use at Harvard for some fifteen years and three later models were built in which, among other improvements, the magnetic clutches were replaced by electrical relays of the type used in communications engineering.

The first of the true electronic computers was the Electronic Numerical Integrator and Calculator (ENIAC), completed in 1946. It was designed and built for the US government by J. W. Mauchly and J. P. Eckert of the University of Pennsylvania. Originally called for to prepare ballistic tables for wartime use, it was in fact not completed until 1946 when it was sent to the Ballistic Research Laboratory in Maryland. Although input and output were still in the form of punched cards – which necessarily limited speed of operation – mechanical parts were eliminated except for switches to control certain sections of the circuitry used for special programming purposes. The moving part was, in effect, a train of electrical impulses generated at the rate of 5000 per second and controlled by electronic gates. The demands of radar

Original ENIAC electronic computer, 1946.

for pulsed signals had already generated much expertise in this field of electrical engineering. Leibniz's stepped wheels or Ohdner's pin-wheels were represented by groups of ten electronic valves. With ENIAC two ten-digit numbers could be multiplied in little more than two-thousandths of a second.

In size, ENIAC was even more formidable than ASCC. It was 100 ft long and contained no less than 18 000 thermionic valves. As it consumed over 100 kilowatts of electricity when in operation the mere dissipation of the heat generated was a serious problem. Nevertheless, it brought computers into the important region in which it was feasible to tackle calculations that would otherwise be too laborious to be worth embarking on at all.

Lady Lovelace, a daughter of Lord Byron, had done much to explain the principles of Babbage's calculator to the general public in the eighteenth century. In particular, she clearly recognized that many calculations involved going through the same sequence of operations many times over, and that provision should be made to enable the machine to embark on these sub-routines automatically. In the 1940s computer engineers again encountered this problem and directed much attention to the memory side of their machines, so that ready-made programs* could be kept in store for use when required. Computer engineers, notably J. von Neumann of the University of Pennsylvania, also went back to a proposal by Leibniz that mechanical calculation could be better performed by the use of a binary rather than a decimal system of notation. This simply means that numbers are built up from only two digits, 0 and 1, rather than ten as in common usage (0–9). Any numbers of digits can, of course, be used to express numbers; the ancient Japanese *soroban,* for example, operates on a quinary system. For an electronic computer the binary notation has the merit that it corresponds to a characteristic of electricity – it is either flowing or not flowing.

The ideas of von Neumann were incorporated in a machine known as EDVAC (Electronic Discrete Variable Automatic Computer), work on which started at the University of Pennsylvania just before ENIAC was finished. It corporated a new form of memory device, based on the circulation of electrically generated sonic pulses through a long tube of mercury. Meanwhile computer research was being undertaken in Britain and elsewhere in Europe, with emphasis on the development of storage capacity. In Britain in 1948, F. C. Williams of Manchester University developed a storage device based on standard cathode-ray tubes; this was incorporated in a number of electronic computers built in Europe and the USA from 1948 to 1956, when it became obsolete. By that time a quite new

---
* In computer science this American spelling has been universally adopted.

UNIVAC 1, the first commercial electronic computer: it was widely publicized in 1952 when it was successfully used to predict General Eisenhower's success in the Presidential election.

storage system, the magnetic core, had been developed, notably by J. Forrester of the Massachusetts Institute of Technology. It depended on the fact that the direction of magnetization of certain magnetic materials, known as ferrites, could be easily and virtually instantaneously reversed – again giving a go/no-go alternative appropriate to a binary notation. Ferrites are so called because they are crystalline compounds containing ferric oxide as a constituent. They had already been known for half a century, but were not seriously investigated until 1933, when the Philips Laboratories in Holland became interested in them. They were being used, among other things, as antenna cores for radio receivers. Magnetic core stores remained in general use until the early 1970s. The magnetic core storage system was first incorporated in the Remington-Rand UNIVAC (Model 1103A) in 1956. It is highly sophisticated but basically depends on the fact that magnetic rings are arranged at the intersection of a matrix of wires carrying electrical pulses. An electrical pulse in one wire is insufficient to reverse the direction of magnetization in the ring, but if two pulses arrive simultaneously reversal occurs. This provided the input. The output was provided by a third set of wires passing through the rings and responding to changes in the direction of magnetization.

By this time computers were experiencing conditions comparable with those developing in the air transport business. There, the higher speeds of aircraft made the journey itself quicker but this was largely offset by delays caused by traffic congestion on the roads leading to and from airports; the need to register well in advance of take-off so that formalities could be completed; delay on arrival while luggage was unloaded and examined; and so on. In the computer field the computing process itself had been enormously accelerated, but the continuing use of punched card or tape for programming and output made it impossible greatly to speed up the process as a whole. The UNIVAC was important in that instead of paper tape it made use of magnetic tape for programming; as we have noted earlier, this had been developed during the war in Germany, and had subsequently been adopted for sound recording, especially in the motion picture industry. The original UNIVAC was designed for the US Census and delivered in 1951. In the following year it was used to predict the outcome of the Presidential election and its correct identification of Eisenhower as the successful candidate was important in directing public attention to the potentialities of computers.

Up to this time electronic computers had been based on thermionic valves; in the late 1950s the replacement of these by transistors represented a major step forward. This takes us rather beyond 1950, but it is an appropriate point at which to close a history of computers nominally ending at mid-century. The magnetic-core storage system, the transistor, and magnetic tape-recording had all been adopted, and the foundation had been laid for a new industry that was to grow at a phenomenal rate and powerfully influence many aspects of modern life.

BIBLIOGRAPHY

Baxandall, D. *Calculating machines and instruments* (revised and updated by Jane Pugh). Science Museum, London (1975).
Booth, A. D. and Booth, K. H. V. *Automatic digital calculators*. Butterworth, London (1953).
Bowden, B. V. *Faster than thought* Pitman, London (1955).
Burck, G. (ed.). *The computer age and its potential for management*. New York (1965).
Couffignal, L. *Les machines à calculer*. Paris (1952).
Eckert, W. J. *Punched card methods in scientific computing*. Columbia Univeristy Press, New York (1940).
Greenberger, M. (ed.). *Management and the computer of the future*. Cambridge (1962).
Hartree, D. R. *Calculating instruments and machines*. Cambridge University Press, London (1950).
Horsburgh, E. H. (ed.). *Modern instruments of calculation*. Bell, London (1914).
— Calculating machines. In *Glazebrook's Dictionary of applied physics*. Cambridge University Press, (1923).

Machines and appliances in government offices. HMSO, London (1947).

Menninger, Karl. *A cultural history of numbers* (translated from the German by
     P. Broneer). MIT Press, Cambridge, Mass. (1969).

Montgomerie, G. A. *Digital calculating machines.* Blackie, London (1956).

Turck, J. A. V. *Origin of modern calculating machines.* Chicago (1921).

28

MEDICINE AND PUBLIC HEALTH

At the beginning of this century the general practitioner treating individual patients at home had few technical aids not possessed by his predecessors a generation earlier. Nevertheless, scientific and technological achievements were enabling him to make available to patients new and effective methods of diagnosis and treatment. By 1950 these achievements had progressed to such an extent that not only had methods of treatment changed radically, but so too, in developed countries, had the whole pattern of medicine. Much of the sophisticated equipment coming into use was expensive and required both medical specialists to interpret its results and skilled technicians to maintain and operate it. Inevitably, the general practitioner had increasingly to refer his patients to hospitals. Nor were such consequences purely national. Medical progress was a major reason for the population explosion in developing countries which today is the source of so many of the world's political, social, and economic problems.

These results cannot be attributed solely to greater success in the treatment of established disease. No less important were developments in preventive medicine. They included not only an extension of already familiar public health services such as pure water supplies and the safe disposal of sewage, but also immunization campaigns against certain diseases and the eradication over large areas of some of the most dangerous insect vectors of disease such as the mosquito. They included, too, a great reduction in the so-called deficiency diseases – such as scurvy and beriberi – as a result of recognition of vitamins, and much progress in the control of diseases, such as diabetes, caused by hormone imbalance.

Though none would dispute the powerful influence of technological innovation on medicine, it is, nevertheless, difficult to distinguish a medical technology as such, for much progress sprang from the adaptation to medicine of results in quite different fields. Indeed, virtually every chapter in this book refers to inventions that have advanced the cause of medicine in one way or another, and in many instances the medical implications have been specifically mentioned. To take a few examples at random: atomic power produced radioactive isotopes of great therapeutic value; progress in chemical technology was essential for the large-scale manufacture of a wide range of synthetic drugs such as the sulphonamides and analgesics; among the new products of the photographic industry were many of importance for the effective use of X-rays; the complex equipment of the modern operating-theatre reflects progress in virtually every branch of engineering – mechanical, electrical, metallurgical, electronic, and chemical. The availability of small thermionic valves – and, just at the end of our period, of the transistor – transformed the hearing-aid. Over and above all this, the practice of medicine was, of course, affected by developments – such as the telephone and fast door-to-door transport – that affected society as a whole.

In the present chapter we can do no more than draw attention, without undue repetition of what has gone before, to some of the major technological developments of direct importance in medicine. In this century, technology became increasingly identified as applied science but we cannot easily distinguish where the role of the scientist ends and that of the technologist begins. There was, for example, great progress in our fundamental knowledge of viruses but much of this was possible only because of the availability of the electron microscope, a highly sophisticated example of electrical engineering. This allows magnification many times greater than is possible with the best optical microscopes, because the wavelength of the radiation involved is much shorter. The first of these instruments was built in the Technical University of Berlin in 1931 by M. Knoll and E. Ruska.

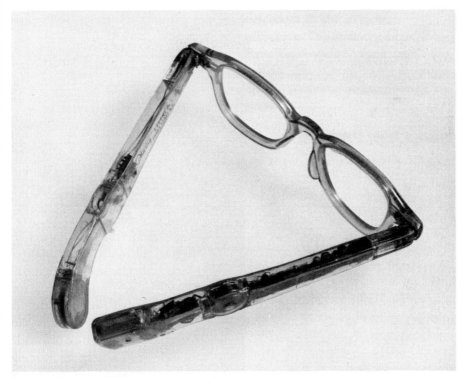

Spectacles with in-built hearing aid, c.1957.

A logical approach to the consideration of advances in medical technology presents difficulties. One approach would be to consider major developments in the main fields of technology as a whole and try to evaluate their medical significance. Since this has to some extent been done in earlier chapters, the alternative has been chosen of selecting the main fields of medicine and considering how far these have been advanced by the advent of new technologies.

I. ANAESTHESIA

Perhaps the greatest advance in nineteenth-century medicine was in surgery. Recognition of the bacterial nature of infection led to the development of sterilization procedures and post-operative care that greatly reduced the risk of infection. At the same time, the introduction of anaesthetics and of improved methods of administering them not only diminished the suffering of the patient but made it possible for the surgeon to embark on more radical operations than were previously possible. Indeed, some medical historians suppose that for a time surgeons became over-enthusiastic in their removal of organs that seemed inessential to the body's working.

In 1900 the administration of an anaesthetic was still generally little more than holding a towel soaked in chloroform or ether over the patient's face; by 1950 the anaesthetist was a medical specialist controlling a formidable array of equipment, including equipment for monitoring the patient's condition. Anaesthesia by inhalation was still widely practised but a succession of new and better agents had been introduced, though not all were retained. They included ethylene (1918), cyclopropane (1930), divinyl oxide (1933), and trichlorethylene (1935). The last of these can dull pain without causing loss of consciousness and was widely adopted in midwifery and dentistry. With greater use of anaesthetics, some of which were expensive, the question of cost became important. In 1924 R. M. Waters pointed out the advantages of closed-circuit anaesthesia in which the anaesthetic is recovered for further use, and in the 1930s the necessary equipment was generally available. In inhalation anaesthesia, the face-piece seriously hampers operations involving the head, a fact which led to the introduction of endotracheal anaesthesia.

Other methods of anaesthesia were introduced, most particularly anaesthesia by injection. From the nineteenth century dentists had made wide use of local anaesthetics such as cocaine (Carl Koller 1884) as pain-killers, though their toxic side-effects could be serious. From the start of the new century a range of synthetic alternatives became available, beginning with procaine in 1905 and followed by others such as nupercaine, the effects of which lasted longer. Following the synthesis of barbital (veronal) by Emil Fischer in 1902, the anaesthetic effects of many barbiturates were investigated, but it was thirty years before barbiturates became popular, after the introduction of evipan in Germany in 1932, and later of thiopentone (pentothal sodium).

Spinal anaesthesia, involving injection into the cerebro-spinal fluid, provided another means of deadening pain locally without loss of consciousness – though not every patient welcomed this. It presented a hazard, however, in that the best available hypodermic needles were liable to break during the injection, with serious consequences. Not until 1950 was a satisfactory needle available, but by then the method was being less widely used because other relaxing drugs such as curare (1942) were being introduced.

II. ANTISEPSIS AND CHEMOTHERAPY

By 1900 the role of bacteria in disease had been clearly recognized and, as we have seen, advances in surgery depended very much on scrupulous attention to cleanliness. Many surgical accessories, such as instruments, could be satisfactorily sterilized by heating in autoclaves, but for general purposes substances with strong anti-bacterial action were necessary, such as carbolic acid or cresol (the latter often emulsified (Lysol)). Such chemicals were, however, too toxic to apply to human tissues, as for example to sterilize the skin before injection or incision or to treat local infections. For such purposes milder agents such as alcohol or hydrogen peroxide were used. To these were added, after the First World War, organic antiseptics such as TCP (tricresyl phosphate) and Dettol (chlorxylenol).

Thus there was no great problem in destroying bacteria if they could, as it were, be directly attacked in the open, for there were plenty of agents to which they rapidly and completely succumbed. This was not true, however, when they lurked in crannies not accessible by ordinary cleaning processes; recognition of this led to important changes in hospital design, especially design of operating-theatres, paving the way to the 'clinical' look they now have.

Bacteria that had invaded human tissues presented a very different problem, however. If the infection was local, as in a boil or infected cut, the natural defences of the body – helped perhaps by antiseptic dressings – would probably overcome it. If bacteria had invaded the body as a whole, and were multiplying and circulating in the blood, the situation was much more dangerous. Again, the body might respond by producing antibodies that would destroy bacteria, and white blood cells would engulf them, but often these natural defences were inadequate; then, in the case of dangerous diseases such as diphtheria and cholera, death might ensue. It had already long been established (Edward Jenner, 1798), of course, that people could be immunized against certain diseases by, in effect, deliberately giving them a mild attack of them by means of a very attenuated culture of the causative organism, and in the first half of this century the manufacture of vaccines, effective against a wide range of infections, was to become an important part of the pharmaceutical industry. Nevertheless, in 1900 it seemed that, once pathogenic bacteria had established a firm hold within the body, medicine could do little about it. The antiseptic substances known to be capable of killing bacteria were far too toxic to introduce into the blood stream. To many, indeed, it seemed that antiseptics attacked all living cells in the same way and there was thus no possibility of differential action.

Fortunately, this pessimistic view was not universally shared, and one who was not convinced by it was Paul Ehrlich, a pioneer in the new science of bacteriology. Working at the Royal Institute of Experimental Therapy in Frankfurt he sought a 'magic bullet' which would destroy invading bacteria but leave human cells unharmed. In 1909 he achieved success with his famous drug 606, so called because it was the 606th variant of a basic arsenical compound, atoxyl, which he had tested against syphilis infection in animals. After further trials the new drug was made available to the medical profession as salvarsan. Clinical trials showed that it achieved considerable success in curing human syphilis, but it was not ideal: injection could be very painful, treatment might have to be prolonged, and there could be toxic side-effects, including jaundice. Ehrlich's neosalvarsan (914) of 1912 was an improvement, mainly because it was more readily soluble.

By modern standards neither salvarsan nor neosalvarsan is a very satisfactory drug, though the latter appeared in the world's pharmacopoeias (except the US) until the 1950s. Their significance lay less in their usefulness than in their influence on medical thought. They dispelled the pessimistic belief that all anti-bacterial agents were, *ipso facto*, toxic to human cells and encouraged research in this new field of chemotherapy. In 1932, Gerhard Domagk, working in the Bayer laboratories at Elberfeld, discovered that a red dye (Prontosil Red) could control streptococcal infections in mice. This observation was not published until 1935 and it was then found – at the Pasteur Institute in Paris – that the anti-bacterial action of Prontosil was exerted by only half the molecule, namely sulphanilamide: this immediately invalidated the Bayer patents relating to Prontosil.

Sulphanilamide (p-amino benzene sulphonamide) is a simple compound and can be cheaply synthesized. It is also very stable, and chemists throughout the world began to synthesize variants of it; by 1942 at least 3600 sulphonamide derivatives had been prepared. The first of these to show clinical promise was sulphapyridine, synthesized in the laboratories of the British chemical manufacturers May and Baker in 1938. Like Ehrlich's 606, it was the last product of a long series of experiments, as indicated by its trade name M & B 693. Although active against a wider range of bacteria than sulphanilamide, it caused more side-effects. These were less apparent in the American sulphathiazole (1939) and sulphadiazine (1941). During the Second World War two other sulphonamide drugs became important, mainly because they are not readily absorbed from the intestine when given by mouth and are therefore useful for treating intestinal infections such as bacillary dysentery, prevalent among the troops. These were sul-phaguanidine (1941) and sulphasuxidine introduced in the same year. By mid-century, therefore, a powerful armament of synthetic drugs was available for the treatment of bacterial infections. To these could be added other synthetic products effective against diseases caused not by bacteria but by other micro-organisms. These included the Bayer product atebrin (mepacrine) which was discovered in 1933 and is at least as effective as the traditional quinine in the treatment of malaria. The supply of quinine was interrupted by the Japanese invasion in the Pacific. In 1943, ICI research workers in Britain discovered another highly effective anti-malarial, introduced as paludrine. This was cheaper than atebrin and did not turn the skin yellow.

Meanwhile, however, events had taken a far more dramatic turn. From the late nineteenth century it had been known that micro-organisms could be mutually antagonistic: if two were grown in a mixed culture, one would destroy the other. In 1889 P. Vuillemin called this phenomenon antibiosis. Over the years there was some interest in the possibility of using antibiosis to treat bacterial infections, but no real progress was made until 1928, when Alexander Fleming, working at St. Mary's Hospital in London, discovered that broth on which the mould *Penicillium notatum* had been grown had a powerful antibacterial activity. This broth he called penicillin, but he failed to appreciate the unique properties of the active substance contained in it, and by 1930 had virtually lost interest in it. The explanation of this is still somewhat controversial, but it is a matter of fact that the development of pencillin as a chemotherapeutic agent was the result of an independent research project initiated by H. W. Florey and E. B. Chain at the Sir William Dunn School of Pathology, Oxford, in 1939. For this they received initially the princely grant of £25 from the Medical Research Council. By 1940 it was clear from experiments with mice that penicillin (the name now given to the active principle and not the broth containing it) had quite exceptional properties: an astonishingly high activity against a wide range of pathogenic bacteria; virtually no toxicity towards animal cells; and no inhibitory effect on leucocytes, the white blood cells that can digest invading microbes. It was then essential to prepare sufficient material for clinical trials with human patients, and in 1941 Florey visited the USA to enlist the aid of the leading pharmaceutical manufacturers there, notably Squibb, Pfizer, Lilly, and Merck. Thereafter, speedy progress was made. The results of clinical trials, first published in *The Lancet* early in 1943, left no doubt that penicillin was a uniquely valuable chemotherapeutic agent. Large-scale manufacture was established in the USA, using deep-culture methods of fermentation not unlike those of the brewing industry and already used for the manufacture of, for example, citric acid. By the time of the Normandy Landings in 1944 sufficient penicillin was available to treat all serious battle casualties. By 1950 pencillin was available in quantities sufficient to meet world demand. Unlike some important natural products, such as vitamin C, penicillin could not be synthesized commercially, and the fermentation process continues to be used.

The immense success of penicillin prompted a worldwide search for other medically useful antibiotics. Although many thousands of examples of antibiosis have been investigated, very few led to useful products. Exceptions were streptomycin (1944), chloramphenicol (1947), and cephalosporin (1953). In addition, useful semi-synthetic penicillins were made by varying the structure of the natural molecule.

III. RADIOLOGY

W. K. Röntgen discovered X-rays in November 1895. Rather surprisingly – for he was a physicist working as professor in the Physical Institute at Wurzburg – they were almost immediately applied in medicine, both for diagnosis and for treatment. Almost simultaneously (1896) A. H. Becquerel discovered that uranium spontaneously emitted a penetrating radiation; this led to the discovery of radium by the Curies, Marie and Pierre, in December 1898. Thus the birth of radiology almost exactly coincided with the start of the twentieth century. Tragically, the new radiation was to prove both a blessing and a hazard. Time showed that excessive exposure could cause fatal illness, notably leukaemia; Marie Curie herself was to be one of the victims. An important element in the use of both X-rays and radium was the development of devices, mainly lead shielding and remote control, to protect the operators.

Diagnostic radiology depended, among other things, on the development of X-ray plate and film. This was stimulated by the medical demands of the First World War – X-rays were an invaluable aid to locating embedded bullets and shrapnel – and Eastman Kodak introduced X-ray film on a cellulose acetate base about 1929. As in photographic film, sensitivity was

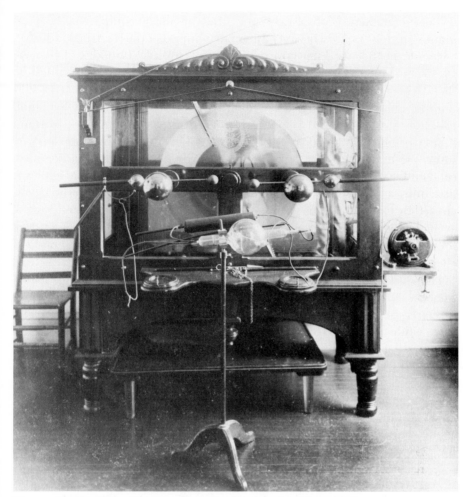

X-ray tube, supplied with power by static electricity machine, 1900.

increased and development times reduced. In 1938, when X-ray examination was routine procedure in all hospitals, the development time was
reduced by two-thirds. As early as 1896, Elihu Thomson of General Electric
invented a stereoscopic X-ray camera that greatly facilitated the location of
objects under examination.

One problem in the use of X-rays is that whereas dense structures, such as
bone, are almost opaque – thus making diagnosis of fractures and malform-
ations easy – soft tissues are almost transparent and therefore difficult to
distinguish. As early as 1897 William Cannon of Boston used a meal
containing bismuth salts to define the gastro-intestinal tract; subsequently
barium was adopted.

The manufacture of X-ray tubes and ancillary equipment, such as induction coils, rapidly became an important part of the electrical industry, particularly after the First World War. In 1916 W. D. Coolidge of General Electric invented a new type of tube containing a target made of tungsten, then being developed in ductile form for use in incandescent electric light bulbs. Generally speaking, three main types of tube were called for: small tubes working at around 70 000 volts for use mostly in dentistry; medium tubes working at 100 000 volts for use in small institutions and some private surgeries; high-voltage tubes for use in large hospitals and radiological centres. By 1930 C. C. Lauritsen, of the California Institute of Technology, had built a 600 000-volt X-ray tube and the million-volt threshold was crossed by General Electric in 1939. The high voltages involved in the use of even small machines created a serious hazard, but in 1921 General Electric introduced a safety unit in which the energizing equipment was immersed in oil in an earthed steel tank.

High-voltage X-ray equipment is used for the treatment of deep-seated tumours where a considerable thickness of normal tissue has first to be traversed before the target is hit. Various devices were developed to ensure that while cancerous tissue remained under attack the exposure of surrounding tissues was much less. One such device adopted a cross-fire technique using several beams. Another rotated either the patient or the X-ray tube while keeping the beam firmly focused on the tumour. Alternatively, seeds or needles containing radium – or, after the 1940s, artificial radioactive isotopes – could be embedded in the cancerous tissue. Seeds containing radon were valuable in that they lost virtually all their radioactivity in about three weeks and it was not necessary to remove them.

IV. ELECTROCARDIOGRAPHY AND ELECTROENCEPHALOGRAPHY

Until 1900 irregularities of the heart were diagnosed largely by the use of the stethoscope, taking the pulse, and measuring the blood pressure. These remain important aids to diagnosis, but in 1903 Willem Einthoven of the University of Leyden built the first electrocardiograph, embodying a string galvanometer, to record the electrical activity of the contracting muscle of the heart. This proved a most valuable aid to the diagnosis of coronary artery disease, especially as a result of work done between 1910 and 1935 by Sir Thomas Lewis in London and F. N. Wilson in the USA on the correlation between clinical data and the tracings of the electrocardiograph, which was second only to X-rays as a physical aid to diagnosis. By 1950, with the introduction of valve amplification and improved methods of analysing the results, the electrocardiograph had become a highly sophisticated and widely used instrument.

A closely allied device, used for the diagnosis of organic diseases of the brain, was the electroencephalograph. The first of these was built by Hans Berger, of Jena, in 1929. It attracted little attention among either psychiatrists or neurologists until the support of the neighbouring firm of Carl Zeiss enabled him to build an improved instrument.

V. WATER SUPPLY AND WASTE DISPOSAL

Apart from their both being important facets of the public health services, water supply and waste disposal are appropriately considered together, as a major problem in supplying water is to avoid contamination by sewage. In the nineteenth century such contamination was a major source of outbreaks of water-borne diseases such as cholera and typhoid. The remarkable drop in the death-rate from typhoid – from 310 per million in 1900 to 2 per million in 1947 in the US – is largely a reflection of improvement in the purity of the water supply. Apart from purity, a major problem in the twentieth century was the sheer volume of water required. Increasing populations, greater personal cleanliness, flush sanitation, and the growing demands of industry all contrived to create problems. One result of this was the initiation of ambitious projects controlling water supply over a large area; these were often combined with hydro-electric schemes, navigational works, and irrigation. One of the best-known examples is the Tennessee Valley Authority established in the USA in 1933. Another very important scheme inaugurated just within our period was the Snowy Mountains scheme in Australia (July 1949). It was designed to provide 3.74 million kW of electricity and 2.5 cubic kilometres of irrigation water per annum.

To a large extent the establishment of a major source of water was a matter of civil engineering, using techniques already discussed in relation to the needs of transport (Ch. 19). Surface water was contained by dams, many of them earth embankments with a core of puddled clay. Increasingly, however, and especially in America, rock-filled dams were used, and by 1900 masonry dams were not uncommon. The most important new material, however, was concrete. Many dams apparently built of masonry are in fact concrete dams faced with stone. The 84-metre Shin Mun Dam, built in 1933–6 to provide water for Hong Kong, was a hybrid: a huge concrete thrust-block was backed by a massive rock-fill. Very often, of course, the local availability of material strongly influenced the type of construction used, and design was increasingly governed by the better understanding of soil mechanics that developed in the 1930s, thanks in particular to the researches of K. Ierzachi, who worked in both Europe and the USA.

Apart from direct extraction from rivers, water had long been extracted by means of wells. In the twentieth century the techniques of the rapidly growing oil industry – particularly the replacement of percussion drilling by rotary drilling – were adopted in the search for water. Diameters of the wells tended to increase, 1 metre being common and 3 metres not exceptional. At the beginning of the century steam-driven reciprocating pumps were used to bring the water to the surface, but these were replaced successively by centrifugal pumps (c.1910) and by submersible electric pumps in the 1930s. Distribution continued to be largely through cast-iron pipes, though asbestos-cement pipes began to be made in Italy in the First World War – with the advantage that they were corrosion-free – and by mid-century polyethylene was coming into use.

However carefully collected, reservoir water must, of course, be treated to make it safe for drinking; in this field there were two major developments in the twentieth century. Up to 1900 the main process was filtration through fine sand. The sand was not in itself sufficient to remove the microscopically small bacteria, but the biological film which quickly developed on it gave it a much more selective action, though it slowed down what was already a slow process. Although fine-sand filtration continued in use, it was increasingly replaced or complemented by coarse-sand filters used in conjunction with a coagulant, usually aluminium sulphate. This produced a flocculent precipitate which had an effect similar to that of the biological film in fine-sand filters.

The other major development was disinfection of the water supply before distribution. Initially bleaching-powder was used, the active ingredient being chlorine. As early as 1896, however, experiments were carried out in America on the disinfection of water with chlorine gas, supplied in cylinders. This became general practice from shortly after the First World War despite considerable opposition, reminiscent of that still encountered in connection with the introduction of fluoride as a means of preventing dental decay: fluoridation was first practised in the USA in 1945. In France, and to a limited extent elsewhere, ozone was used as an alternative to chlorine for disinfection, one advantage being that it does not affect the taste of the water.

In the nineteenth century the growing use of the water-closet resulted in disposal problems. The common solution was to discharge sewage into the drains carrying off surface water and ultimately emptying into rivers or the sea. This caused pollution – especially in the hot dry summer months – which was not only unpleasant but dangerous, for much drinking water was drawn from rivers. There were, therefore, obvious advantages in adopting the so-called separate system, in which sewage is separately collected and processed before discharge. The main, and usually overriding, argument against this was the expense. Nevertheless, in the twentieth century the separate system was increasingly used for new developments and when replacing old ones. Up to about 1930 main sewers continued to be built in brickwork, and many such still exist, but thereafter greater use was made of concrete, either poured *in situ* or in the form of pre-cast pipes and ancillary units such as inspection chambers. For smaller drains, earthenware pipes were used up to the First World War, but afterwards asbestos-cement pipes – similar to those used for water distribution – were introduced, as well as pitch-fibre pipes.

For reasons of health as well as amenity, it was desirable to treat raw sewage before discharge into rivers or the sea, though the standard set depended on local circumstances. Thus waste discharge from small urban communities into large fast-flowing rivers – as in North America – naturally presented less of a problem than the discharge from some of the great conurbations of Europe situated on relatively slow and sluggish rivers. From the 1890s until the 1920s the main purification system was biological degradation during filtration through beds of coke, gravel, or similar relatively coarse material promoting fairly rapid flow. This process was still

Mogden Sewage Treatment Works, near London, 1936.

in use in 1950, but after the First World War increasing use was made of the faster activated-sludge process, in which aeration was combined with sedimentation in tanks.

BIBLIOGRAPHY

Berger, K. W. *The hearing aid: its operation and development*. National Hearing Aid Society, Detroit (1970).

Brecher, R. and Brecher, E. *The rays: a history of radiology in the U.S. and Canada*. Williams and Wilkins, Baltimore (1969).

Curuchet, P. D. *The origin, evolution and modification of surgical instruments*. Buenos Aires (1964).

Davis, L. (ed.). *Fifty years of surgical progress*. Franklin H. Martin Memorial Foundation, Chicago (1955).

Dickinson, H. W. *Water supply of Greater London*. Newcomen Society, London (1954).

Duncum, Barbara M. *The development of inhalation anaesthesia*. Wellcome Historical Medical Museum: Oxford University Press (1947).

Duthie, E. S. *Molecules against microbes*. Sigma, London (1946).

Johnson, S. L. *The history of cardiac surgery 1896–1955*. Johns Hopkins Press, Baltimore (1970).

Macfarlane, Gwyn. *Howard Florey: the making of a great scientist*. Oxford University Press, Oxford (1979).

Major, Ralph H. *A history of medicine*. Thomas, Springfield, Ill. (1954).

Marton, L. *Early history of the electron microscope*. San Francisco Press (1968).

Metropolitan Water Board. *London's water supply, 1903–1953*. Staples Press, London (1953).

Parish, H. J. *Victory with vaccines: the story of immunisation*. E. & S. Livingstone, Edinburgh (1968).

—— *A history of immunization*. E. & S. Livingstone, Edinburgh (1965).

Sidgwick, J. M. and Murray, J. E. A brief history of sewage treatment. *Effluent and Water Treatment Journal*, **16**, 65, 193, 295, 403, 515, 609 (1976).

Smith, N. A. F. *A history of dams*. Peter Davies, London (1971).

Snowy Mountains Hydro-Electric Authority. *Snowy Mountains scheme, Cooma, NSW* (1963).

29

MILITARY TECHNOLOGY

To define military technology, especially in the context of the twentieth century, is by no means easy. Clearly any definition must comprise the main instruments of war, such as rifles and artillery, battleships and submarines, aircraft and tanks. But in the present century, and especially in the Second World War, nations were involved as a whole and not merely through their fighting services. In total war victory depended not merely on success in the field but increasingly also on the ability and will of whole nations to survive under siege conditions. Military technology came to mean also fertilizers to maintain food supplies; canned food for the fighting services; synthetic rubber to keep transport on the road; drugs to treat not only military and civilian casualties but also to maintain the health of the nation as a whole; special alloy steels for armour-plating in warships, gun shields, and tanks. In a sense, therefore, the whole of this book is relevant to military technology. In the present chapter we will restrict ourselves to weaponry as generally understood, though even this must be taken as including such a commonplace civilian material as the barbed wire which, from the 1880s, had been used in vast quantities to enclose the cattle ranches of America and Australia. For many, barbed wire and mud epitomized the battlefields of the First World War.

We must recognize, too, that even at the level of weaponry military technology had only a limited independence. In the main, existing technologies were adapted to military needs, and certainly in the First World War, and to a considerable extent in the Second, this adaptation was slow and reluctant, and more often the result of outside pressure rather than demand from the military services concerned. There is some truth in the observation that each new war began with the weapons that had won the last: certainly the political and military minds of the European powers and of the USA were nothing if not conventional. The reasons for this we cannot pursue here, but probably we need not look much further than the educational and social systems in vogue in the early decades of this century. It is true that science and technology had been increasingly introduced into schools and universities as their commercial value to industry was recognized. The political leaders and the military élite, however, were not drawn from those with this sort of training but were still largely the product of a traditional classical education: worse, they were antipathetic to forms of

In the twentieth century steel had a special role in warfare. Here armour plate is being rolled in the mills of the English Steel Corporation (1940). Power is provided by 12,000-hp engines.

education which they regarded as unsuitable for gentlemen. Generally speaking, the military and political hierarchy was very slow to grasp any but the most direct implications of new technology. However, as the great powers were alike in this, the balance of power was not disturbed. Nevertheless, by mid-century events had shown that in the long run technological superiority was a prerequisite for victory, a concept that is now firmly embedded in military strategy.

Although it was the Second World War that first saw a policy of total national involvement adopted by the principal contestants, the beginnings were apparent in the First. In Britain, for example, it was clear by the spring of 1915 that Winston Churchill's maxim 'Business as Usual' could not be realized: a long war of attrition stretched ahead. It thus became essential to reorganize industry both to supply the armed services on a scale far greater than expected and to secure the basic needs of the civilian population in the face of an intensifying blockade by submarines that, in the spring of 1917, was to become critical. It was in this situation that Lloyd George, in 1915, set up the Ministry of Munitions. Effectively, this assumed control of all sections of industry concerned with the war effort; by the end of the war it had a staff of 65 000 and controlled three million workers.

Although a temporary expedient adopted to meet a relatively short-lived emergency, the Ministry of Munitions had far-reaching effects: it conditioned industry to a degree of government participation in its affairs that would have been unacceptable before 1914. In the inter-war years, therefore, government involvement with technological development became firmly established – as, for example, through the Department of Scientific and Industrial Research in Britain and the TNO organization in the Netherlands.

Leaving such generalities aside, there is no simple way of making a general survey of military technology, for some topics are relevant to all forms of warfare, whereas others are more limited. Thus firearms have much in common whether they are the infantryman's rifle, an air pilot's machine-gun, or the big guns of a battleship. The development of radio opened up new possibilities for swift long-distance communication for all the services. By contrast, submarines, aeroplanes, and tanks are peculiar to specific branches of the armed services. For want of an ideal solution we will, therefore, consider the technological aspects of war on land, sea, and air and then try to draw some general conclusions.

I. GROUND WARFARE

In 1900 the key weapons for land forces were the rifle – with bayonet for fighting at close quarters – the machine-gun, and artillery. Although cavalry still had a distinctive role to play where the circumstances of the terrain and the nature of the opposing forces were favourable – as in the Boer War (1899–1902)—their role was diminishing. Not until the Second World War, with the tragic deployment of Polish cavalry against German tanks, was it finally realized that cavalry had had its day. Nevertheless, up to 1918 the horse still dominated the transport field, both for moving supplies and for drawing heavy guns.

Although a variety of rifles were in use, they were all similar – a breech-loading repeater with a magazine. To some extent their effectiveness depended on the discipline and training of the troops using them; it is said that when the Germans first encountered British riflemen in 1914 the rate of fire was so rapid that they thought they were machine-gunners. This weapon had a great advantage over the old muzzle-loader, still widely used in more primitive parts of the world, in that it could be loaded in the prone position. P. M. E. Vieille's smokeless powder (1886) further changed the situation by making it difficult to locate an enemy marksman. It was claimed, not without reason, that a single rifleman had become equivalent to a platoon of musketeers.

As early as 1866 the US Ordinance had accepted R. J. Gatling's machine-gun, but at the turn of the century the most widely used weapon of this kind was the Maxim, in which the power of the recoil loaded the next round. In 1887 Hiram G. Maxim, at a demonstration in England, fired 666 rounds in one minute; the Maxim's predecessor, the Hotchkiss (1872), fired only 33. With such a gun, one man could equal a former battalion: in the Matabele War (1893) four Maxim machine-gunners successfully fought off 5000 African tribesmen.

Artillery was similarly improved. In the nineteenth century the rate of fire was slow because, owing to the recoil, the gun had to be relaid after every shot. This difficulty was overcome in the famous French 'Seventy-five' which came into production at the beginning of this century. In this, the recoil was absorbed by a hydraulic mechanism, and twenty aimed shots could be fired in a single minute by a skilled crew.

These developments, limited though they were in modern terms, emphasized the growing importance of economic factors. It is reported that when the Chinese first saw a demonstration of machine-gun fire, their reaction was that they could not afford the ammunition. Half a century later, the atom bomb was to be limited, at least for a time, to the very few wealthy nations which could afford the huge investment necessary for its production.

Tanks. Such were the weapons with which the Second World War began on land in 1914; such, with relatively few exceptions, were the weapons with which it ended in 1918. The most important exceptions were gas and the tank. Although the tank as a major military weapon is entirely a twentieth-century development, the idea was by no means new. Descriptions of armoured vehicles in which soldiers might safely approach their objectives – especially siegeworks – abound in Medieval and Renaissance literature; Leonardo (1572) and Ramelli (1588) were among those who made proposals for mobile 'battle cars'. All were necessarily impracticable, however, in the absence of a suitable source of motive power. The advent of the steam-engine gave new life to such proposals and nineteenth-century literature contains many proposals for steam-powered fighting vehicles; moreover a number of them embodied the very important concept of 'caterpillar' traction – an endless chain round the wheels which, in effect, laid a continuous track over rough ground. It was, however, the internal combustion engine that finally made the modern tank – like the aeroplane – a practical possibility. It is interesting to note, however, that Winston

Churchill, in a memorandum to Asquith written in January 1915, visualized steam as the motive power. He referred to:

A number of steam tractors with small armoured shelters in which men and machine guns could be placed, which would be bullet proof . . . The caterpillar system would enable trenches to be crossed quite easily and the weight of the machine would destroy all wire entanglements. Forty or fifty of these machines . . . could advance quite certainly into the enemy's trenches with their machine gun fire, and with grenades thrown out of the top.

The germ of this idea was probably a proposal made by Major (later Major-General) E. D. Swinton, as early as October 1914, for a heavily armoured car with caterpillar traction to crush barbed-wire entanglements: the War Office had been actively interested in such vehicles since at least 1908.

Both gas and tanks were mishandled. When the Germans first used gas in the spring of 1915 they were unprepared for its success and thus unable to exploit it. By the end of that year, the Allies had improvised gas masks filled with adsorbent charcoal and also were able to retaliate, especially as the prevailing wind was in their favour: in the event neither side gained any clear advantage. Equally, the British evaluation of the role of tanks – a code word used in official reports to ensure secrecy – was misconceived. In September 1916 some twenty were deployed against the enemy but, although they were highly effective, their success was too limited to be exploited; their main achievement was to give the Germans warning of a new weapon. Not until comparatively late in the war, at Cambrai in 1917, were tanks deployed without preliminary artillery bombardment in large numbers (378), but again their success could not be exploited. Nevertheless, as a battle Cambrai was a major success and vindicated the claims made for the tank as a means of breaking the deadlock on the Western Front. In the spring and summer of 1918, however, things were very different. Ludendorff's great offensive, which brought his artillery within earshot of Paris, was broken by Haig's attack on the Somme supported by massed tanks on 8 August. A major breakthrough was effected on what Ludendorff described as 'the blackest day of the German Army in the history of the war'. Nevertheless, casualties, including mechanical casualties, were very heavy: out of 400 tanks in action only 40 were serviceable three days later. In 1914 the Germans had entered the war with 93 cavalry regiments, but it is surprising that even at this late stage British cavalry were used in a futile attempt to exploit the massive break made by the tanks in the enemy lines.

In the First World War tanks were predominantly a British weapon, some 2000 being built in all. The prototype – Little Willie – was built by William Foster and Co. of Lincoln, and first ran in September 1915. It is indicative of

military thinking that responsibility for their production was assigned to the
Admiralty, on the rather specious plea that they were land battleships. It is
true, however, that tanks were at least partially inspired by the armoured
cars of the Royal Naval Air Service, which had the all-important caterpillar
traction. It was, too, an apt phrase, because the most important future of the
tank was that it conferred on land forces a degree of mobility comparable
with that enjoyed by warships. Even in 1918 there were engagements
between opposing tanks, and in the Second World War tank battles were to
be fought that had much in common with naval ones. Although they
produced rather fewer tanks than the British, the French also used them on a
considerable scale, especially the light Renault. In July 1918 they deployed
some 500 tanks at Soissons. By contrast, the Germans made little use of tanks
in this war: their total force was probably no more than 40, reinforced by a
few captured vehicles.

Even during the war considerable development of the tank occurred: in
Britain, Mark I was ordered at the end of 1915, and when the Armistice was
signed in November 1918 a Mark VIII version was in production, though
never employed. The initial role of the tank was conceived to be that of a
machine-gun destroyer: its long caterpillar tracks could bridge intervening
trenches and its armour-plating gave protection against small-arms fire.
Improvements included simplification of the steering (1917) so that only a
driver was required, instead of the driver, brakesman, and two gearsmen
needed for the Mark I version; heavier (12 mm) armour, proof against
German K-type bullets; and a completely rotatable turret, introduced by
the French for their light Renaults in 1916 but not used in British tanks until
after the war. Speeds were increased also, from 4 m.p.h. to 8 m.p.h., but it
was realized that much higher speeds were desirable. Fast free-ranging
tanks could not only overrun machine-gun nests near the front line but
create havoc in the more lightly defended rear areas.

The great success of tanks in 1918 when properly handled in favourable
terrain made a great impression. Had the war continued into 1919, Britain
alone planned to put 10 000 tanks into the field, half medium and half
heavy. This programme would have included vehicles with a speed of 20
m.p.h. and a range of 200 miles. The USA had an equally ambitious
programme, based mainly on the British Mark VIII and the French
Renault, and contracts were placed for 1450 of the former type and 4400 of
the latter. Additionally, plans were made to produce an American Ford
tank at the rate of 100 per day from the beginning of 1919. As the Germans
had virtually no tanks, even in the last year of the war, nor any longer the
means to make them, these might well have proved the final instrument of

Allied victory. In retrospect, it seems likely that, had the concept of the tank – advocated within a few months of the outbreak of war – been quickly and effectively realized, years of carnage might have been avoided. Nevertheless, these novel ideas were not easily developed in the face of conventional military thinking.

With the end of the war in November 1918 all these plans were, of course, abandoned, and in the years that followed economic stringency restricted developments largely to the drawing-board and prototypes. There emerged the concept of two different kinds of tank: heavily armoured infantry tanks to be used in close co-operation with infantry in attack and fast mobile cruiser tanks designed to make forays deep into enemy territory. For the latter, new forms of suspension had to be developed, of which the most successful was the Christie. This was incorporated in the 10-ton American Christie cruiser tank of 1931, which had a speed of 40 m.p.h.; it was subsequently adopted for the British cruiser tank of 1939, with the same speed, and the Russian T.34 of the same year. The last, somewhat unusual in being of welded construction, was a utility tank lacking many of the refinements usual elsewhere.

Although Germany failed to develop tanks in the First World War the lesson had been learnt. When the Nazis came to power in 1933, and a massive 'Guns not Butter' rearmament programme was embarked upon, the concept of blitzkrieg was developed. This relied on a high degree of mechanization for the transport of troops and guns, and extensive use of tanks. At the outbreak of war in 1939 Germany had some 3000 tanks. The most important were the *Panzer* cruiser tanks, weighing 20–25 tons. As the war developed these were followed by the 45-ton Panthers (1942) – which proved mechanically troublesome – and the giant 67-ton Royal Tigers, which were cruiser and infantry tanks respectively. The Royal Tigers were the heaviest operational tanks of that war. When defeated in 1945, the Germans were found to have plans for far heavier tanks, up to 180 tons, but whether these would ever have materialized is doubtful. Vehicles of this weight present great problems: for example, they are not easily transported to the scene of action and few bridges will bear their weight.

While most tanks of 1939–45 came into the category of infantry or cruiser types, a number of specialist types were also evolved. One of the most important of these was the flail tank, which literally flailed the ground with heavy iron chains as it advanced in order to detonate land-mines, one of the most effective anti-tank weapons. Others produced the same effect by means of a heavy roller. To cross deep and wide depressions some tanks were equipped with portable bridges giving spans up to 10 metres. Some other tanks again, such as the British Valentine and the US Sherman, were made amphibious; they were used with great effect in the Normandy landings and in the crossing of the Rhine.

Artillery. Developments in artillery were concerned more with technique than with direct technology. Indirectly, new technology had important consequences; in particular, motorized artillery was in almost all circumstances more easily and quickly moved and manœuvred than its horse-drawn counterpart. Radio communication – dispensing with the need for field telephones with their vulnerable cables – made fire-control vastly easier. The British, American, and Russian armies relied heavily on the concentration of artillery fire against enemy targets. The Germans had a relatively low artillery establishment for their infantry divisions, and in their first successes – especially in Europe in 1940 – their commitment to a mobile war relying mainly on tanks and aircraft was fully justified. After 1942, however, when they increasingly lost control of the air, their weakness in artillery proved a serious disadvantage. Concentrated artillery fire was a major factor in the British successes at El Alamein in 1942, perhaps the turning-point of the war.

The early success of German tanks was partly due to the inadequacy of opposing anti-tank weapons. The British 2-pounder of 1939 failed to penetrate heavy German armour, but the story was different when from 1942 6-pounders, and later self-propelled 17-pounders, became available. At the same time, as we have noted, Allied tanks began to match those of Germany.

The tank was not the only new weapon that had become of major importance since 1918: the role of aircraft too had totally changed by 1939, and they became an important artillery target. At the outbreak of war, however, none of the combatants had adequately prepared their anti-aircraft defences. The guns were inadequate and so were the sonar tracking devices; overall, the scene had changed little since the Armistice in 1918. In Britain the heavy weapons – as opposed to machine-guns used for defence against low-level attacks – were largely 3-inch guns. Fortunately, the British Expeditionary Force preferred their essentially obsolete weapons to the superior 3.7-inch guns then becoming available, and thus after the fall of France the latter were still largely available, though in painfully small numbers, for the defence of Britain. As an immediate stopgap large numbers of rocket projectiles were produced. In defence against night attacks searchlights initially proved more of a hindrance than a help, indicating target areas without greatly assisting the gunners. Later, however, especially with the advent of radar, they greatly assisted fighter pilots by pinpointing their targets.

Heavier armour was countered by more penetrating missiles. This picture
shows a 2-cm tungsten-carbide armour-piercing projectile embedded in 11-
cm tank armour plate.

In the event, anti-aircraft artillery made a substantial contribution to the
Battle of Britain, destroying 337 German aircraft; by the end of the war the
total had risen to 833. In addition, when Germany launched her flying-
bomb attacks towards the end of the war (June 1944), anti-aircraft fire
brought down nearly 2000 out of a total of just over 4000 destroyed. At the
height of this attack a decision was taken to move the whole of London's AA
defences to the south coast. This was completed in four days and involved
nearly three million lorry-miles of transport, another example of the crucial
role of motor vehicles.

The above figures, for Britain, are indicative of the general situation. In
all theatres of war, including the war at sea, anti-aircraft guns played a role
which was scarcely imaginable at the end of hostilities in 1918. A very
important development was the proximity fuse, which in effect was a
miniature radar transmitter and receiver which would detonate an anti-
aircraft shell at the point at which it would do maximum damage. It was
successfully developed for the US Navy's 5-inch anti-aircraft gun – first used
only in the Pacific to avoid risk of one falling into enemy hands – and
improved by a factor of ten the performance of pre-set fuses. It was used in
Europe in 1944 and very materially contributed to the defeat of the German
V-1 assault against Britain. German attempts to produce such a fuse failed.

II. WARFARE AT SEA

Like many other technologies, naval technologies followed two lines. On the one hand was what may be termed an evolutionary line: the development of existing weapons by the application of new techniques. On the other hand, was the introduction of wholly new weapons. An example of evolutionary development was the British *Dreadnought* of 1906. With its massive armament, both offensive and defensive, this was in a class by itself; no other existing warship could match it. The most important of the new weapons, which had no direct military ancestors, were the submarine and the aeroplane.

Submersible self-propelled craft date back at least to 1620, when a submarine propelled by twelve oarsmen operated for several hours in the Thames at London at a depth of a dozen feet; it was built by the Dutch engineer Cornelius van Drebbel. In the War of American Independence an American submarine designed by David Bushnell unsuccessfully attempted to attack the British warship *Eagle* at anchor off New York. In 1864, during the Civil War, the *Housatonic*, blockading Charleston, was sunk by a Confederate submarine which was itself lost in the attack. These were sporadic episodes, however, and at the beginning of the twentieth century the submarine was of trifling importance. In 1901 the French navy had a dozen 30-ton submarines, electrically propelled, based on the prototype *Gymnote* of 1888. The US Navy had one – the 120-ton *Holland*, so called after its designer J. P. Holland – and five more under construction. The British Navy had five Holland-type craft under construction at Barrow. Germany and Italy each had two submarines. Thus the entire world's submarine fleet at that time totalled only about thirty vessels.

Although insignificant as a naval force, these early submarines embodied most of the essential features of their successors. While the earliest French submarines relied for propulsion entirely on electric motors supplied by storage batteries which could be recharged only in port, a dual system soon became the general rule. When running on the surface submarines used conventional power units – at first steam but later internal combustion engines – but electric motors, which required no air-intake or exhaust gas disposal, were employed during submersion. The engines were used to recharge the batteries at sea, an operation that could be carried out only on the surface. As the charging process resulted in copious production of explosive hydrogen gas, careful attention had to be given to ventilation. Not until 1955, when the nuclear-powered USS *Nautilus* was commissioned, was a power unit available which enabled a submarine to remain submerged and mobile for indefinitely long periods and thus realize its full military potential.

While there is advantage simply in being submerged and unsuspected, an operational submarine must have some means of observation. In the very earliest types this was effected by means of an observation dome fitted with portholes; the submarine had just to break surface – and thus reveal its presence – for the commander to take a sight. As early as 1902, however, the optical periscope had appeared; although giving a narrow field of view, say 15°, it was fully rotatable. Initially it was designed only to scan the horizon, but as aircraft presented a new threat in the Second World War the sky, too, had to be searched.

From the outset, the principal weapon of the submarine was the torpedo, invented by Robert Whitehead in 1866 and much improved by him and his son in the latter years of the nineteenth century. The problem of precise directional control was not fully solved until 1896, when a gryo-compass was fitted. The Whitehead torpedo was licensed to the principal naval powers and demonstrated its effectiveness in the Japanese attack on the Russian fleet at Port Arthur in 1904. Up to the outbreak of the First World War few submarines carried guns, but during the war this became normal practice, not so much for attacking other warships as for intimidating weaker vessels. Again, the new circumstances of the Second World War demanded the addition of anti-aircraft guns.

In the inter-war years the great naval powers developed two main classes of submarine: a relatively small (600-ton) type equipped with 4–6 torpedo tubes and a larger (1200-ton) type fitted with 6–8 torpedo tubes. Each tube could fire two torpedoes, one carried in the tube itself and the other standing by in the torpedo room. All vessels carried guns of approximately 4-inch calibre. Some submarines were equipped for laying mines. In the Second World War midget submarines were developed with a crew of two or four men, but although they had a few spectacular successes – such as the disabling of *Tirpitz* in Alta Fjord in 1943 – their total contribution was very small.

One reason for the slow development of the submarine up to 1914 was the Geneva Convention which provided that no merchant ship was to be sunk without warning and that the crew should be obliged to take to the boats only within reasonable distance of land. It was difficult for a submarine to observe these provisions without unduly endangering itself. The German way out of this dilemma was simply to inaugurate, in January 1915, a policy of sink-at-sight, which enabled the Germans to make very effective use of their new fleet of submarines, greatly increased from the twenty-eight with which they entered the war, as indiscriminate commerce raiders. Such was the success of this campaign that early in 1917 essential Allied imports were so greatly impaired that victory hung in the balance. However, this marked a turning-point in the hostilities, and thereafter the campaign was much less effective. In all, Allied shipping losses totalled some 20 million tons. The Second World War followed a similar course. Germany gave clear warning of her intentions with the sinking of the liner *Athenia* on the first night of hostilities; by 1942 her submarine fleet had increased from 60 to 250. As in 1914–18, the Allies were forced to adopt a convoy system. For their part, of course, Allied submarines were active in all theatres of war. Within the space of less than half a century the submarine had risen from literally nothing to be a major naval weapon. In this time perhaps the only major innovation was the schnorkel, a Dutch invention, quickly adopted by the Germans (1944–5). This was an inconspicuous double-ducted tube, extending a short distance above the surface, which enabled the batteries to be charged while the vessel remained submerged.

In the latter part of the First World War aeroplanes were increasingly used to locate and attack targets at sea, and as early as August 1915 a British seaplane torpedoed an enemy ship. Later, aircraft (often seaplanes) were carried on board ship where they were of great value for scouting purposes and reporting the fall of shot. They were launched by catapult devices but their recovery was hazardous and uncertain: the pilot had either to land his aircraft in what might be a rough sea or hope to find a landing place ashore. Moreover, ditched planes could be recovered only when the ship was stationary and therefore vulnerable. In September 1918, however, a notable advance was made with the commissioning into the British Navy of *Argus*, the first aircraft-carrier in the modern sense, with through-deck facilities for transporting aircraft from their hangars to a flight-deck which they could use for both take off and landing.

Argus never saw action, and it is thus fair to say that the aircraft-carrier played no part in the First World War. The situation was very different in 1939–45, however, when all the major naval powers relied heavily on custom-built carriers: Britain had 11, the USA 6 with 6 more under construction, and Japan 7. The importance of the aircraft-carrier was convincingly demonstrated by the Japanese attack on Pearl Harbor in 1941, which at one blow immobilized a great part of the US Pacific Fleet. Thereafter the war in the Pacific involved a series of major engagements between aircraft and warships working in close co-operation.

The details of these highly sophisticated vessels are quite beyond the scope of this work and we can note only some major principles of their construction. Aircraft-carriers of the USA differed from those of other countries in that their operational aircraft were kept on the flight-deck; others, seeking to assist pilots by providing as unencumbered a landing space as possible, kept all planes below the deck except at take-off or landing. Attempts to clear the landing-deck completely by eliminating bridge, funnel, etc. proved unsatisfactory and virtually all operational carriers had an 'island' structure. The necessarily short landing-strip (600–900 feet) presented the pilot with a grave risk of overshooting and landing in the sea. To avoid this hazard, tensioned ropes were stretched across the deck in order quickly to reduce the speed of landing aircraft; safety nets provided an additional precaution.

The defence of aircraft-carriers presented a special problem, in that no evasive action could be taken while aircraft were taking off and landing: at other times the carriers' high speed, normally in excess of 30 knots, was a valuable safety factor. Moreover, they carried formidable batteries of anti-aircraft guns and were powerfully escorted.

At sea, in both world wars, mines were important defensive weapons, and we have already referred to the use of submarines for laying them. Most, however, were laid by surface craft. Even in 1900 they were no novelty, having been used at least as early as the Crimean War. In the First World War they were largely of the contact type; that is to say, they detonated only when the projecting 'horns' were struck by a passing vessel. In the Second World War much more sophisticated devices were used, in particular the magnetic mine, introduced by the Germans in 1940, which was detonated by the magnetic field of a steel hull; this was quickly countered by a degaussing technique in which this field was exactly neutralized by an electromagnetic field constantly generated by the ship's own dynamos. Acoustic mines were also used. It is estimated that during the Second World War half a million mines were laid by the combatants, and all concerned had to divert a great deal of effort to sweeping enemy minefields.

III. AERIAL WARFARE

Aircraft proved their military importance in the First World War. In the Second, they proved indispensable: in general, victory in any theatre depended on command of the air. This the Germans failed to achieve in 1940, and as a direct consequence they had to abandon their plan to invade Britain and thus secure to themselves the whole of western Europe. By 1944, Allied superiority in the air was undisputed and thus the Normandy landings became feasible. The main technological developments have already been discussed in the context of air transport (Ch. 22), for the shape of civil aviation was profoundly influenced by military programmes.

The use of military aircraft suffered from one major constraint that did not apply to tanks on land or to ships at sea: the overriding importance of keeping the weight as low as possible virtually precluded protection by armour-plating. Aircraft had to depend for their defence on their own

Supermarine Spitfire, perhaps the most famous of fighter aircraft. It was of stressed-skin construction with a liquid-cooled piston engine.

weapons or evasive tactics. Fighter aircraft had to rely primarily on themselves, with such supporting fire as other aeroplanes in the same formation could give in the confusion of aerial combat. Bombers had some facilities to defend themselves, and were supported by escorting fighters so far as the latters' range permitted. The later American bombers of the Flying Fortress type, favouring daylight rather than night raiding, each carried a formidable array of guns and flew in close formation. The total fire that could potentially be directed against an attacking fighter force was enormous.

The main developments lay in the scale of operations, the operational techniques, and – above all – in the application of all kinds of new technology to the problems of war in the air. Radar, for example, not only vastly improved navigation for aircraft of all kinds but also made it possible for fighter aircraft to find their bomber targets, and bombers their ground targets, with a precision inconceivable in 1939. Such devices generated their own counter-measures. For example, the Germans confused British radar by scattering clouds of metallic ribbon cut into lengths that would interfere with the radar wavelength. The availability of larger aeroplanes, and thus of larger bomb-loads, influenced the whole strategy of the war. In particular, it led to the concept – the validity of which is still argued – that an enemy could be brought to defeat by massive bombing of industrial and military targets far behind the front lines. In pursuit of this policy, reliance was no longer placed solely on the destructive effects of high explosives: it was

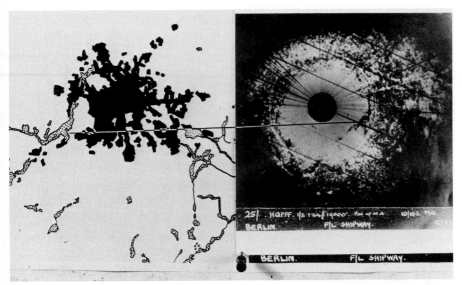

Wartime H$_2$S picture of Berlin.

realized that cascades of lightweight incendiary bombs – a sort of twentieth-century version of Greek Fire – could wreak havoc, especially when earlier systematic bombing had disorganized the fire-fighting services. Random bombing attacks gave way to highly organized ones in which the pattern of bombing was determined by statistical analysis in order to achieve the greatest destructive effect.

How far the weakening of the German economy by bombing (particularly the disruption of oil supplies) was responsible for Germany's ultimate defeat in 1945 may be arguable, but it is a matter of fact that the capitulation of Japan was an immediate and inescapable response to the atom bomb attacks on Hiroshima and Nagasaki on 6 and 9 August 1945. At that stage of the war the ultimate defeat of Japan by concentrated attacks with conventional weapons was clearly inevitable, but nevertheless a fierce raid on Tokyo in March 1945, which caused more deaths than either of the atom bomb attacks, seemed to effect no diminution of the Japanese will to continue to support what must already have seemed to many a lost cause.

The principles of the atom bomb, the technology of its production, and the mechanisms developed to detonate it have been described elsewhere (Ch. 5), but in order satisfactorily to round off this section on aerial warfare we must extend our history a little beyond mid-century. After major wars the combatants are normally content to disband their fighting forces and restore their industrial economies before devoting very much effort to modernizing their weaponry. This was not so after the end of hostilities in

1945. The partnership between the western Allies and Russia had always been an uneasy one, the main bond being a common enemy; the unquestionable dominance given to the USA by possession of the atom bomb left Russia at the end of hostilities in an unacceptable position of military inferiority which, as events were to prove, was not matched by technological inferiority. It was a position, therefore, that she had no necessity to accept.

Two immensely important new lines of development had been opened up by the V-2 rocket and the atom bomb. As early as March 1947 the Russians, having closely examined captured V-2s, determined in principle to develop an intercontinental ballistic missile (ICBM) – that is to say, a rocket capable of reaching the USA from Europe. The Russian T-3 – with a range of over 5000 miles – was successfully tested in August 1957 and any doubts about the validity of their claims were dispelled with the launching of Sputnik I only a few weeks later. The American Thor, with an intermediate range of about 1200 miles, appeared in 1958, and the Atlas D, comparable in performance with T-3, in the following year. As the American monopoly of atomic weaponry had been broken by the Russians in 1947 – thus permitting a nuclear warhead – a new era of military strategy was opened, the scope of which was still further extended by the launching of the first US nuclear-powdered submarine *Nautilus* in 1955, followed by a Russian version in 1962. In that same year the first American submarine capable of launching an ICBM was launched. In barely a decade the role of the heavy bomber, which had been dominant at the end of the Second World War, had been very substantially diminished in favour of unmanned missiles. Its utility was further diminished by the development of a range of anti-aircraft missiles – such as the Nike Hercules (1958) – which actively sought out their targets.

IV. SOME GENERAL CONSIDERATIONS

In its narrowest sense military technology concerns itself with what weapons were developed and how they were made. Clearly, however, far broader issues than this have to be taken into consideration. Commitment to a major strategic policy – such as massive bomber attacks against enemy industrial areas – in itself absorbs so much in the way of resources that other options no longer remain open. The transport of large numbers of troops, tanks, artillery, aircraft, and other weapons to a distant theatre of war – such as North Africa – automatically precludes not only major offensives elsewhere but also weakens the defences of the homeland. Technologically, the D-day landings in 1944 – when the Allies decisively moved again to the offensive – were a matter of landing-craft, aircraft, gliders, artillery, and the like. But assembling this vast force without disclosing its intention to the enemy; collecting the necessary stores and loading them in such a manner that they were available in the right order on landing; supplying fuel through a specially laid underwater pipeline (Pluto); creating instant harbour facilities (Mulberry); ensuring the availability of prefabricated (Bailey) bridges—all demanded managerial skills of the highest order.

In wartime, the need to survive is all-important. In peacetime, economic considerations become of far greater importance. The cost of the sophisticated and exceedingly expensive weaponry that was developed immediately after the Second World War had to be balanced against the strain it put on national resources. Democratic countries were at a disadvantage compared with totalitarian ones, in that governments had to persuade, rather than merely inform, their peoples that such massive expenditure was in the national interest. Persuasion had to be undertaken in the face of powerful pacifist and environmental lobbies which, quite legitimately though perhaps not wisely, believed that their own causes were of higher priority than national defence. Moreover, those responsible for national

Prefabricated Bailey bridges were widely used by the Allied forces in the Second World War.

defence, and the development of essential new military technologies, had no special gift of prophecy and had to guess as best they could at the future trend of power politics.

BIBLIOGRAPHY

Barnes, G. M. *Weapons of World War II*, New York (1947).

Bethell, H. A. *Modern guns and artillery*. London (1907).

Brodie, Bernard and Brodie, Fawn. *From crossbow to H-bomb*. New York (1962).

Charbonnier, P. *Histoire de la ballistique*. Paris (1928).

Christmas, J. K. Manufacture of high-speed tanks. In USA *Army Ordnance*, Washington, DC (1939).

Dommett, W. E. *Submarine vessels including mines etc*. London (1915).

Dornberger, W. *V-2*. New York (1954).

Emme, E. M. (ed.). *The impact of air power*. New York (1959).

— *The history of rocket technology*. Detroit (1964).

— *Aeronautics and astronautics: an American chronology of science and technology in the exploration of space 1915–1960*. Washington, DC (1961).

Equevilley, R. d' *Bateaux sous-marins*. Paris (c.1900).

W. Foster and Co. *Tank; birth and development*. Lincoln (1919).

Fuller, J. F. C. *Armament and history*. New York (1945).

Hovgaard, G. W. *Submarine boats*. London (1887).

Hutchinson, G. S. *Machine guns: their history and tactical employment*. London (1938).

Low, A. M. *Musket to machine gun*. London (1943).

— *Submarine at war*. London (1941).

Kissinger, Henry A. *Nuclear weapons and foreign policy*. New York (1957).

Mahan, A. T. *Mahan on naval warfare*. Westcott (1919).

Martel, G. Le Q. *In the wake of the tank: the first fifteen years of mechanization in the British Army*. London (1931).

National Aeronautics and Space Administration. *Documents in the history of NASA: an anthology*. NASA History Office, Washington, DC (1975).

Newbolt, H. J. *History of submarine war*. London (1918).

Newman, J. R. *The tools of war*. New York (1942).

Schwiebert. *A history of the U.S. Air Force ballistic missiles*. New York (1965).

Stewart, O. *Strategy and tactics of air fighting*. London (1925).

Sueter, M. F. *Evolution of the submarine from the 16th century*. London (1937).

Swinton, E. D. *Eyewitness*. London (1932).

Tunis, E. *Weapons: a pictorial history*. Cleveland (1954).

Woodward, E. L. *Great Britain and the German Navy*. Oxford University Press, London (1935).

Van Creveld, Martin. *Supplying war: logistics from Wallenstein to Patton*. Cambridge University Press, London (1980).

30

TECHNOLOGY IN THE HOME

Many of the technological developments described in previous chapters had a direct influence on domestic life: as examples we may cite electric light, wireless, the telephone, and canned and frozen food. While it is impossible to identify anything that can properly be called domestic technology, the impact of technology on daily life in the home was so great as to deserve separate consideration. This can be prefaced by three generalizations. Firstly, most of the appliances used in the first half of the twentieth century were developments of ones already in use. Secondly, the centre of innovation was the USA, which as a nation was inclined towards labour-saving devices simply because labour was scarce and expensive. In Europe, where domestic servants were readily available up to the First World War, the incentive was correspondingly less. Thirdly, there was a clearly discernible trend away from manually operated devices towards ones driven by small electric motors. This was made possible by the increasing availability of electricity in the home for lighting and heating.

I. HEATING AND LIGHTING

At the beginning of this century open fires were commonplace, especially for living-rooms. They were inefficient because most of the heat went up the chimney, but fuel was cheap; they had to be cleared and replenished, but labour was plentiful; they were popular, because they were traditional and cheerful. By 1900, however, the supremacy of the open fire for space heating was being challenged. Gas, which throughout the nineteenth century had been predominantly used for lighting, was increasingly being promoted for use in heating, mainly in response to the growing challenge of electricity for lighting with consequent loss of the main market. In Britain, the change of balance is indicated by the Gas Regulation Act of 1920 which required gas undertakings to sell gas on the basis of its calorific value rather than on its illuminating quality, which had been mandatory for nearly a century. The relatively new petroleum industry was producing quantities of low-boiling fractions (paraffin, kerosene) very suitable for small domestic heaters which – like oil-lamps – were particularly useful in the large areas still without either gas or electricity.

Electric fires began to be significant in the 1890s but one of the problems – apart from the then high cost of electricity – was the difficulty of finding a

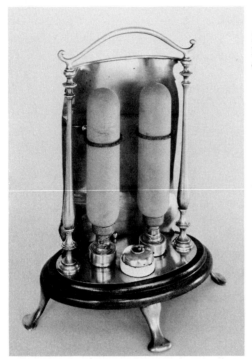

Electric fire by GEC *c.*1914. The two
heating elements consist of glass bulbs
containing carbon filaments.

suitable material for the heating elements. Iron wires lost their strength and
oxidized readily at red heat, but in 1906 A. L. Marsh patented an alloy of
nickel and chromium (nichrome) that was very much more satisfactory.
From then on electric fires based on nichrome wire (often spirally coiled) on
a ceramic base came into general use, frequently with a metallic reflector to
radiate the heat. In the USA, in the 1920s, silicon carbide rods were used as
heating elements. The Bastian fire (1909) embedded the heating wire in a
tube of quartz, a device that was revived after the Second World War.
Another post-war development, prompted by the invention of the cylindrical (tangential) fan by Bruno Eck in 1953, was the fan-heater. In this air
was blown over the heating element, thereby keeping its operating
temperature down and distributing warm air throughout the room. At
about the same time, gas heaters operating on a similar principle became
more generally used; they had first appeared in the 1860s. Air was drawn
over the heated surfaces, without mixing with the flue gases, and returned to
the room.

The coal-fired range typical of the nineteenth-century kitchen succumbed to gas and electricity. Two important developments were common
to both. Firstly, there were great improvements in design. From the 1920s
the old black cast-iron stoves, which had to be polished with black lead,

rapidly gave way to enamelled models which could be wiped clean with a damp cloth and a little cleaning powder. Secondly, automatic temperature-control devices (thermostats) were introduced which enabled cooking to proceed satisfactorily without the constant attention of the housewife. In Britain, the first such cooker was the Davis New World of 1923, fitted with a Regulo. Ten years later the Creda electric cooker appeared with its Credastat.

In the nineteenth century hot water was a luxury: for baths or washing it had to be carried in cans from the kitchen to the bedrooms. By the First World War, and increasingly later, a general supply of hot water was common in middle-class houses and this was achieved in two ways. Firstly, it could be supplied from a central boiler which might also supply a number of hot-water radiators. The boiler was usually fired by solid fuel (often cheap coke) or sometimes gas; oil was not much used in the home for this purpose before the Second World War. Hot water might also be supplied from a central cylinder fitted with an electric immersion heater, thermostatically controlled. Secondly, hot water might be generated where it was actually required, often from a gas-fired geyser of the type invented by B. W. Maughan in 1868. Other makers, such as Ewart, entered this market but the major development came in the 1930s with the advent of heaters fitted with

Gas cooker with lagged oven and thermostatic control, 1923. Enamelled side-panels allow easier cleaning.

Gas-fired geyser, 1914. This advertisement typifies the end of a social era.

powerful aerated burners. They were German in origin, and became familiar in Britain as Ascots or the short-lived Progas.

Up to the Second World War domestic hot-water systems operated on convection principles and demanded relatively large-bore pipes to ensure adequate circulation. After the war small electric pumps became available that could be relied on to function for very long periods without attention, and this made it possible to construct central-heating systems with small-bore tubing.

One of the problems of the electric supply industry is that demand is not continuous but is concentrated in peak periods: during the night it is very small. It is, therefore, feasible to offer cheaper electricity in off-peak periods if a demand can be created. This led to the development of storage heaters, in which heavy blocks of concrete or brick are heated up during the night and release their heat to warm the house during the day. The system was developed in Sweden during the 1920s. It was not applicable to gas since gas, unlike electricity, can be stored at source.

Small quantities of hot water were generally taken from a kettle, and all gas and electric cookers had rings for heating these. Electric kettles were offered at an early date, however, and appeared in Crompton's catalogue as early as 1894. Early models were relatively inefficient, because the heating element was attached to the bottom of the kettle on the outside. In 1921, however, kettles were introduced in which the heating element was inside and completely immersed. The risk with these was that if forgotten the water would boil away and the element would burn out; by mid-century, however, this could be obviated by an automatic cut-off that came into operation when the water boiled. Crompton's 1894 catalogue also listed electric irons; the main improvements in these have been the introduction of thermostats, to enable the temperature to be fixed according to the nature of the fabric being ironed, and – after the Second World War – water reservoirs enabling the fabrics to be automatically damped. Gas-heated steam presses had, however, long been in use in the clothing industry, being introduced about 1900 by a New York tailor, A. J. Hoffman.

The direct use of coal in the home steadily diminished in favour of gas and electricity, but this was no great loss to the mining industry, for coal was necessary to generate both electricity and gas – save in North America, where the use of natural gas was well established by the beginning of the century. Nevertheless, one important new solid-fuel appliance was introduced. This was the Aga cooker, invented in 1924 by a Swede, Gustaf Dalen, who had won a Nobel Prize for physics in 1912. Its ingenious design,

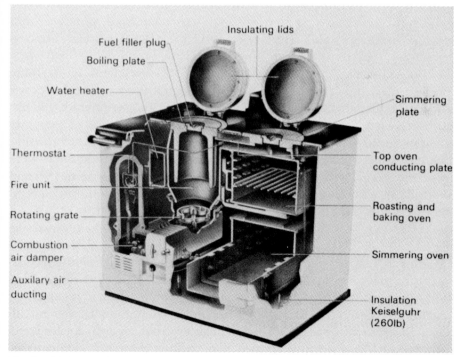

'Aga' solid-fuel cooker, 1946 model.

with careful attention to insulation, made it extremely efficient and economical; it immediately evokes comparison with the then equally revolutionary new stoves designed for domestic use by Count Rumford at the end of the eighteenth century. Originally designed for solid smokeless fuel, some later Aga models could burn oil or gas.

II. REFRIGERATION

The preservation of food by chilling it is of great antiquity: large estates could afford well-insulated ice-houses kept cool by blocks of ice put in during the winter or brought down from the mountains later in the year. Such luxuries were not for ordinary citizens, however; they had to rely on cool larders or cellars. In the 1850s refrigerating machinery began to be developed on a large scale and was used, for example, to import chilled meat in bulk from Australia and South America. Small refrigerating plants were set up in the larger towns and it became possible for fishmongers, hotel-keepers, and other traders to buy blocks of ice cheaply. So, too, could householders, and regular deliveries were possible; many middle-class homes therefore began to regard ice-boxes as standard kitchen equipment.

Not until the 1920s, however, was refrigerating plant built on a small enough scale for it to be used in the home. Domestic refrigerators were first made in the USA, and by 1923 some 20 000 were in use there. Like their industrial counterparts, they worked on the compression principle, and were powered by small electric motors. In most, ammonia was the refrigerant, but from the late 1920s synthetic products (Freons) were introduced; these were less disastrous if a leak developed. In 1922 the Swedish Electrolux firm brought out domestic refrigerators working on a different principle (absorption) which required no motor and thus no moving parts to wear out. All that was necessary was a source of heat; this could be electricity, but gas or paraffin could equally well be used in areas where electricity was not available. In Europe, domestic refrigerators were by no means rare, but they became almost universal only after the Second World War, a development which went hand in hand with the advance of the frozen food industry and the availability of deep-freeze lockers.

III. CLEANING

Cleaning is one of the most tedious of household chores and it is not surprising that much attention was given to methods of mechanizing it. One of the first and biggest successes was the vacuum cleaner, invented in London in 1901 by H. C. Booth who immediately founded the British Vacuum Cleaner Company. At first the plant was clumsy and mounted on a cart; the suction tubes were passed into the house through the doors or

Vapour-compression
refrigerator, 1932; note large
cooling element.

Washing machine with
wooden tub; Beatty Bros.,
Canada, 1920. Water was
externally heated but agi-
tation and mangling was
effected by an electric
motor.

Booth's vacuum cleaner, 1904.

windows. Very soon, however, the firm began to manufacture much smaller machines which could be brought into the house and plugged into the domestic electricity supply. Booth's success brought many rivals, but his patent position was strong. In 1909 Hiram Maxim (better known for his machine-gun) brought out a hand-operated vacuum cleaner, and many other machines of this type followed. Nevertheless, they involved much more physical effort than those fitted with electric motors. Indeed, they did not offer a great deal more than the cheaper and simpler carpet-sweeper which had been developed in the USA in the 1860s and was little changed a century later.

The washing of clothes was another important and regular household task, even though heavier items might be sent out to laundries. Large country houses and institutions had crude mechanical devices in the nineteenth century, but for the ordinary housewife it was a chore that had to be done by hand. Common practice was to heat the water in a coal-fired copper, but in the 1920s gas-fired boilers began to appear. At the same time, in the USA and Canada, machines for domestic use were introduced which mechanically agitated the hot water by means of an electric motor. At first, however, the water had to be heated separately, and it was not until after the Second World War that machines appeared which both heated the water and swirled the clothes around. Many were also equipped with small mangles with rubber rollers – a welcome alternative to the traditional mangle with its heavy iron frame and big wooden rollers. Spin-driers were in use in the USA in the 1920s, but did not appear in Europe until some forty years later.

Dish-washing is an even more frequent task than laundering. Although primitive dish-washers using jets of hot water were manufactured in the USA from the 1860s, their use was limited to large establishments such as hotels and restaurants. Domestic dish-washers were still rare in 1950.

BIBLIOGRAPHY

Bohle, H. The electrical kitchen for private houses. *Electricity* (July 1924).

Booth, H. C. Origin of the vacuum cleaner. *Newcomen Society Transactions*, **15**, 93, 1935.

Frederick, C. *Housekeeping with efficiency*. New York (1913)

— The new housekeeping. *Ladies Home Journal* **29**, 16, 1912.

— *The new housekeeping: efficiency studies in home management*. New York (1913).

Giedion, Siegfried. *Mechanization takes command*. Oxford University Press, New York (1948).

Hull, H. B. *Household refrigeration*. Chicago (1912).

Morton, F. N. The evolution of the gas stove. *Public Service*, July 1908.

Summerton, A. *A treatise on vacuum cleaning*. London (1912).

What women want in their kitchens of tomorrow: a report on the Kitchen of Tomorrow contest. *McCall's Magazine*, p. 155, 1944.

Witte, Irene. *Heim und Technik in Amerika*. Berlin (1928).

Wright, Lawrence. *Clean and decent*. Routledge and Kegan Paul, London (1960).

— *Home fires burning*. Routledge and Kegan Paul, London (1964).

EPILOGUE

In the introductory chapter we considered some of the major factors that influenced the progress of technology in the first half of this century. These included expansion and changes in national systems of education; new attitudes among, and relations between, management and workers; government intervention expressed in such ways as the encouragement of research, restrictive or protective legislation, and modification of the patent systems; and the incentives conducive to innovation. In succeeding chapters we recorded the progress of events in the main fields of technology and the factors that governed it. Now, in conclusion, it is appropriate to consider briefly how these changes, without historical parallel in their extent or speed, affected the quality of human life.

Such an evaluation poses immediate and obvious difficulties, for it takes us from a field of factual description where, broadly speaking, all knowledgeable chroniclers will tell much the same tale, into a controversial and emotive one where there are great differences of opinion. Moreover, as our survey has been on a global scale, the consequences of new technology necessarily vary enormously from place to place and time to time. For a variety of reasons the world's great conurbations tend to be affected first and most diversely by technological change; at the other end of the scale – in Papua New Guinea, for example, or on the Upper Amazon – there still exist people whose way of life has changed little for centuries. Again, different kinds of community within a single nation may have different views as to what constitutes technological progress: in the 1920s, for example, the cinema was making a considerable impact on life in the towns, but in the countryside the advent of motor transport, giving an unprecedented degree of mobility, was generally more significant. Reaction to new technology will be conditioned, too, by local economic and social circumstances. Europe, with its abundance of labour, was less disposed to adopt labour-saving devices than the USA, where labour was not so readily available. In developing countries, with vast numbers of unskilled labourers seeking employment even at low wages, the incentive was even less. It was just such countries, however, which most needed the fruits of modern technology: materials and machinery for irrigation schemes; fertilizers and plant-protection chemicals to improve the yield of crops; drugs to control the most prevalent diseases;

and so on. Technology transfer became one of the major problems of the twentieth century.

Nevertheless, despite such considerable reservations, certain trends can be seen. First, perhaps – and one that brought important consequences in its train – was a very great increase in man's material wealth. For many, this encouraged the hope that the long-sought brotherhood of mankind would be brought demonstrably nearer; the elimination of poverty and disease, it was argued, would remove the main causes of strife and create a milieu favourable to universal peace. In the event no such thing happened: nine years of our half-century were given over to two World Wars and the peace between and after was uneasy in both the military and the industrial sphere. The new wealth was not evenly distributed: if anything the rich became richer and the poor became poorer, aggravating existing social tensions. Moreover the very richness of the rich permitted them to indulge in strife that could not be afforded in less prosperous circumstances. Understandably, new technology brought not only the fear but also the reality of unemployment, and led to bitter disputes not only between unions and employers but also between unions themselves; such disputes were often more prolonged because, in a rich society, both sides tended to have greater reserves and thus greater staying-power. While it can be argued convincingly that new technology created rather than diminished employment, in the sense that at mid-century a very large number of people worked in industries that in 1900 did not even exist, individuals made redundant were often unfitted by age or training to grasp the new opportunities, which were the prize of the next generation.

Equally important was the fact that wealth provided a degree of security that made it possible to stand back and question the means by which it had been created. Technology was no longer satisfying merely basic needs – food, clothing, shelter – for a growing population, but had created new ones. Abundant and sophisticated foods out of season; non-stop entertainment by radio and, later, television; cheap travel, in particular the overseas package holiday by air, were all pleasant new adjuncts to life but certainly not necessities. Paradoxically, perhaps, the first half of this century witnessed the beginning of an anti-technology movement that was quickly to represent a political force too powerful to be ignored.

For many, the first manifestation of this movement – but certainly not its genesis – was the publication of Rachel Carson's *Silent Spring* in 1962, inveighing against the destruction of wild life by the excessive use of synthetic insecticides. That she had a case is undeniable, but unfortunately the chemical industry was slow to respond and to point out the enormous

benefits conferred by these same products. One of the first major successes of
DDT was to stop in its tracks a louse-borne typhus epidemic in Naples in the
winter of 1943/4; DDT made it possible to deinfest 73 000 people in one day.
By contrast, during the First World War there were more deaths from
typhus than from any other cause, including battle injuries. Again, in 1971
the World Health Organization, initiated in Geneva in 1948, was able to
announce that three-quarters of the estimated 1800 million people living in
areas of the world originally subject to malaria were now free of this menace
as a result of mosquito eradication campaigns. Literally tens of millions of
lives were saved and as many more people were spared the severely
debilitating effect of the disease. While land-drainage schemes and other
public health measures played an important part, synthetic insecticides
played a major part in this outstandingly successful campaign. Such
successes must be weighed in the balance against the undeniably harmful
side reactions of insecticides based on chlorinated hydrocarbons, which in
any case were soon superseded by better and more acceptable products.

The large-scale eradication of malaria and other insect-borne diseases is
only one of the contributions of technology to the war against disease.
Sulphonamides, and later penicillin and other antibiotics, so successfully
reduced mortality as to create serious new problems. The paradox was
succinctly stated by A. V. Hill in his Presidential Address in 1952 to the
British Association for the Advancement of Science: 'The conquest of disease
has led to a vast increase in the world's population. The result may be
starvation, unrest, and even the end of civilization. If ethical principles deny
our right to do evil that good may happen, are we justified in doing good
when the foreseeable consequence is evil?'

Symptomatic of the reaction against technology and its consequences was
the tendency – small overall but nevertheless significant – to return to small
communities practising a simple way of life. In this, of course, there is
nothing very new: from time immemorial educated people have by choice,
and not from necessity, retreated from the world, as witness the foundations
of the great religious orders throughout the world. The difference today is
that the motive is generally not the positive one of religious belief but the
negative one of escapism.

In this context much has been made of the so-called alternative
technology, one using simple devices to make life easier but eschewing the
complexities of the most advanced industries. To some extent, however, the
escape is symbolic rather than real, for much alternative technology can in
reality exist only in the shadow of its big brother. For example, to use a
windmill to generate electric power for lighting seems, at first sight, a return
to the simple life – until one remembers that some of the essential

components are quite complex. Windmills themselves have been used in millions from ancient times, but local craftsmen cannot manufacture dynamos, storage batteries, or electric lamp bulbs; the making of the last is a particularly complex process. Moreover, these new primitives find it hard to refuse for themselves – and particularly for their families – the kind of help that modern technology alone can provide in the event of a medical emergency. Such help includes the telephone to summon aid; the motor ambulance for swift conveyance to hospital; a complex array of healing drugs and equipment; and – most important but easily forgotten – the generation of sufficient wealth to make welfare systems possible.

For the hundreds of millions of true primitives – the populations of the developing countries – such temptations do not exist. Their worry is not whether crops have been grown with the aid of synthetic chemicals or wheat ground between steel rollers rather than millstones; too often their anxiety is whether the day will produce food at all. Equally, there are still millions for whom the chance of skilled medical attention at any time in their lives is virtually nil.

Basically, perhaps, the question is one of selectivity: people like technology *à la carte*, choosing what they like and rejecting what they do not. Choose the medical aids, the nylon stockings, the convenience foods; reject the nuclear power stations, the agricultural chemicals, the water supply schemes that change the landscape. The difficulty is that it is the nature of technology to advance on a broad front. It is difficult, if not impossible, to set desirable objectives – supposing that general agreement could be reached on what these are – without at the same time becoming involved in quite different ventures. Areas where progress has become slow are suddenly vitalized as a consequence of seemingly unrelated developments elsewhere; in the present volume we have seen innumerable examples of this. In medicine one of the most important of recent developments has been the introduction of highly active radioisotopes, but these are available only because of the atomic energy programme as a whole. The adoption of the thermionic valve, and later the transistor, transformed the computer and the field of information retrieval, with far-reaching social consequences. Synthetic fertilizers, with their profound effects on agricultural economy, derive from the high-pressure technology that also produces nitric acid for military explosives. Radar and the jet engine, developed primarily in response to military needs, are an essential part of post-war developments in civil aeronautics.

Patently, and disappointingly, progress in technology has not been matched by progress in social relations. To take civil aeronautics as our example once again, it can justly be claimed that by mid-century the technology had been fairly satisfactorily mastered, though succeeding decades were to show plenty of room for progress. But those same decades were marked by a steady decline in other respects. While the traveller was less subject to delays caused by mechanical faults or bad weather, he suffered increasingly from strikes among essential operating staff and elaborate security measures necessary to frustrate terrorists. The same may be said of medical services. The full benefit of new techniques could be limited by disputes and restrictive practices among those on whom their utilization depended.

To draw attention to these realities of modern society is not to imply that the scientists and technologists got it right and the rest of society failed to support them. Nor, as some would have us believe, is it the case that scientists and technologists got it all wrong and thoughtlessly drove an innocent society down roads to disaster. Rather, it is an indication that society generally – whether technically oriented or not – failed to adapt itself to a world in which both material and moral values were changing far more rapidly than ever before in human experience. Taking the position as it was at mid-century, and as it still is, we may appropriately conclude with a further sentence from A. V. Hill's address to the British Association: 'The forces of good and evil depend not on the scientist but on the moral judgment of the whole community.'

INDEX OF PERSONAL NAMES

GENERAL INDEX